11th International Conference on Numerical Modeling of Space Plasma Flows (ASTRONUM 2016)

Journal of Physics: Conference Series Volume 837

Monterey, California, USA
6 – 10 June 2016

ISBN: 978-1-5108-4183-3
ISSN: 1742-6588

Printed from e-media with permission by:

Curran Associates, Inc.
57 Morehouse Lane
Red Hook, NY 12571

proceedings.com

Some format issues inherent in the e-media version may also appear in this print version.

Printed by Curran Associates, Inc. (2017)

Institute of Physics
Dirac House, Temple Back
Bristol BS1 6BE UK

Phone: 44 1 17 929 7481
Fax: 44 1 17 920 0979

techtracking@iop.org

Additional copies of this publication are available from:

Curran Associates, Inc.
57 Morehouse Lane
Red Hook, NY 12571 USA
Phone: 845-758-0400
Fax: 845-758-2633
Email: curran@proceedings.com
Web: www.proceedings.com

Table of contents

Volume 837

11th International Conference on Numerical Modeling of Space Plasma Flows: ASTRONUM-2016

6–10 June 2016, Monterey, California, US

Accepted papers received: 27 May 2017
Published online: 30 May 2017

Astrophysical Flows

Space Plasma Flows

Numerical Methods

IOP Conf. Series: Journal of Physics: Conf. Series **837** (2017) 011001 doi:10.1088/1742-6596/837/1/011001

PREFACE

The Center for Space Plasma and Aeronomic Research (CSPAR) at the University of Alabama in Huntsville (UAH) and Maison de la Simulation at the French Alternative Energies and Atomic Energy Commission (Commissariat a l'energie atomique et aux energies alternatives, CEA) organized the 11th annual International Conference on Numerical Modeling of Space Plasma Flows (ASTRONUM-2016) on June 6—10, 2016 in Monterey, California, USA.

The Program Committee consisted of Tahar Amari (CNRS Ecole Polytechnique, France), Edouard Audit (CEA/CNRS Maison de la Simulation, Gif-sur-Yvette, France, co-chair), Amitava Bhattacharjee (Princeton University, USA), Phillip Colella (Lawrence Berkeley National Laboratory, USA), Anthony Mezzacappa (University of Tennessee, Knoxville, USA), Ewald Müller (Max-Planck-Institute for Astrophysics, Garching, Germany), Nikolai Pogorelov (University of Alabama in Huntsville/CSPAR, USA, chair), Kazunari Shibata (Kyoto University, Japan), James Stone (Princeton University, USA), Jon Linker (Predictive Science, Inc., USA), and Gary P. Zank (University of Alabama in Huntsville, USA).

The conference attracted 92 scientists representing different branches of the plasma simulation community. The distinctive feature of this conference is a combination of diverse research topics, all of which are essential for performing high-resolution, continuum mechanics and particle, simulations of physical phenomena in space physics and astrophysics. Among such topics were software packages for modeling and analyzing plasma flows; advanced numerical methods for space and astrophysical flows; large-scale fluid-based, kinetic, and hybrid simulations; turbulence and cosmic ray transport; and magnetohydrodynamics. The applications discussed included cosmology and galaxy formation, supernova explosions, physics of the Sun-heliosphere-magnetosphere environments, the interstellar medium and star formation, stellar physics, experimental plasma physics, astrophysical accretion, numerical methods for ideal and non-ideal, relativistic and nonrelativistic MHD, etc. The proceedings volume is structured so that it covers all of these topics.

A distinctive feature of the ASTRONUM conference series is based on the idea that modelers working in seemingly distant fields should have an opportunity to share their scientific achievements with the broad community of computational scientists performing numerical experiments. As in previous ASTRONUM meetings, we were interested in physical systems that are coupled across a multiplicity of spatial and temporal that incorporate diverse physical processes.

The contributors to this volume are both young researchers and renowned experts in space physics and astrophysics, applied mathematics, and computational physics. This book describes the application of numerical methods and the algorithms themselves, allowing us to discuss the challenges that theory imposes on numerical schemes for solving partial differential equations describing collisional and collisionless processes in space and astrophysical plasmas.

We would like to thank the participants who submitted their papers to Proceedings of ASTRONUM-2016 and especially to those who reviewed manuscripts thus ensuring the high quality of this publication. We also are grateful to Adele Corona (International Conferencing & Networking Solutions) for the excellent management of the conference.

The book will be useful to graduate and postgraduate students majoring in space physics, astrophysics, numerical, engineering, and applied mathematics. It is also aimed at specialists in applied mathematics, and various fields of physics that involve flows of partially ionized plasmas at both the collisional and collisionless levels.

Nikolai V. Pogorelov, Edouard Audit, and Gary P. Zank

April 21, 2017

IOP Conf. Series: Journal of Physics: Conf. Series **837** (2017) 011002 doi:10.1088/1742-6596/837/1/011002

Peer review statement

All papers published in this volume of *Journal of Physics: Conference Series* have been peer reviewed through processes administered by the proceedings Editors. Reviews were conducted by expert referees to the professional and scientific standards expected of a proceedings journal published by IOP Publishing.

Particle acceleration in MHD turbulence

Andrey Beresnyak
Naval Research Laboratory, Washington, DC 20375

Abstract. MHD Turbulence describes dynamics of astrophysical plasmas on large scale. It is characterized by energy transfer between different scales and the exchange of energy with nonthermal population – typically cosmic rays. Recent progress in theory regarding almost all basic regimes of turbulence – from the dynamo to the decaying case and the asymptotic scaling laws, allowed us to proceed with more observationally motivated questions. One of them is why almost all strongly magnetized environments are indeed observable, e.g. why such environments are infused with high-energy particles, their distributions stretching to energies orders of magnitude higher than thermal. It turns out that there are generic mechanisms of acceleration in turbulence, both second and first order in v/c, the latter was somehow unnoticed. These generic mechanisms may explain the ubiquity of non-thermal tails in various magnetized astrophysical environments, e.g. solar chromosphere, pulsar magnetospheres, jets from supermassive black holes, gamma-ray bursts, etc.

1. Introduction

Magnetically-dominated environments are fairly common in Astrophysics and a sizable fraction of astrophysical objects are made of rarefied plasma. These objects are only visible because they contain a non-thermal particle population which dominates emission across most of the spectrum. Following the idea of [1] that collisionless particles can get energy by scattering in fluid motions, especially converging motions, the diffusive shock acceleration (DSA) mechanism has been proposed [2, 3, 4] and has become rather popular in explaining non-thermal electromagnetic emission, as well as cosmic rays – energetic charged particles, detected at Earth. In the DSA, however, the acceleration rate is related to the diffusion coefficient, with the latter being importantly bound by the so-called Bohm limit[1] which assumes that scattering rate must be lower than Larmor gyration frequency. This makes acceleration progressively slower at higher energies.

On the other hand, the observations of many variable astrophysical objects suggested extremely fast acceleration timescales, incompatible with DSA. Few blazar jets powered by supermassive black holes exhibit ~ 10 min variations in TeV emissions, e.g., [6, 7]. Such fast time variabilities, along with other emission region constraints, have been suggested as evidence for mini-jets generated by reconnection [8]. Recent observations of gamma-ray flares from Crab [9] revealed that the impulsive nature of the energy release and the associated particle acceleration might need an alternative explanation as well [10]. It has also been suggested that reconnection plays a crucial role in producing high energy emissions from gamma-ray bursts [11].

[1] In the case of parallel shocks. Whether the perpendicular or oblique shocks can be more efficient in accelerating particles is an interesting question which is beyond the scope of our paper, see e.g. [5]

In the environments which are known to be magnetically dominated, e.g., the solar corona or the pulsar wind nebula, it is natural to expect the magnetic energy to be the main energy source. The global energetics of the powerful X-ray flares confirms this assumption. Reconnection and associated phenomena has been a active field of study, see, e.g. [12] for a review.

Here we describe a generic mechanism to transfer magnetic energy into particles and to conceptually understand the nature of recent numerical results that demonstrated that in both MHD fluid simulations [13] and ab-initio plasma simulations [14, 15, 16] of reconnection there is a regular acceleration of particles

Particle acceleration is often classified into "first order Fermi" mechanism where particles are gaining energy regularly, e.g., by colliding with converging magnetic mirrors and "second order Fermi" where particles can both gain and lose energy [1]. These two are not mutually exclusive and represent two different terms in the equation for the evolution of the distribution function: the terms describing advection and diffusion in energy space correspondingly. In practical terms, the first order mechanism usually dominates, if present, as the acceleration rate is proportional to the first order of u/v, where u is the typical velocity of the scatterer and v is the particle velocity. The outcome of the first order acceleration can be described in terms of the rate at which particle gain energy, acceleration rate, r_{acc} and the rate of particle escaping from the system, r_{esc}. If escape is negligible and r_{acc} is constant with energy, the energy of each particle grows exponentially. Also, if r_{esc}/r_{acc} does not depend on energy, the stationary solution for the particle distribution is a power-law, with the power law index determined by $-1-r_{esc}/r_{acc}$, see, e.g. [17]. Various environments, such as supernova shocks, were thought to satisfy this condition and produce power-law distributed cosmic rays, which become consistent with observations after being modified by diffusion. Acceleration within many orders of magnitude in energy was regarded as a result of a large-scale physical layout of the acceleration site, e.g., the planar shock can be thought of as a set of large-scale converging mirrors. The very same picture could also be applied to the large-scale reconnection site, where the two sides of the inflow effectively work as converging mirrors [18, 19]. Such acceleration lacks characteristic energy scale because there is only a single spacial scale – the scale of the system, which determines the maximum achievable energy.

In this paper, however, we will try to find regular acceleration over large energy ranges in systems that do not necessarily possess global regular structure – however they could still be scale-free in statistical sense, such as turbulent systems. Normally, turbulent environments are expected to be regions of second-order acceleration, see, e.g., [20, 21]. In this paper we point to the mechanism of regular or first order acceleration that was overlooked in the literature. This mechanism is inherently related to a certain statistical measure of energy transfer in turbulence and, therefore, does not rely on a particular geometry and is very robust. As we will show below, the direction of energy transfer from magnetic field to kinetic motions and the sign of curvature drift acceleration are inherently related, so that in systems with the average positive energy transfer from magnetic energy to kinetic motions there is an average positive curvature drift acceleration, while in the opposite case, there is an average curvature drift cooling.

One of the commonly considered cases of magnetically-driven flows is magnetic reconnection. A significant effort was put into understanding on non-ideal plasma effects that could both cause reconnection and create non-zero parallel component of the electric field [22, 23]. In this paper we apply our model to reconnecting layers which are large-scale in a sense that the current layer is many orders of magnitude thicker than the ion skin depth d_i. In this case the reconnecting layer will have multiple X-points and while non-ideal effects are indeed required for the individual field lines to break and reconnect, their influence is limited to fairly small scales, typically around the ion skin depth d_i. In this paper, we instead focus on larger scales and ignore non-ideal effects. In order to understand high energy particle acceleration, one normally has to consider plasma dynamics on scales much larger than d_i, that is on MHD scales. While modern simulations such

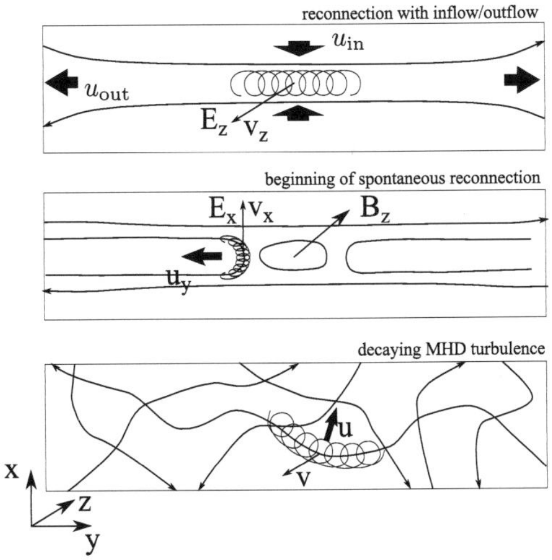

Figure 1. Three acceleration scenarious in magnetized environments. (upper) – reconnection with inflow, acceleration is due to gradient drift, the rate is determined by $v_z E_z$. (middle) – acceleration driven by contracting field lines which drive fluid motion and at the same time cause the curvature drift of particles. The direction of drift is typically perpendicular to the fluid motion. The dominant term is associated with E_x and the field curvature of B_y component. The bottom panel – same mechanism as in the middle panel but in a homogeneous and isotropic setup of decaying MHD turbulence, which also has contracting field lines (all vector components contribute equally to acceleration).

as [14, 15, 16], are able to reach box sizes of several hundred d_i, some theory work is needed to disentangle acceleration from MHD and the non-ideal effects.

2. MHD flows and energy transfer

Well-conductive plasmas can be described on large scales as inviscid and perfectly conducting fluid (ideal MHD). The ideal MHD equations allows for exchange between kinetic and magnetic energies. The Lorentz force density, multiplied by the fluid velocity, $\mathbf{u}\cdot[\mathbf{j}\times\mathbf{B}]/c$ is the amount of energy transferred from magnetic to kinetic energy. While macroscopic (i.e. kinetic plus magnetic) energy is expected to be conserved in the ideal MHD, it is not the case in real systems which have non-zero dissipation coefficients. This is qualitatively explained by the nonlinear turbulent cascade that brings macroscopic energy to smaller and smaller scales until it dissipates into thermal energy. One of the important examples of this is the spontaneous reconnection where the thin current layer becomes turbulent and starts dissipating magnetic energy at a constant rate. The small scales of these turbulent flows resemble "normal" MHD turbulence which has equipartition between magnetic and kinetic energies [24], it is also true that kinetic and magnetic part of the cascade each contribute around half of the total cascade rate. Therefore, if we assume that the turbulent cascade is being fed with magnetic energy, approximately half of the magnetic energy has to be transferred into kinetic energy before equipartition cascade sets in. It follows that the $\mathbf{B}$ to $\mathbf{v}$ energy transfer must be *positive on average* and could be approximated by one half of the volumetric energy dissipation rate ϵ, the main parameter of turbulence. The term $\mathbf{u}\cdot[\mathbf{j}\times\mathbf{B}]/c$ is the Eulerian expression for the work done by magnetic tension upon the fluid element. This term can be rewritten as the sum of $-(\mathbf{u}\cdot\nabla)B^2/8\pi$, advection of magnetic energy density by the fluid, and $\mathcal{T}_{bv}=\mathbf{u}\cdot(\mathbf{B}\cdot\nabla)\mathbf{B}/4\pi$, the actual energy transfer between $\mathbf{B}$ and $\mathbf{v}$. For the purpose of future calculations we will separate the $\mathcal{T}_{bv}$ in the following way:

$$\mathcal{T}=\frac{1}{4\pi}\mathbf{u}\cdot(\mathbf{B}\cdot\nabla)\mathbf{B}=\frac{1}{4\pi}(\mathbf{u}\cdot\mathbf{B})(\mathbf{b}\cdot\nabla)B+\frac{B}{4\pi}\mathbf{u}\cdot(\mathbf{B}\cdot\nabla)\mathbf{b}=\mathcal{X}+\mathcal{D}, \tag{1}$$

The term $\mathcal{X}$ contains cross helicity density $\mathbf{u}\cdot\mathbf{B}$. We argue that in those systems where $\langle\mathbf{u}\cdot\mathbf{B}\rangle=0$, which include many physically relevant cases, such as spontaneous reconnection, the whole $\mathcal{X}$ term could average out (see also Fig. 2). The second term $\mathcal{D}$ contains magnetic field curvature $(\mathbf{B}\cdot\nabla)\mathbf{b}$ and will be important for subsequent calculation of curvature drift.

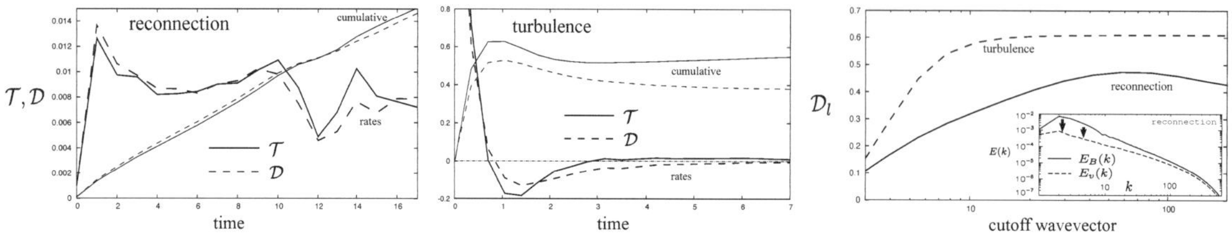

Figure 2. A case study of terms related to curvature drift acceleration in spontaneous reconnection and decaying MHD turbulence. (left): case A, averaged energy conversion rate $\mathcal{T}$ and curvature acceleration rate $\mathcal{D}$ in MHD simulation with turbulent current layer produced by spontaneous reconnection with setup described in [24]. (middle): case B, the same for decaying turbulence generated by random initial field. On both panels we also show cumulatives $\int\{\mathcal{T},\mathcal{D}\}dt$. The reconnection case (left panel) is characterized by approximately constant turbulent dissipation rate and it also show stable rate of energy conversion $\mathcal{T}$, while $\mathcal{D}$ closely follow $\mathcal{T}$. The decaying turbulence (middle panel) shows a burst of energy conversion rate within a few dynamical times (Alfvénic times). (right): the scale-dependency of the $\mathcal{D}$ term, by plotting $\mathcal{D}_l$ which is obtained by calculating $\mathcal{D}$ with a coarse-grained dynamical quantities $\mathbf{v},\mathbf{B}$ (Gaussian low pass filter with a cutoff wavenumber $k=2\pi/l$). $\mathcal{D}_l$ has a physical meaning of all energy transfer B to v accumulated down to scale l, for which reason it asymptotes to a constant at small l or large cutoff wavenumbers. The inset in the right panel shows kinetic and magnetic spectra in case A to demonstrate the range of scales within which magnetic energy is transferred to kinetic – down to $k\approx 30$ with thick arrows depicting the energy transfer.

3. Curvature drift

To explore the implications of magnetic energy transfer in non-thermal particle acceleration, it is instructive to consider the particles motion in slowly-varying electric and magnetic field, which can be described in the so-called drift approximation. The leading drift terms are known as electric, gradient and curvature drifts. While electric drift, proportional to $[\mathbf{E}\times\mathbf{B}]$, can not produce acceleration, the other two drifts can. For example, imagine the configuration of the reconnection with the inflow, Fig 1 top panel. The gradient drift $\sim[\mathbf{B}\times\nabla B]$ is along $-z$, and so is the electric field in the ideal case $E=-[\mathbf{u}\times\mathbf{B}]/c$. Their product will be positive and will result in acceleration which is due to particles being compressed by the converging inflow. This mechanism does not work in the initial, most energetic stages of spontaneous reconnection which has negligible inflow, Fig. 1 middle panel [24]. It is this initial stage that has the highest volumetric dissipation rate, however. Fig 1 illustrates why curvature drift acceleration is important in this configuration. It also turns out that in any magnetically-driven turbulent environment, such as depicted on the bottom panel of Fig 1, curvature drift acceleration will accelerate particles on average.

Let us look carefully at the term which is responsible for acceleration by curvature drift,

$$\frac{d\mathcal{E}}{dt}=-2\frac{\mathcal{E}_{\|}}{B}[\mathbf{u}\times\mathbf{B}]\cdot[\mathbf{b}\times(\mathbf{b}\cdot\nabla)\mathbf{b}],\tag{2}$$

see, e.g., [25], where $\mathcal{E}_{\|}=v_{\|}p_{\|}/2=\gamma m v_{\|}^2/2$ is a particle's parallel kinetic energy. With some manipulation, this expression could be equivalently transformed as $2(\mathcal{E}_{\|}/B)\mathbf{u}\cdot(\mathbf{B}\cdot\nabla)\mathbf{b}$. It now becomes clear that this term is related to the transfer rate between magnetic and kinetic energies, in particular it is a fraction of $\mathcal{D}$:

$$\frac{d\mathcal{E}}{dt}=\mathcal{E}_{\|}\frac{8\pi}{B^2}\mathcal{D}.\tag{3}$$

The quantity $\mathcal{D}$ depends on the coordinate and can be negative or positive, however we can average the Equation 3 in a volume sufficient to obtain average energy transfer between v and

B, e.g. in a domain of characteristic size of the outer scale of turbulence L. In this volume, we can estimate $\mathcal{E}_{\|} \approx \mathcal{E}/3$, this implies averaged particle distribution to be isotropic, note that it can still be locally anisotropic. The acceleration rate, as the rate of growth of the total energy will be determined by $8\pi\mathcal{D}/3B^2$, not including the contribution from the $\mathcal{X}$ term. The above quantity is a $1/3\tau_{\rm tr}$, where $\tau_{\rm tr}$ is the characteristic energy transfer timescale.

4. A case study

We can test the general ideas outlined above in two physical cases that feature turbulent energy transfer from magnetic to kinetic energies. Using spontaneous reconnection and the decaying MHD turbulence simulated numerically, we can directly calculate the discussed terms and compare them. The spontaneous reconnection numerical experiment was started with thin planar current sheet and small perturbations in $\mathbf{u}$ and $\mathbf{B}$ and was described in detail in [24], while the decaying MHD turbulence was similar to our previous incompressible driven simulations in [26], except that there was no driving and the initial conditions were set as a random magnetic field with wavenumbers $1 < k < 5$ and zero velocity. Both simulations developed magnetically-driven flows, from which we calculated the average $\mathcal{T}$, $\mathcal{D}$, and $\mathcal{X}$ terms and presented them in Fig. 2.

The spontaneous reconnection case had fairly stable reconnection rate, this also corresponded to the approximately constant $\mathcal{D}$ integrated over the volume. The $\mathcal{X}$ term didn't seem to be sign-definite and contributed relatively little. Gradient drift acceleration was also negligible, possibly due to the absence of global compression. Keeping in mind that all the energy had to come from magnetic energy, and given that the dissipation rate was approximately constant, it was no surprise that the average $\mathcal{D}$ term evolved relatively little. It should be noted, however, that in the spontaneous reconnection experiment the width of the reconnection region was growing approximately linear with time [24], so the $\mathcal{D}$ term magnitude, pertaining to the reconnection region itself was much higher than that of an averaged $\mathcal{D}$ over the total volume. Given the reconnection layer thickness $l(t)$, the local $\mathcal{D}$ could be estimated as $(1/2)v_r(B^2/8\pi)/l(t)$, where v_r is a reconnection rate, which was $v_r \approx 0.015v_A$ in [24] and the $1/2$ comes from only half of magnetic energy being transferred into kinetic energy before physically dissipating on very small dissipation scale $\eta \ll l(t)$. This would correspond to acceleration rate of $(1/4)v_r/l(t)$ and can be very high, because the $l(t)$ could be as small as the Sweet-Parker current sheet width or the ion skin depth.

The decaying MHD turbulence experienced two regimes: 1) the initial oscillation when excessive amount of magnetic energy was converted into kinetic energy and the bounce back and partial inverse conversion afterwards; 2) the self-similar decay stage of MHD turbulence. The first stage had the strongest conversion term which was dominated by $\mathcal{D}$. All terms integrated over time were mostly accumulated within the first 1-2 Alfvénic times.

5. Reconnection Simulations Setup

Earlier we simulated spontaneous reconnection in resistive MHD with Lundquist numbers up to 6×10^4, in which which the turbulent current layer is expanding in time [24]. The setup of these simulations on Fig 3. Additionally to this, we obtained a range of statistical measures of current layer turbulence [24]. Most interestingly: a) the power spectra of turbulent perturbations resembled decaying magnetic turbulence with the magnetic contribution dominating over the kinetic contribution on large scales and with an approximate energy equipartition on small scales; b) the energy spectral slope was around $-1.5 \div -1.7$, roughly consistent with the [27] scaling, such spectral slopes are indicative of local-in-scale turbulence [28]; c) the turbulence was observed to have scale-dependent anisotropy, again, in accord with [27].

As we noted above, locality implied that, if the ion Larmor radius (r_L) and the ion skin depth (d_i) are both below the equilibrium layer width, $0.015L$, the bulk quantities, such as the average

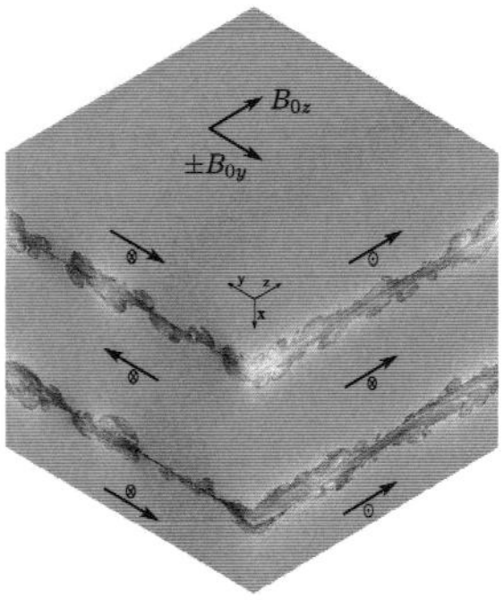

Figure 3. The all-periodic box reconnection setup with two current layers between the regions of reconnecting field $\pm B_{y0}$ (left). Simulations were seeded with small initial perturbations. The time evolution of the current layer width Δ (right) and the inferred reconnection speed $v_r = d\Delta/dt$ (inset). From [24].

reconnection rate or the average dissipation rate, are not affected by plasma processes. This has already been established for reconnection due to ambient turbulence [29], but we extended this result to the case where reconnection develops spontaneously without an external agent. Unlike the 2D case, in which large plasmoids couple large and small scales directly, the 3D turbulence that we observe appears to be local-in-scale. This implies that MHD reconnection rates are universal [24].

6. The particle-MHD approach

Recently we developed a novel kinetic particle solver with a broad range of physical capabilities and coupled it fully dynamically with a Hall-MHD code. This code allows us to study environments which have significant mutual interaction between particles and the MHD fluid, such as reconnection with efficient particle acceleration. The code was developed from the ground up with a focus on scalability. Furthermore, it implements the most general prescription for particle motion. In particular, apart from the Lorentz force there is the possibility of imposing various stochastic diffusion terms intended to reproduce the subgrid diffusion, i.e., we solve for trajectories of pseudoparticles whose distribution satisfies a Fokker-Plank type of equation. Due to this, particle splitting and merging comes naturally in this code and it allows for better coverage of the particle energy space by splitting highly energetic particles. In addition, for all particles we use a flexible precision-controlled solver, which significantly improves accuracy and stability in situations when particle energies vary by orders of magnitude.

The MHD part of the code solves the incompressible MHD equations with source terms from pseudoparticles. The pseudoparticle part solves stochastic differential equations (SDE), which are essentially equivalent to the Fokker-Planck equation [30]. The MHD equations are modified by adding $-(\mathbf{J}^{cr} \times \mathbf{B})/c + en^{cr}(\mathbf{U} \times \mathbf{B})/c$ to the RHS of the momentum equation, where $\mathbf{J}^{cr}, n^{cr}$ are the particle current and density, obtained from volume-sampled pseudoparticle distribution first and zeroth moments. The first term reflects the fact that only the fluid part of the current should be included in the Lorentz force, while the second part is the Lorentz force due to extra electrons in the fluid to maintain quasi-neutrality. The term $(c\eta\nabla \times \mathbf{J^{cr}})/4\pi$, where η is the magnetic diffusivity, could be added to the RHS of the induction equation and is due to the fact that only the fluid current is subject to resistivity. In most non-relativistic flows only the $-\mathbf{J}^{cr} \times \mathbf{B}$ term is relevant and other terms could be omitted. The pseudoparticles are evolved using equations

$$mcd\mathbf{u}/dt = e(\mathbf{E} + [\mathbf{u} \times \mathbf{B}]/\sqrt{1+\mathbf{u}^2}) + d\mathbf{p}_w/dt, \quad d\mathbf{x}/dt = \mathbf{u}c + d\mathbf{x}_w/dt, \tag{4}$$

where $\mathbf{E}$ and $\mathbf{B}$ are the MHD fields and $\mathbf{u}$ is a space part of the particle 4-velocity. The stochastic terms $d\mathbf{p}_w/dt$ and $d\mathbf{x}_w/dt$ are numerically calculated by increments in the frame associated with the local mean magnetic field, where the diffusion tensor has a diagonal

IOP Conf. Series: Journal of Physics: Conf. Series **837** (2017) 012001 doi:10.1088/1742-6596/837/1/012001

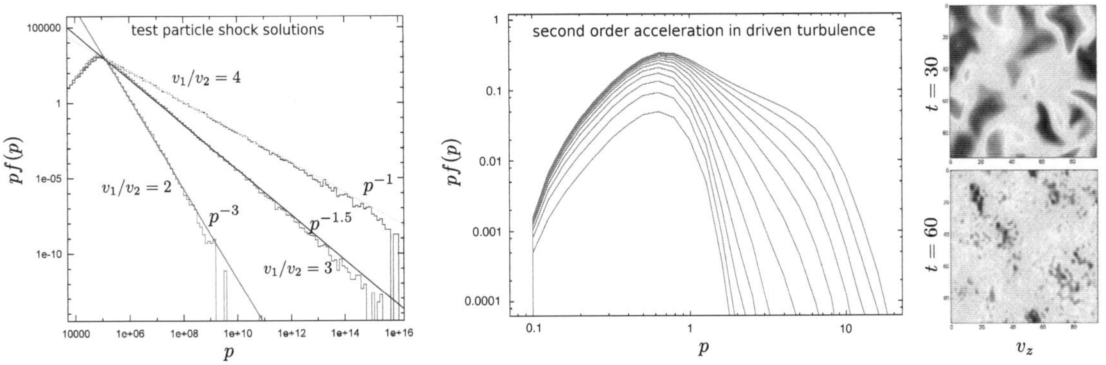

Figure 4. Test and pilot simulations with a new PIC-MHD code *Hephaestus*. *Left:* momentum distribution function obtained in a simulation with prescribed velocity field, constant magnetic field and ad-hoc diffusion after reaching stationary state. The velocity was prescribed a jump of a factor of 2, 3 and 4 in the middle of the box. The solutions reproduced well theoretical power spectra. *Middle:* momentum distribution function obtained in a driven fully self-consistent 3D simulation in a periodic box (evolution in time shown, no stationary state exists). *Right:* one velocity component for the xy slice in a driven simulation. Earlier times show large scales produced by driving which later times overlaid by kinetic micro-turbulence occurring from interaction of particles and the fluid.

formΔ $\hat{\mathbf{x}}_w = (g_1\sqrt{2D_{\|x}}, g_2\sqrt{2D_{\perp x}}, g_6\sqrt{2D_{\perp x}})\sqrt{\Delta t}$, and are then transformed to the laboratory frame[2]. Here g_* are Gaussian random numbers andΔ t is a time step.Δ $\hat{\mathbf{p}}_w$ have similar form with momentum diffusion tensor instead of spacial diffusion tensor. Each pseudoparticle also carries "weight", which is accounted for, when calculating the RHS side of the MHD equations. Due to the weight, pseudoparticles could be "split", e.g., two pseudoparticles are created in place of one with the weight split between them. Such split particles follow different trajectories due to the stochastic part. The splitting is usually required in the acceleration scenario when higher energy particles are represented by fewer numbers, and splitting is necessary to maintain good statistics for high energies.

Physically, stochastic terms represent evolution due to the unresolved, subgrid-scale electromagnetic perturbations. The so-called Monte Carlo simulations do not solve for the fluid flow, but use the prescribed velocity structure. In this case stochastic terms represent all magnetic perturbations, which are conjectured to exist but are not calculated explicitly by the model. Ordinarily Monte Carlo calculations use a simplistic prescription for the diffusion terms, such as Bohm scattering, implicitly assuming that scattering is very efficient. Our approach does not imply such strong assumptions and solves a coupled MHD-pseudoparticle system, which naturally includes particle-fluid interactions. The diffusion terms could be either set *ad hoc* as functions of energy or obtained from earlier local box simulations. In the reconnection experiments we describe below the particle diffusion terms has been set to zero.

7. Test results

Due to the flexibility of our code it can be used in a variety of circumstances, from a simple Monte Carlo solver to calculate particle dynamics in a prescribed flows to a fully self-consistent, dynamically relevant interplay between particles and the fluid. This flexibility also makes validation easier, as the separate parts of the code can be independently tested. The particle

[2] This assumes homogeneous diffusion. We did not implement inhomogeneous diffusion terms yet.

IOP Conf. Series: Journal of Physics: Conf. Series **837** (2017) 012001 doi:10.1088/1742-6596/837/1/012001

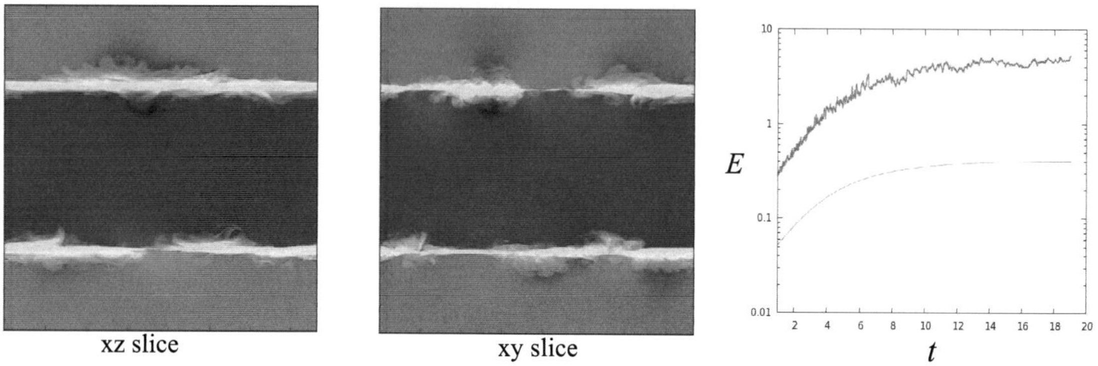

Figure 5. Results from 1536^3 simulations with 5×10^7 particles of spontaneous reconnection and acceleration. The reconnection setup is the same as in Fig. 3. Particles were only injected in the lower part of the datacube to estimate the dynamical effect of particles on reconnection by comparing upper and lower current layers. Left and middle depicts B_y (reconnecting) component of the magnetic field in slices xz (parallel to the mean field B_z) and xy (perpendicular to B_z). Note small-scale perturbations created in the lower half of the datacube by particles. Plot on the right shows evolution of the maximum energy (red) and average energy (green) of the particles. The particle distribution function also confirms that in this scenario all particles in the current layer are accelerated with the same rate, consistent with the theory presented in this paper.

code can be tested in a problem with a prescribed flow, the MHD code can be tested by introducing zero particles, and the test particle case can be studied by setting the pseudo-particle "weights" to zero. We already completed a variety of such tests, e.g. confirming the momentum conservation and energy conservation for the coupled system. Figure 4 show two tests. The first is the Monte Carlo test verifying that our particle part indeed solves Fokker-Plank equations and gives standard diffusive shock acceleration solutions for the test particle case. Note that this case feature spectra with 10 orders of magnitude range of energies. This is not unusual for the Monte Carlo with efficient particle splitting, however our code actually solves for trajectories. The second test features fully dynamic interplay between the fluid turbulence and particles, where energy is provided by driving the fluid, while particles are accelerated by the 2nd order mechanism and provide back-reaction to the fluid in the form of small-scale micro-turbulence, which in turn greatly enhances particle scattering.

8. Reconnection Simulations with particles

Figure 5 shows results of our simulations of spontaneous reconnection with particles. Note that coupled PIC-MHD system have four extra dimensionless numbers, characterizing the system, in addition to the MHD numbers. These are: m_p, mass load of particles, the ratio of particle mass to the fluid mass, e_p, energy load, the ratio of particle energy density to characteristic kinetic plus magnetic energy density of the fluid, r_p, the ratio of Larmor radius of the particles to the dissipation/grid scale of the fluid. v_p, the ratio of characteristic particle speed and characteristic fluid speed. Of these four numbers, r_p is naturally set to values somewhat below one, to ensure that particles are accelerated from "thermal pool" with subgrid Larmor radii. Also v_p is set to a large number (unless it limits numerical efficiency), as we are dealing with particles much faster that the fluid, e.g. relativistic particles. We typically inject particles with $\gamma \sim 1$ that are further accelerated to relativistic energies. Finally, m_p and e_p are free parameters dependent on the problem. In simulations described on Figure 5 we chose a very small mass load of 10^{-4} and a sizable initial energy load of 1.3×10^{-2}, which increased by an order of magnitude during the simulation. Our results have demonstrated that even when particles have negligible momentum compared to the fluid they still back-react to reconnection considerably.

9. Discussion

In this paper we showed that magnetic configurations that relax to the lower states of magnetic energy will also regularly accelerate particles, on timescales which are, typically, Alfvénic, but can be much shorter, e.g., in the beginning of spontaneous reconnection. This mechanism of acceleration of collisionless non-thermal particles by MHD electric field should not be confused with the acceleration of the bulk of the plasma by magnetic tension. Indeed, for particles with low energies the drift terms could be neglected, i.e. Eq. (2) RHS will be zero. In the bulk fluid acceleration the energy gained by each particle does not depend on its initial energy, while for drift acceleration scenario it is proportional to the particle energy.

Particle acceleration during reconnection is a topic under intense study, but the mechanism discussed in this paper is distinctly different from the direct acceleration by the reconnecting electric field at the X-line. In fact, we completely ignore non-ideal effects which produce local $\mathbf{E}\|\mathbf{B}$. Also, our mechanism is not tied to a special X-point, but instead volumetric.

The regular acceleration due to converging field lines have been suggested in [18], later it was pointed by [31] that an outflow cooling should also be included. In this paper we do not rely on simple transport equation, such as Parker's, therefore we relax the above approach's requirement that particle distribution need to be quasi-isotropic. In fact, in cases with significant acceleration, such as described in Section 8 we confirmed directly that particle distribution have significant deviations from isotropy locally.

The acceleration in turbulent reconnection has been further numerically studied in [13]. Plasma PIC simulations has also been increasingly used to understand particle acceleration. The emphasis was mostly on the non-ideal effects near X-line regions and interaction with magnetic islands [32, 22, 23, 33, 34, 35, 36]. The change of energy due to curvature drift in a single collision of a particle with magnetic island was estimated analytically in [33]. An important question that was left out in that study was whether this term will result in acceleration or cooling, on average. Without understanding this, it was not clear whether this process results in acceleration or deceleration or the second-order effect. In this paper we unambiguously decide this by relating the answer to a certain well-known statistical measure in turbulence – the direction of transfer between magnetic and kinetic energies. We also showed that the curvature drift acceleration does not require particles to be trapped in contracting magnetic islands, so their energy is not limited by this requirement, meaning that the energy cutoffi s not related to the island size, instead it is related to the system size.

PIC simulations are limited in the range of scales and energies they cover. Recent simulations in [14, 15] demonstrated acceleration up to 100 MeV in electron energies, which is below maximum energy in most astrophysical sources. Theory, therefore, is necessary to supplement conjectures based on the observed PIC distribution tails, explaining the underlying physical mechanism and making predictions for astrophysical systems which often feature gigantic scale separation between plasma scales and the size of the system. The feedback from simulations, nevertheless, was very useful, in particular the recent simulations [15] that reached MHD scales and confirmed the prediction that the curvature drift acceleration will dominate compared to the non-ideal electric field acceleration.

References

[1] Fermi E 1949 *Physical Review* **75** 1169–1174
[2] Krymskii G F 1977 *Soviet Physics Doklady* **22** 327
[3] Bell A R 1978 MNRAS **182** 147–156
[4] Malkov M A and O'C Drury L 2001 *Reports on Progress in Physics* **64** 429–481
[5] Caprioli D and Spitkovsky A 2014 Astrophys. J. **783** 91 (*Preprint* 1310.2943)

[6] Aharonian F *et al.* 2007 A&A **464** 235–243 (*Preprint* astro-ph/0611813)
[7] Aleksić J *et al.* 2011 ApJ **730** L8 (*Preprint* 1101.4645)
[8] Giannios D 2013 MNRAS **431** 355–363 (*Preprint* 1211.0296)
[9] Abdo A *et al.* 2011 *Science* **331** 739–742
[10] Clausen-Brown E and Lyutikov M 2012 *Monthly Notices of the Royal Astronomical Society* **426** 1374–1384
[11] Zhang B and Yan H 2011 Astrophys. J. **726** 90 (*Preprint* 1011.1197)
[12] Uzdensky D A 2015 *ArXiv:1510.05397* (*Preprint* 1510.05397)
[13] Kowal G, de Gouveia Dal Pino E M and Lazarian A 2012 *Physical Review Letters* **108** 241102
[14] Sironi L and Spitkovsky A 2014 ApJ **783** L21 (*Preprint* 1401.5471)
[15] Guo F, Li H, Daughton W and Liu Y H 2014 *arXiv:1405.4040* (*Preprint* 1405.4040)
[16] Uzdensky D A and Rightley S 2014 *Reports on Progress in Physics* **77** 036902 (*Preprint* 1401.5110)
[17] Drury L O, Duffy P, Eichler D and Mastichiadis A 1999 A&A **347** 370–374 (*Preprint* astro-ph/9905178)
[18] de Gouveia dal Pino E M and Lazarian A 2005 A&A **441** 845–853
[19] Lazarian A, Beresnyak A, Yan H, Opher M and Liu Y 2009 SSR **143** 387–413 (*Preprint* 0811.0826)
[20] Schlickeiser R 2002 *Cosmic Ray Astrophysics* (Berlin: Springer)
[21] Cho J and Lazarian A 2006 Astrophys. J. **638** 811–826 (*Preprint* arXiv:astro-ph/0509385)
[22] Pritchett P 2006 *Journal of Geophysical Research: Space Physics (1978–2012)* **111** 10212
[23] Fu X, Lu Q and Wang S 2006 *Physics of Plasmas (1994-present)* **13** 012309
[24] Beresnyak A 2013 *ApJ, submitted, 2016, arXiv:1301.7424* (*Preprint* 1301.7424)
[25] Sivukhin D V 1965 *Reviews of Plasma Physics* **1** 1
[26] Beresnyak A and Lazarian A 2009 Astrophys. J. **702** 1190–1198 (*Preprint* 0812.0812)
[27] Goldreich P and Sridhar S 1995 Astrophys. J. **438** 763–775
[28] Eyink G L 2005 *Physica D Nonlinear Phenomena* **207** 91–116
[29] Eyink G L, Lazarian A and Vishniac E T 2011 Astrophys. J. **743** 51 (*Preprint* 1103.1882)
[30] Gardiner C W *et al.* 1985 *Handbook of stochastic methods* vol 3 (Springer Berlin)
[31] Drury L O 2012 MNRAS **422** 2474–2476 (*Preprint* 1201.6612)
[32] Zenitani S and Hoshino M 2001 *The Astrophysical Journal Letters* **562** L63
[33] Drake J F, Swisdak M, Che H and Shay M A 2006 Nature **443** 553–556
[34] Oka M, Phan T D, Krucker S, Fujimoto M and Shinohara I 2010 *The Astrophysical Journal* **714** 915
[35] Huang C, Lu Q and Wang S 2010 *Physics of Plasmas (1994-present)* **17** 072306
[36] Dahlin J T, Drake J F and Swisdak M 2014 *Physics of Plasmas* **21** 092304 (*Preprint* 1406.0831)

IOP Conf. Series: Journal of Physics: Conf. Series **837** (2017) 012002 doi:10.1088/1742-6596/837/1/012002

Effects of driving scale on astrophysical turbulence and a modified Chandrasekhar-Fermi method

Jungyeon Cho

Department of Astronomy and Space Science, Chungnam National University, Daejeon, Korea (ROK)

E-mail: jcho@cnu.ac.kr

Abstract. Many astrophysical fluids are in turbulent state. To maintain turbulence, energy must be injected into the fluid. In turbulence simulations, it is customary to drive the fluids on a scale comparable to the size of the computational domain. In this paper, we show how some statistics of turbulence, especially turbulence statistics projected on the plane of the sky, is changed when we do not follow the conventional approach. As an example of such a statistics, we discuss in detail how the Chandrasekhar-Fermi method is affected by the small-scale driving, which is a simple and powerful technique for estimating the strength of the mean magnetic field projected on the plane of the sky.

1. Introduction

Driving is required to initiate and maintain turbulence. When hydrodynamic turbulence is driven on a scale L_f, the injected energy cascades down to small scales and, when the energy cascade reaches the dissipation scale L_d, the energy is lost through viscous damping. The energy spectrum of turbulence peaks at the driving wavenumber k_f and shows a power-law scaling in the range between k_f and the dissipation wave number k_d. The spectrum drops rapidly for wavenumbers larger than k_d.

In turbulence simulations, the driving scale L_f is usually very close to the computational box size L_{sys}. The reason for this is to maximize the inertial range. Let us assume $L_{sys} = 2\pi$ throughout the paper. In most turbulence simulations, the dissipation wavenumber k_d is determined by the numerical resolution: k_d is slightly smaller than the maximum wavenumber $N_x/2$, where N_x is the number of grid points in one direction. Therefore, in order to maximize the inertial range, it is necessary to drive turbulence near $k \sim 1$, which means $k_f \sim 1$ and $L_f \sim L_{sys} = 2\pi$.

In astronomy, observed quantities are usually integrated along the line-of-sight (LOS). Since observed quantities are two-dimensional (2D) data, obtaining three-dimensional (3D) quantities from observations is not easy. Nevertheless, there are techniques developed to derive 3D turbulence properties from 2D observations. However, those techniques have an intrinsic shortcoming: it is difficult to verify them. To overcome this difficulty, people use computer simulation data to verify their techniques.

One such example is the Chandrasekhar-Fermi (CF) method [1], which is a powerful technique for estimating the strength of the plane-of-the-sky component of the mean magnetic field, $B_{0,sky}$. The CF method makes use of polarized emission from magnetically aligned dust grains. In the conventional CF method, we can obtain an estimate of $B_{0,sky}$ from

$$B_{0,sky} = \xi\sqrt{4\pi\bar{\rho}}\frac{\delta v_{los}}{\delta\phi}, \tag{1}$$

where ξ is a constant of order unity, $r\bar{h}o$ is average density, δv_{los} is the LOS velocity dispersion (or, typical width of an optically thin emission line), and $\delta\phi$ is the variation of the polarization angle. Researchers used numerical simulations to verify the technique and obtain the value of ξ [2, 3, 4].

In this paper, we demonstrate that we should be careful in using simulations to verify a theoretical method: if we do not take into account the driving scale, it is possible to make a mistake. We focus on the CF method. In section 2, we show that the conventional CF method is prone to such a mistake. To be specific, we show that the conventional CF method overestimates $B_{0,sky}$ by a factor of $\sqrt{N_{eddy}}$, where N_{eddy} is the number of independent eddies along the LOS. Note that $N_{eddy} \approx L_{sys}/L_f$ in simulation data. In Section 3, we introduce our modified CF method (see [5] for details). In Section 4, we provide discussions and summary.

2. The conventional Chandrasekhar-Fermi (CF) method

2.1. Derivation of the conventional CF method

Consider a turbulent medium with a mean magnetic field ($\mathbf{B}_0^{3D}$) and a fluctuating magnetic field ($\mathbf{b}^{3D}$). For simplicity, let us assume that the mean magnetic field is along the x direction and the LOS is along the z direction. In this setup, the strength of the plane-of-the-sky mean magnetic field ($B_{0,sky}^{3D}$) is just B_0^{3D}:

$$B_{0,sky}^{3D} = B_0^{3D}, \quad \text{if } \mathbf{B}_0^{3D} \perp \text{ LOS}. \tag{2}$$

Let us consider a fluid filled with Alfvén waves or Alfvénic turbulence. In Alfvénic disturbances, the r.m.s. fluctuation of magnetic field (δb^{3D}) and the r.m.s. velocity (δv^{3D}) are related by

$$\frac{\delta b^{3D}}{\sqrt{4\pi\bar{\rho}}} \sim \delta v^{3D} \quad \text{or} \quad 1 \sim \sqrt{4\pi\bar{\rho}}\frac{\delta v^{3D}}{\delta b^{3D}}, \tag{3}$$

where $\bar{\rho}$ is average density. Since we are dealing with Alfvén waves, we assume the density is constant. If we multiply both sides by the mean plane-of-the-sky magnetic field $B_{0,sky}^{3D}$, we obtain

$$B_{0,sky}^{3D} \sim \sqrt{4\pi\bar{\rho}}\frac{\delta v^{3D}}{\left(\delta b^{3D}/B_{0,sky}^{3D}\right)}. \tag{4}$$

It is relatively easy to estimate the numerator on the right-hand side (i.e. δv^{3D}) from observations. If we observe the turbulent medium using an optically thin molecular emission line, then the width of the emission line profile ($\equiv \delta v_{los}$) is approximately $\sim \delta v^{3D}$:

$$\delta v^{3D} \sim \delta v_{los}. \tag{5}$$

Note that δv^{3D} roughly corresponds to the typical length of the velocity vectors in the 3D space, which is $\sim \sqrt{3}$ times larger the typical length of the LOS velocity vectors in the 3D space (see red arrows in Figure 1(a)).

Estimation of the denominator on the right-hand side of Equation (4) is rather tricky. Suppose that we have coherent Alfvén waves (see Figure 1(b)). In this case, it is trivial to show that

$$\frac{b_y^{2D}}{B_{0,sky}^{2D}} = \frac{\int b_y^{3D}dz}{\int B_{0,sky}^{3D}dz} = \frac{b_y^{3D}}{B_{0,sky}^{3D}}, \tag{6}$$

where $B^{2D}_{0,sky}$ is the plane-of-the-sky mean magnetic field, which is proportional to $B^{3D}_{0,sky}$ integrated along the LOS, and b^{2D}_y is the y component of the plane-of-the-sky magnetic field, which is proportional to b^{3D}_y integrated along the LOS. Note that, if we ignore the fluctuating magnetic field in the x direction, $B^{2D}_{0,sky}$ and b^{2D}_y are proportional to the x and y components of the plane-of-the-sky magnetic field, respectively, measured by a distant observer located on the z axis. Therefore, the ratio $b^{2D}_y/B^{2D}_{0,sky}$ determines the direction of the observed magnetic field on the plane of the sky. From (6), we can write

$$\frac{\delta b^{3D}_y}{B^{3D}_{0,sky}} = \delta\left(\frac{b^{3D}_y}{B^{3D}_{0,sky}}\right) = \delta\left(\frac{b^{2D}_y}{B^{2D}_{0,sky}}\right) = \frac{\delta b^{2D}_y}{B^{2D}_{0,sky}} = \delta(\tan\phi) \sim \delta\phi, \tag{7}$$

where ϕ is the angle between the plane-of-the-sky magnetic field and the mean plane-of-the-sky magnetic field ($\propto \mathbf{B}^{2D}_{0,sky}$) and $\delta\phi$ is its variation.

Since the polarized emission from magnetically aligned dust grains is perpendicular to the direction of the observed magnetic field on the plane of the sky, the angle ϕ in Equation (7) is virtually identical to the polarization angle with respect to the average polarization direction. Therefore, if we observe the variation of polarization angle (i.e. $\delta\phi$), we can obtain an estimate for the denominator in Equation (4).

In summary, the conventional CF method can be written by

$$B^{3D}_{0,sky} \sim \xi\sqrt{4\pi\bar{\rho}}\frac{\delta v_{los}}{\delta\phi}, \qquad (\leftarrow \text{for coherent waves}) \tag{8}$$

where ξ is a constant of order unity. Note that we have assumed that Alfvén waves are coherent (see Figure 1(b)) to obtain this relation.

2.2. A problem with the conventional CF method

In the previous subsection, we have assumed that Alfvén waves are all in phase. In this case, the conventional CF method should work fine. But, if either Alfvén waves have different phases or they form turbulence, then the conventional CF method needs modification.

Suppose that we have Alfvénic turbulence. Then, several large-scale eddies can exist along the LOS (see Figure 1(c)). It is natural to assume that physical quantities do not change much inside each large-scale eddy. For example, we may assume that the value of b^{3D}_y is roughly constant in each large-scale eddy. However, the direction of b^{3D}_y may change from eddy to eddy. Therefore, if we integrate b^{3D}_y along the LOS, we have

$$b^{2D}_y \propto \int b^{3D}_y dz \sim \delta b^{3D}_y L_f \sqrt{N_{eddy}}, \tag{9}$$

where N_{eddy} ($\approx L_{sys}/L_f$) is the number of large-scale eddies along the LOS and L_f is the driving scale of turbulence. Note that the typical size of the large-scale eddies is similar to the driving scale L_f.

Since the direction of the 3D mean magnetic field does not change along the LOS, integration of $B^{3D}_{0,sky}$ is different from that of b^{3D}_y:

$$B^{2D}_{0,sky} \propto \int B^{3D}_{0,sky} dz = B^{3D}_{0,sky} L_{sys} = B^{3D}_{0,sky} L_f N_{eddy}. \tag{10}$$

From Equations (9) and (10), we have

$$\frac{b^{2D}_y}{B^{2D}_{0,sky}} \sim \sqrt{N_{eddy}}\frac{b^{3D}_y}{B^{3D}_{0,sky}}, \tag{11}$$

IOP Conf. Series: Journal of Physics: Conf. Series **837** (2017) 012002 doi:10.1088/1742-6596/837/1/012002

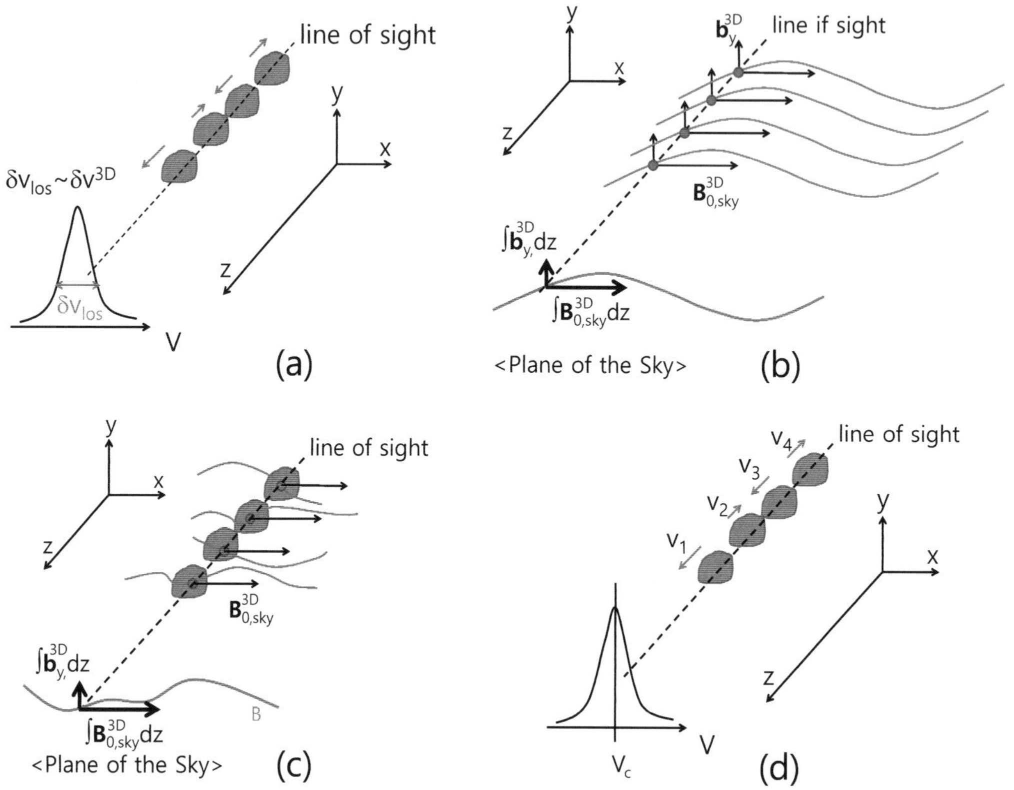

Figure 1. Observations of velocity and magnetic field. The line of sight (LOS) is along the z direction and the sky plane is parallel to the xy plane. For simplicity, we assume density is constant. (a) The width of an optically thin line is equal to the LOS velocity dispersion δv_{los}. (b) When Alfvén waves are all in phase, we have $b_y^{2D}/B_{0,sky}^{2D} = b_y^{3D}/B_{0,sky}^{3D}$, where b_y^{2D} and $B_{0,sky}^{2D}$ are magnetic fields integrated along the LOS. (c) When Alfvén waves are out of phase, we have $b_y^{2D}/B_{0,sky}^{2D} < b_y^{3D}/B_{0,sky}^{3D}$ due to random-walk-like behavior of b_y^{3D}. (d) The centroid velocity V_c is an average velocity. The area of a line profile on the left-hand side of V_c and that on the right-hand side of V_c are same.

which is obviously different from the relation for coherent waves in Equation (6). If we insert this relation into Equation (4), we get

$$\sqrt{N_{eddy}} B_{0,sky}^{3D} \sim \sqrt{4\pi\bar{\rho}} \frac{\delta v_{los}}{\delta\phi}, \qquad (\leftarrow \text{ for turbulence}) \tag{12}$$

where we have replaced δv^{3D} with δv_{los} and used relations in Equation (7). Equation (12) implies that the conventional CF technique (see the right-hand side of the equation) overestimates $B_{0,sky}^{3D}$ by a factor of $\sqrt{N_{eddy}}$.

Another way to look at this problem is that, when there are many independent eddies along the LOS, $\delta\phi$ becomes reduced due to averaging effects, which makes the conventional CF method to overestimate $B_{0,sky}^{3D}$. This averaging effect, in fact, has been discussed by earlier researchers (see, for example, [6, 7, 8]).

2.3. Numerical tests of the conventional CF method

We use direct numerical simulations to test the conventional CF method (see [5] for details). We solve the following compressible MHD equations in a periodic box of size 2π using an Essentially Non-Oscillatory scheme (see [9]):

$$\partial\rho/\partial t + \nabla\cdot(\rho\mathbf{v}) = 0, \tag{13}$$

$$\partial\mathbf{v}/\partial t + \mathbf{v}\cdot\nabla\mathbf{v} + \rho^{-1}\nabla(C_s^2\rho) - (\nabla\times\mathbf{B})\times\mathbf{B}/4\pi\rho = \mathbf{f}, \tag{14}$$

$$\partial\mathbf{B}/\partial t - \nabla\times(\mathbf{v}\times\mathbf{B}) = 0, \tag{15}$$

with $\nabla\cdot\mathbf{B}=0$ and an isothermal equation of state $P=C_s^2\rho$, where C_s is the sound speed and ρ is density. Here $\mathbf{v}$ is the velocity, $\mathbf{B}$ is the magnetic field, and $\mathbf{f}$ is the driving force. We use 512^3 grid points. In our simulations, $C_s = 0.1$, $\bar{\rho} = 1$, and $B_0/\sqrt{4\pi\bar{\rho}} = 1$. In all simulations, the r.m.s. velocity v_{rms} is between $\sim$0.7 and $\sim$0.8, and the sonic Mach number is $M_s \equiv v_{rms}/C_s \sim 7$. Since the Alfvén speed of the mean field ($V_A = B_0/\sqrt{4\pi\bar{\rho}}$) is 1, the Alfvén Mach number is $M_A \equiv v_{rms}/V_A \sim 0.7$, which means that turbulence considered here is sub-Alfvénic.

Table 1. Simulations

Runs	Resolution	Sonic Mach number(M_s)	$B_{0,sky}^{3D}/\sqrt{4\pi\bar{\rho}}$	k_f	N_{eddy} $(= L_{sys}/L_f)$
KF3	512^3	$\sim$7	1	3	~ 3
KF5	512^3	$\sim$8	1	5	~ 5
KF10	512^3	$\sim$7	1	10	~ 10
KF20	512^3	$\sim$7	1	20	~ 20

We perform 4 different simulations with different driving wavenumbers (see Table 1). The driving wavenumbers are k_f=3, 5, 10, and 20, respectively and the corresponding driving scales are L_f=$L_{sys}/3$, $L_{sys}/5$, $L_{sys}/10$, and $L_{sys}/20$, respectively. Note that the number of independent eddies along the LOS is $\sim 1/k_f$.

Except the driving scale, other conditions are virtually same (see Table 1). Using the MHD turbulence data and numerical method in Ref. [10] (see also [4]), we obtain synthetic polarization maps arising from magnetically-aligned dust grains at a far-infrared/sub-mm wavelength. Using the polarization maps, we calculate $\delta\phi$. We also calculate δv_{los} using the turbulence data. After getting $\delta\phi$ and δv_{los}, we calculate the quantity

$$\sqrt{4\pi\bar{\rho}}\frac{\delta v_{los}}{\delta\phi}, \tag{16}$$

which is proportional to $B_{0,sky}^{3D}$ in the conventional CF method. Since $B_0/\sqrt{4\pi\bar{\rho}} = 1$ in all our simulations, the above expression should be constant if the conventional CF method is correct. We plot the results in the left panel of Figure 2. As we can see in the figure, the conventional CF method overestimates $B_{0,sky}^{3D}$ as N_{eddy} ($\sim 1/k_f$) increases.

3. A Modified CF Method

3.1. Estimation of N_{eddy} and our modified CF method

Since the conventional CF method overestimates $B_{0,sky}^{3D}$ by a factor of $\sqrt{N_{eddy}}$, we need to know N_{eddy} to correct it. In this subsection, we show that variation of centroid velocity δV_c is

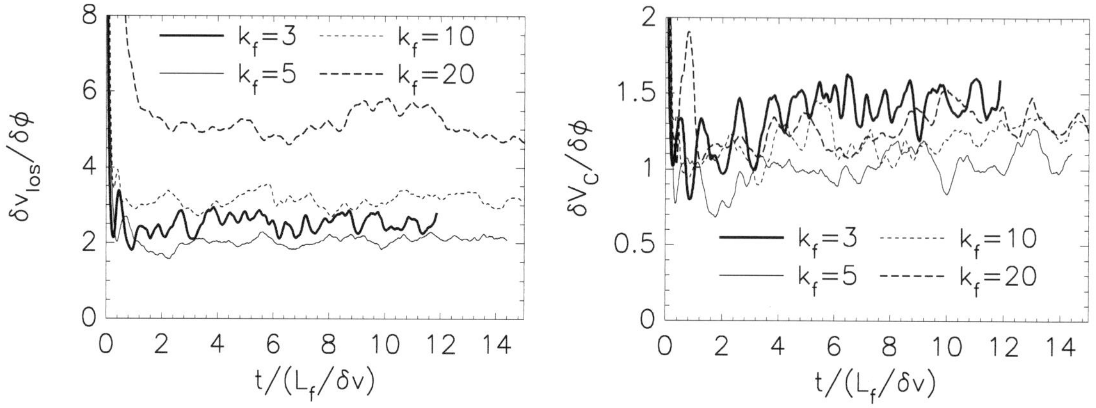

Figure 2. Estimates of $B^{3D}_{0,sky}/\sqrt{4\pi\rho}$, which is 1 in our simulations, from the conventional CF method (left panel) and our modified CF method (right panel). (Left) The conventional CF method overestimates $B^{3D}_{0,sky}$ if k_f is large. Note that, if turbulence is driven near k_f, there are $\sim k_f$ independent eddies along the LOS. (Right) Our modified CF method returns better estimates of $B^{3D}_{0,sky}$ even if k_f is large. From [5].

proportional to $\sqrt{N_{eddy}}$ (see [5] for details). The centroid velocity is a kind of average velocity. When we have an emission line with intensity profile $I(v)$, the centroid velocity is defined by

$$V_c = \int v_{los} I(v_{los}) dv_{los} \Big/ \int I(v_{los}) dv_{los}. \tag{17}$$

Suppose that there are N_{eddy} independent eddies along a LOS (see Figure 1(d)). Each eddy has typical LOS velocity $v_{los,i}$, which we denote as v_i for simplicity. Although v_i's are not directly observable, we can consider the following sum for the LOS:

$$v_1 + v_2 + ... + v_{N_{eddy}}. \tag{18}$$

Now, let us consider similar summations for many *different* LOS's. Since we have now many summations, we can calculate the standard deviation of the summations. Note that v_i's show a random-walk-like behavior and the summation in Equation (18) corresponds to the net displacement of a one-dimensional random walk. Therefore, the standard deviation of the summations follows

$$\text{Standard deviation of } (v_1 + v_2 + ... + v_{N_{eddy}}) \sim \sqrt{N_{eddy}} \delta v_{los}, \tag{19}$$

which becomes

$$\text{Standard deviation of } (v_1 + v_2 + ... + v_{N_{eddy}})/N_{eddy} \sim \frac{\delta v_{los}}{\sqrt{N_{eddy}}}. \tag{20}$$

The left-hand side is equal to the standard deviation of the (arithmetic) average velocity. Centroid velocity is also a kind of average velocity. Therefore, we expect that the standard deviation of centroid velocity also follows a similar scaling:

$$\delta V_c \sim \frac{\delta v_{los}}{\sqrt{N_{eddy}}}. \tag{21}$$

From Equations (12) and (21), we have

$$B^{3D}_{0,sky} \sim \sqrt{4\pi\bar{\rho}}\frac{\delta v_{los}}{\delta\phi}\frac{1}{\sqrt{N_{eddy}}} \sim \sqrt{4\pi\bar{\rho}}\frac{\delta v_{los}}{\delta\phi}\frac{\delta V_c}{\delta v_{los}} \sim \sqrt{4\pi\bar{\rho}}\frac{\delta V_c}{\delta\phi}. \quad (22)$$

Therefore, our modified CF method becomes

$$B^{3D}_{0,sky} = \xi'\sqrt{4\pi\bar{\rho}}\frac{\delta V_c}{\delta\phi}, \quad (23)$$

where ξ' is a constant of order unity that can be determined by numerical simulations.

3.2. Numerical tests of the modified CF method

The right panel of Figure 2 shows our main result, in which we plot estimates of $B^{3D}_{0,sky}/\sqrt{4\pi\bar{\rho}}$ from our modified CF method:

$$\frac{\delta V_c}{\delta\phi}. \quad (24)$$

In the panel, we can see that the estimates do not depend on the driving scale strongly and are fluctuating between ~1.0 and ~1.5. Therefore, since $B^{3D}_{0,sky}/\sqrt{4\pi\bar{\rho}} = 1$ in our simulations, the constant ξ' in Equation (23) is between ~0.7 and ~1.

4. Discussions and Summary

In this paper, we have considered the effects of driving scale on the estimates of the mean plane-of-the-sky magnetic field $B^{3D}_{0,sky}$ from the Chandrasekhar-Fermi (CF) method. We have shown that the conventional CF method tends to overestimate $B^{3D}_{0,sky}$ if the driving scale is smaller than the system size and proposed a modified CF method that can correct the effect. The method we propose in Equation (23) with $0.7 < \xi' < 1.0$ does not require new observations [5]. That is, the method is readily applicable for present observational data. Apart from numerical constants, the only difference between our method and the conventional CF method is that our method requires the standard deviation of velocity centroids δV_c, while the conventional method requires average width of the emission line profiles δv_{los}. The standard deviation of velocity centroids δV_c can be easily obtained from existing optically-thin emission line profiles. If such emission line profiles ($I(v_{los})$'s) are available for n_{obs} lines of sight, then we need the following two steps to obtain δV_c:

(i) We calculate the centroid velocity V_c (see Equation (17)) for each line of sight. Let $V_{c,i}$ be the centroid velocity for line of sight i:

$$V_{c,i} = \int v_{los} I_i(v_{los}) dv_{los} \bigg/ \int I_i(v_{los}) dv_{los}, \quad (25)$$

where $I_i(v_{los})$ is the optically-thin emission line profile for the line of sight.

(ii) We calculate δV_c from the formula

$$\delta V_c^2 \equiv \frac{1}{n_{obs}}\sum_{i=1}^{n_{obs}} V_{c,i}^2 - \left(\frac{1}{n_{obs}}\sum_{i=1}^{n_{obs}} V_{c,i}\right)^2. \quad (26)$$

In this paper we have demonstrated that the conventional CF method indeed overestimates the mean plane-of-the-sky magnetic field $B^{3D}_{0,sky}$ by a factor of $\sqrt{N_{eddy}}$, where N_{eddy} is the number of independent eddies along the line of sight. We have found that the standard deviation of centroid velocities divided by the average line-of-sight velocity dispersion ($\delta V_c/\delta v_{los}$) is proportional to $1/\sqrt{N_{eddy}}$ (Equation (21)). Therefore Equation (23) with $\xi' = 0.7 \sim 1$ provides a better estimate for $B^{3D}_{0,sky}$.

References

[1] Chandrasekhar S and Fermi E 1953 *Astrophys. J.* **118** 113
[2] Ostriker E C, Stone J M and Gammie C F 2001 *Astrophys. J.* **546** 980–1005 (*Preprint* astro-ph/0008454)
[3] Padoan P, Goodman A, Draine B T, Juvela M, Nordlund Å and Rögnvaldsson Ö E 2001 *Astrophys. J.* **559** 1005–1018 (*Preprint* astro-ph/0104231)
[4] Heitsch F, Zweibel E G, Mac Low M M, Li P and Norman M L 2001 *Astrophys. J.* **561** 800–814 (*Preprint* astro-ph/0103286)
[5] Cho J and Yoo H 2016 *Astrophys. J.* **821** 21 (*Preprint* 1603.08537)
[6] Myers P C and Goodman A A 1991 *Astrophys. J.* **373** 509–524
[7] Zweibel E G 1996 *Polarimetry of the Interstellar Medium* (*Astronomical Society of the Pacific Conference Series* vol 97) ed Roberge W G and Whittet D C B p 486
[8] Houde M, Vaillancourt J E, Hildebrand R H, Chitsazzadeh S and Kirby L 2009 *Astrophys. J.* **706** 1504–1516 (*Preprint* 0909.5227)
[9] Cho J and Lazarian A 2002 *Physical Review Letters* **88** 245001 (*Preprint* astro-ph/0205282)
[10] Fiege J D and Pudritz R E 2000 *Astrophys. J.* **544** 830–837 (*Preprint* astro-ph/0005363)

Galactic cosmic ray propagation models using Picard

Ralf Kissmann

Leopold Franzens Universität Innsbruck, Austria

E-mail: `ralf.kissmann@uibk.ac.at`

Abstract. I will give an overview on the numerical approach for the solution of the Galactic cosmic-ray propagation problem used within the recently introduced code Picard. The code solves the transport equation for each cosmic-ray species resulting in a prediction of the cosmic ray flux everywhere in the Galaxy. Picard is a prominent example for the current transition from two-dimensional azimuthally symmetric models of Galactic cosmic-ray transport to those that allow higher degrees of realism. In Picard this is achieved by use of contemporary numerical solvers that allow efficient computation of very high resolution models. I will address the new physics that can be incorporated in such three-dimensional models, e.g. example problems like the transition from axisymmetric cosmic-ray source distributions to spiral-arm cosmic-ray source distributions, including consequences for the various observables related to cosmic rays.

1. Introduction

The majority of cosmic rays up to an energy of 10^{15} eV are thought to be of Galactic nature. They are accelerated by different sources and mechanisms all throughout the Galaxy (see (1) for a discussion of different sources). Here, our focus is on the transport of these particles throughout the Galaxy – to the observer at Earth. The transport of Galactic cosmic rays is determined by the interplay of several processes. Cosmic rays are subject to diffusion in space and also in energy due to scattering of the high-energy particles off irregularities in the interstellar magnetic field. They suffer from different kinds of energy-losses, where electrons are most severely affected. Additionally, heavy cosmic-ray nuclei can fragment in nuclear interactions with the interstellar gas. This is a loss process for the heavy nuclei and an additional secondary source process for lighter particles. In total, Galactic cosmic-ray transport is described by the following transport equation (see also 2; 3):

$$\frac{\partial \psi_i}{\partial t} = q(\vec{r}, p) + \nabla \cdot \mathcal{D} \nabla \psi_i + \frac{\partial}{\partial p} p^2 D_{pp} \frac{\partial}{\partial p} \frac{1}{p^2} \psi_i - \nabla \cdot \vec{v} \psi_i - \frac{\partial}{\partial p} \left\{ \dot{p} \psi_i - \frac{p}{3} (\nabla \cdot \vec{v}) \psi_i \right\} - \frac{1}{\tau_f} \psi_i - \frac{1}{\tau_r} \psi_i. \quad (1)$$

This transport equations determines the cosmic-ray flux ψ_i for each cosmic-ray species i. In this equation, $q(\vec{r}, p)$ describes the spatial distribution and the energy dependence of the cosmic-ray sources. $\mathcal{D}$ is the diffusion tensor, that can depend on energy, position and also direction in space. Similarly, D_{pp} is the momentum diffusion coefficient. Cosmic rays can be advected together with the interstellar gas moving with advection velocity $\vec{v}$. Additionally, $\dot{p}$ describes the energy losses and τ_f and τ_r are the time scales for fragmentation- or radiative losses, respectively.

Mathematically, the Galactic cosmic-ray transport equation is a four-dimensional diffusion-convection equation. Typically, the energy-dependence of the different terms in the cosmic-ray transport equation can be described as a power law in energy. This pertains in particular to the energy dependence of the cosmic-ray source strength, the energy dependence of spatial- and momentum diffusion, and also the energy dependence of the different energy-loss rates. As a result the spectra of the different particle species are also of power-law form. However, the power-law index differs for different species and also in different energy regimes – depending on the relative importance of the transport parameters. Together with the wide range of energies relevant for Galactic cosmic-ray transport (ranging from hundreds up to some 10^9 MeV) the solution of this transport equation is numerically very demanding. In the following we will discuss available solution approaches in section 2 and, more in particular, the approach used in the PICARD code (4) in section 3. Finally, we present a few results in section 4.

2. Numerical solution of the cosmic-ray transport equation

For the numerical solution of the transport equation (1) there exist a range of numerical methods. These include grid-based methods, where the differential equation is solved by a discretisation in space, energy, and time, and a range of methods that propagate pseudo particles according to the prescription of the transport equation. The latter include Monte-Carlo-based methods as, e.g., discussed by (5) and also the use of so called stochastic differential equations (SDEs) (6; 7; 8). SDEs are particularly efficient in computing the cosmic-ray flux at a given position in space due to the possibility to perform a computation backwards in time – from the observer to the sources of cosmic rays. When, however, the distribution of cosmic rays everywhere in the Galaxy is of relevance, as for example in a study of Galactic diffuse emission (9), this method becomes computationally expensive. With a focus on methods suitable for the latter, we will only discuss grid-based solution methods of the cosmic-ray transport equation.

There are several Galactic cosmic-ray propagation codes available that rely on a grid-based solution method. These included, but are not limited to, DRAGON (10), GALPROP (11), and PICARD (4). The former two of these codes rely on a very similar numerical scheme: they use a dimensional splitting method for all spatial and the momentum dimension. For each dimensions the transport equation is discretised – typically using a Crank-Nicolson method – and the resulting tri-diagonal system of equations is solved via a standard method. The solution is then advanced in time according to a user-defined configuration of the given time-integration method.

Depending on the highest energy taken into account the characteristic time scales for Galactic cosmic-ray transport can become rather short - on the order of tens of years for cosmic-ray electrons. Additionally, the time until a converged solution is reached at <GeV energies, can be hundreds of millions of years. To avoid an extremely large number of time steps both schemes start the implicit time-integration scheme with a very large time step. Time-step size is then successively reduced until a smallest time step is reached that should reflect the characteristic time scale of the problem. For instance in GALPROP, however, the particular setup of the time-integration method is not controlled by the code but rather by the configuration file. While standard parameters for this setup usually yield satisfactory results, often the users are not aware of possible problems caused by an unsuitable choice of time-integration parameters (12). We found that sometimes also standard time-integration parameters can result in an error of up to ten percent in the cosmic-ray flux.

3. Steady-state solution scheme

Since the majority of studies relating to Galactic cosmic-ray transport focus on steady-state problems, i.e. problems where all transport parameters are time independent, the solution method applied by PICARD explicitly uses:

$$\frac{\partial \psi_i}{\partial t} = 0 \tag{2}$$

No time-integration is used in PICARD. Instead the transport equation is discretised in space and momentum only. This does not allow a dimensional splitting method as used in DRAGON and GALPROP any more, but leads to a system of algebraic equations with a band-diagonal matrix. In contrast to a tri-diagonal matrix there is no direct solution method available with a numerical cost of order $\mathcal{O}(N)$, where N is the number of coupled algebraic equations. Instead PICARD uses iterative methods to solve the system of equations. This does not yield an exact solution of the system of equations, but the iterations are stopped only when the residual is smaller than a threshold value that can be imposed by the user. The residual values reached by the code are smaller than the discretisation error. In the following we will describe the specifics of the solution approach in PICARD.

There are two types of steady-state solvers available in PICARD – depending on the transport-parameter setup. In Galactic cosmic-ray transport at least three types of propagation models are distinguished. In a so-called plain-diffusion model only spatial but no momentum diffusion is taken into account. In a reacceleration model also diffusive reacceleration or Fermi II acceleration is taken into account. In this case D_{pp} is non zero, effectively resulting in a diffusion in momentum space with an additional term describing an energy gain. In both these models spatial convection is often not taken into account. Thus, the third kind of models are such, where convection is additionally considered.

Apparently, in a plain-diffusion model without spatial convection, there is no term in the transport equation that can describe a possible energy gain. Cosmic rays are accelerated at the sources, resulting in a pre-defined source spectrum, and are thereafter only subject to spatial diffusion and energy losses. Also in a plain-diffusion model with non-zero convection fulfilling the condition $\nabla \cdot \vec{v} > 0$ everywhere there are no energy gains. This implies that the cosmic-ray flux at a given energy is independent from the flux at lower energies. This motivates the first numerical scheme used in PICARD, where the transport equation is solved successively at the different energy levels starting from the highest energy. At a given energy, the flux at the next-higher grid point in energy is used as input or initial condition and then only the spatial problem needs to be solved. By this the four-dimensional transport equation can be solved as a successive number of three-dimensional problems.

Regarding the numerical implementation this procedure is similar to a time-integration scheme, only that in PICARD we integrate from high to low energy. Ignoring spatial convection the transport equation in this particular case is:

$$-\nabla \cdot \mathcal{D} \nabla \psi_i + \frac{\partial}{\partial p} \{\dot{p}\psi_i\} + \frac{1}{\tau}\psi_i = q(\vec{r}, p), \tag{3}$$

where the time scales for fragmentation and radiative decay were combined into the decay time scale τ. Here, PICARD offers two principle sub-solvers. The first is analogous to a time-forward implicit discretisation of first order. For this an upwind finite-difference discretisation is used on the energy grid. The grid points are distributed as a power-law in energy, resulting in an equidistant representation when plotted logarithmically. After the discretisation in energy, at each energy level n a three-dimensional transport equation of the form:

$$-\nabla \cdot \mathcal{D} \nabla \psi_i\big|_n - \frac{\dot{p}\psi_i|_n}{p^{n+1} - p^n} + \frac{\psi_i}{\tau}\bigg|_n = q(\vec{r}, p^n) - \frac{\dot{p}\psi_i|_{n+1}}{p^{n+1} - p^n} \tag{4}$$

is solved. For this Eq. 4 is discretised in space (currently, configuration space is discretised using finite differences on a Cartesian grid with constant grid spacings along each dimension) and then solved using an iterative method for band-diagonal matrices. In PICARD two solution methods are available for such band-diagonal matrices: a multigrid method (13) and a BiCGStab solver (14). For a model using spatially constant isotropic diffusion, the multigrid method is more efficient, but for spatially-variable diffusion possibly with additional off-diagonal elements the BiCGStab method reaches a converged solution more quickly. The modular framework of the code also allows adding of additional solvers with relative ease.

For the multigrid solver different versions are implemented. For spatially constant, isotropic diffusion a simple red-black Gauss-Seidel method is very efficient. For anisotropic and possibly spatially variable diffusion an alternating plane solver is the preferred option.

The second sub-solver is second order in momentum. This is achieved by integrating Eq. (3) from p^n to p^{n+1}, which results in the alternative form for the energy-discrete transport equation:

$$-\nabla\cdot\mathcal{D}\nabla\psi_i\big|_n - \frac{2\dot{p}\psi_i|_n}{p^{n+1}-p^n} + \left.\frac{\psi_i}{\tau}\right|_n = \tilde{q}, \tag{5}$$

with

$$\tilde{q} = \nabla\cdot\mathcal{D}\nabla\psi_i\big|_{n+1} + q(\vec{r},p^n) + q(\vec{r},p^{n+1}) - \frac{2\dot{p}\psi_i|_{n+1}}{p^{n+1}-p^n} - \left.\frac{\psi_i}{\tau}\right|_{n+1}. \tag{6}$$

Here, the momentum integrals of all but the energy-loss term in Eq. (3) are evaluated via the trapozoidal rule, leading to the desired second-order behaviour even on the non-linear momentum grid. Eq. (5) is of the same principle form as Eq. (4) and can, thus, be solved using the same numerical scheme. This shows that the second order scheme is nearly identical in numerical cost, while offering higher precision – or the option to use fewer grid points in momentum.

Whenever energy gains are possible in a given configuration the plain-diffusion solver discussed above is not applicable any more. In that case a fully four-dimensional solver is used. Currently, the most efficient option in PICARD is using a four-dimensional multigrid solver. In this case the spatial derivatives are discretised in the same form as in the other schemes. Momentum derivatives are computed via finite differences on the power-law grid. For the usual three-point stencil this leads to first order behaviour for the energy-loss terms because central differencing is not suitable here, and an upwind discretisation needs to be used. As an alternative, PICARD also allows for a second-order discretisation of the energy-loss terms, by using a wider stencil.

At each multigrid level all three-dimensional sub problems at even momentum grid points and then those at odd momentum grid points are solved. In that regard the approach is similar to the red-black Gauss-Seidel solver. For the solution of the corresponding three-dimensional sub-problems at given momenta we apply the same three-dimensional solvers as for the plain-diffusion form of the transport equation.

Here, we described the specific solvers for the transport equation of a single cosmic-ray species. Due to radioactive decay and fragmentation reactions different species can interact with each other. This interaction is realised by the loss-terms and the source term in Eq. (1). To acknowledge the coupling of the different cosmic-ray species, the transport equation for the heaviest nucleus with the highest charge is solved first. Then the transport equations for the lighter nuclei are solved successively. Whenever a nucleus is treated that can decay into one that was addressed earlier in this nuclear reaction network, and can thus contribute to its source function, the transport equation for the earlier nucleus is solved again. Thus, the part of the nuclear network between the two coupled species is iterated several times. In (12) we discuss further details of this implementation and show that usually two such iterations are sufficient to lead to a converged result.

All solvers have been verified by being applied to a range of analytically solvable problems in (4). Additionally, results produced with other codes have successfully been reproduced using PICARD, as, e.g., shown in (12). PICARD is fully MPI parallel and has been run on distributed- and shared-memory machines successfully. By construction and due to its efficiency, PICARD is particularly suited for the solution of three-dimensional propagation problems, where in the

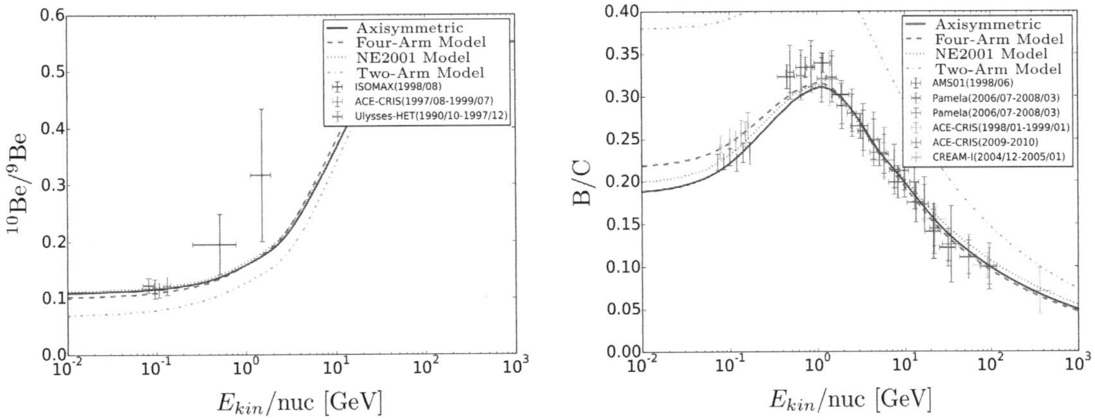

Figure 1. Secondary to primary ratios for a range of spiral-arm source-distribution models in the context of Galactic cosmic-ray transport. On the left we show results for the ^{10}Be to ^{9}Be ratio and on the right for the total Boron to total Carbon ratio.

past often axisymmetric setups were used for the Galaxy (see, e.g. 3) applying a two-dimensional setup using cylindrical coordinates.

4. Example results

Due to the optimisation for the solution of spatially three-dimensional problems, PICARD has been applied to the Galactic cosmic-ray transport problem under the assumption that the sources of cosmic rays are concentrated along the Galactic spiral arms. This has lately also been addressed by other groups (5; 8; 15; 16; 7). Such a source distribution is motivated by cosmic rays being supposedly accelerated predominantly at young astrophysical objects like, e.g., supernova remnants and pulsar-wind nebulae.

In (17) we investigated the impact of different such localised source distributions on the proton and electron flux at different energies, which are the most relevant species for the Galactic diffuse gamma-ray emission (9). In a follow-up study (12) we verified that a set of propagation parameters can be found by which also models using a localised source distribution yield a good fit to cosmic-ray fluxes observed at Earth.

Corresponding results for the resulting secondary-to-primary ratios are shown in Fig. 1 where propagation parameters as discussed in (12) are used. In this case we investigated an axisymmetric source distribution model taken from (9), a model with a logarithmic four-arm source distribution model extracted from (18, the four-arm Model), a source distribution model based on the parametrisation by (19, the NE2001-Model) and a two-arm model with a Galactic bar based on (20). The ^{10}Be/^{9}Be ratio and the B/C ratio shown in Fig. 1 are the best measured of the secondary-to-primary ratios, allowing to judge the quality of a fit – especially since they do not show a power-law distribution in energy. Another interesting constraint for the propagation physics is the measured antiproton flux, but reproducing these measurements can still be problematic (see, e.g., 21). For this, an investigation of the impact of the more recent presciption for the antiproton production cross section by (22) will be interesting. This cross section has been a recent addition to the PICARD code.

For the four-arm Model and the NE2001 model we found a satisfactory agreement with observations at Earth. For the two-arm Model, however, it was so far not possible to find a set of transport parameters that could describe the observations with convincing values for the different parameters (see also 23). It turns out, however, that the quality of the fit is mostly determined by the sources nearest to the Earth, where the two-arm model suffers from the nearest spiral arms not being taken into account in that case. However, when adding the local spiral-arm segment near Earth to the model, also the two-arm model should be able to reproduce the observations (see 12).

This shows that very different source-distribution models can lead to a good fit to observational data - despite very different spatial distributions of the cosmic-ray flux. As mentioned above, this is because the observable flux at Earth is predominantly determined by the nearest cosmic-ray sources. This also means that the diffuse gamma-ray emission should be rather different in the different models, since it originates everywhere in the Galaxy – this is currently being studied using PICARD. Apart from that, the results also show the applicability of PICARD to studies of fully three-dimensional transport problems of Galactic cosmic rays.

5. Conclusions

Here, we discussed the specific implementation of the numerical solution scheme used in the PICARD code to solve the Galactic cosmic-ray transport problem. We focussed on the implementation of the different steady-state solvers in PICARD, although there is also a time-evolution scheme for time-dependent cosmic-ray transport problems. PICARD has successfully been applied to transport problems with localised source distributions, where a few results have been discussed, here. Current developments in PICARD include the application of spatially variable diffusion using a full diffusion tensor and the implementation of a new interstellar radiation field. The latter will be particularly important in a study of the Galactic center physics. For this relevant constraints will mostly come from diffuse gamma-ray emission, because the cosmic-ray flux at Earth is mostly determined by near-by sources.

References

[1] Aharonian F, Akhperjanian A G, Bazer-Bachi A R, Beilicke M, Benbow W, Berge D, Bernlöhr K, Boisson C, Bolz O, Borrel V, Braun I, Breitling F, Brown A M, Chadwick P M, Chounet L M, Cornils R, Costamante L, Degrange B, Dickinson H J, Djannati-Ataï A, Drury L O, Dubus G, Emmanoulopoulos D, Espigat P, Feinstein F, Fontaine G, Fuchs Y, Funk S, Gallant Y A, Giebels B, Gillessen S, Glicenstein J F, Goret P, Hadjichristidis C, Hauser M, Heinzelmann G, Henri G, Hermann G, Hinton J A, Hofmann W, Holleran M, Horns D, Jacholkowska A, de Jager O C, Khélifi B, Komin N, Konopelko A, Latham I J, Le Gallou R, Lemière A, Lemoine-Goumard M, Leroy N, Lohse T, Martin J M, Martineau-Huynh O, Marcowith A, Masterson C, McComb T J L, de Naurois M, Nolan S J, Noutsos A, Orford K J, Osborne J L, Ouchrif M, Panter M, Pelletier G, Pita S, Pühlhofer G, Punch M, Raubenheimer B C, Raue M, Raux J, Rayner S M, Reimer A, Reimer O, Ripken J, Rob L, Rolland L, Rowell G, Sahakian V, Saugé L, Schlenker S, Schlickeiser R, Schuster C, Schwanke U, Siewert M, Sol H, Spangler D, Steenkamp R, Stegmann C, Tavernet J P, Terrier R, Théoret C G, Tluczykont M, Vasileiadis G, Venter C, Vincent P, Völk H J and Wagner S J 2006 *ApJ* **636** 777–797 (*Preprint* astro-ph/0510397)

[2] Parker E N 1965 *Planet. Space Sci.* **13** 9–49

[3] Strong A W, Moskalenko I V and Ptuskin V S 2007 *Annual Review of Nuclear and Particle Science* **57** 285–327 (*Preprint* arXiv:astro-ph/0701517)

[4] Kissmann R 2014 *Astroparticle Physics* **55** 37–50 (*Preprint* 1401.4035)

[5] Benyamin D, Nakar E, Piran T and Shaviv N J 2014 *ApJ* **782** 34 (*Preprint* 1308.1727)

[6] Zhang M 1999 *ApJ* **513** 409–420

[7] Kopp A, Büsching I, Potgieter M S and Strauss R D 2014 *New Astronomy* **30** 32–37

[8] Effenberger F, Fichtner H, Scherer K and Büsching I 2012 *A&A* **547** A120 (*Preprint* 1210.1423)

[9] Ackermann M, Ajello M, Atwood W B, Baldini L, Ballet J, Barbiellini G, Bastieri D, Bechtol K, Bellazzini R, Berenji B, Blandford R D, Bloom E D, Bonamente E, Borgland A W, Brandt T J, Bregeon J, Brigida M, Bruel P, Buehler R, Buson S, Caliandro G A, Cameron R A, Caraveo P A, Cavazzuti E, Cecchi C, Charles E, Chekhtman A, Chiang J, Ciprini S, Claus R, Cohen-Tanugi J, Conrad J, Cutini S, de Angelis A, de Palma F, Dermer C D, Digel S W, Silva E d C e, Drell P S, Drlica-Wagner A, Falletti L, Favuzzi C, Fegan S J, Ferrara E C, Focke W B, Fortin P, Fukazawa Y, Funk S, Fusco P, Gaggero D, Gargano F, Germani S, Giglietto N, Giordano F, Giroletti M, Glanzman T, Godfrey G, Grove J E, Guiriec S, Gustafsson M, Hadasch D, Hanabata Y, Harding A K, Hayashida M, Hays E, Horan D, Hou X, Hughes R E, Jóhannesson G, Johnson A S, Johnson R P, Kamae T, Katagiri H, Kataoka J, Knödlseder J, Kuss M, Lande J, Latronico L, Lee S H, Lemoine-Goumard M, Longo F, Loparco F, Lott B, Lovellette M N, Lubrano P, Mazziotta M N, McEnery J E, Michelson P F, Mitthumsiri W, Mizuno T, Monte C, Monzani M E, Morselli A, Moskalenko I V, Murgia S, Naumann-Godo M, Norris J P, Nuss E, Ohsugi T, Okumura A, Omodei N, Orlando E, Ormes J F, Paneque D, Panetta J H, Parent D, Pesce-Rollins M, Pierbattista M, Piron F, Pivato G, Porter T A, Rainò S, Rando R, Razzano M, Razzaque S, Reimer A, Reimer O, Sadrozinski H F W, Sgrò C, Siskind E J, Spandre G, Spinelli P, Strong A W, Suson D J, Takahashi H, Tanaka T, Thayer J G, Thayer J B, Thompson D J, Tibaldo L, Tinivella M, Torres D F, Tosti G, Troja E, Usher T L, Vandenbroucke J, Vasileiou V, Vianello G, Vitale V, Waite A P, Wang P, Winer B L, Wood K S, Wood M, Yang Z, Ziegler M and Zimmer S 2012 *ApJ* **750** 3

[10] Evoli C, Gaggero D, Grasso D and Maccione L 2008 *J. Cosmology Astropart. Phys.* **10** 18 (*Preprint* 0807.4730)

[11] Strong A W and Moskalenko I V 1998 *ApJ* **509** 212–228 (*Preprint* arXiv:astro-ph/9807150)

[12] Kissmann R, Werner M, Reimer O and Strong A W 2015 *Astroparticle Physics* **70** 39–53 (*Preprint* 1504.08249)

[13] Trottenberg U, Osterlee C and Schuller A 2001 *Multigrid* (Orlando, FL, USA: Academic Press, Inc.) ISBN 0-12-701070-X

[14] Sleijpen G L and Fokkema D R 1993 *Electronic Transactions on Numerical Analysis* **1** 2000

[15] Jóhannesson G, Moskalenko I V and Porter T 2013 *Proceedings of the 33rd International Cosmic Ray Conference*

[16] Jóhannesson G, Moskalenko I V, Orlando E, Porter T and Strong A W

[17] Werner M, Kissmann R, Strong A W and Reimer O 2015 *Astroparticle Physics* **64** 18–33 (*Preprint* 1410.5266)

[18] Steiman-Cameron T Y, Wolfire M and Hollenbach D 2010 *ApJ* **722** 1460–1473

[19] Yusifov I and Küçük I 2004 *A&A* **422** 545–553 (*Preprint* astro-ph/0405559)

[20] Dame T M and Thaddeus P 2011 *ApJ* **734** L24 (*Preprint* 1105.2523)

[21] Jóhannesson G, Ruiz de Austri R, Vincent A C, Moskalenko I V, Orlando E, Porter T A, Strong A W, Trotta R, Feroz F, Graff P and Hobson M P 2016 *ApJ* **824** 16 (*Preprint* 1602.02243)

[22] Kachelriess M, Moskalenko I V and Ostapchenko S S 2015 *ApJ* **803** 54 (*Preprint* 1502.04158)

[23] Urquhart J S, Figura C C, Moore T J T, Hoare M G, Lumsden S L, Mottram J C, Thompson M A and Oudmaijer R D 2014 *MNRAS* **437** 1791–1807 (*Preprint* 1310.4758)

IOP Conf. Series: Journal of Physics: Conf. Series **837** (2017) 012004 doi:10.1088/1742-6596/837/1/012004

Dissipation and particle energization in moderate to low beta turbulent plasma via PIC simulations

Kirit Makwana[1], Hui Li[2], Fan Guo[2] and Xiaocan Li[3]

[1] Center for mathematical Plasma Astrophysics, KU Leuven, Belgium
[2] Los Alamos National Laboratory, New Mexico, USA
[3] University of Alabama-Huntsville, Alabama, USA

E-mail: kirit.makwana@kuleuven.be

Abstract. We simulate decaying turbulence in electron-positron pair plasmas using a fully-kinetic particle-in-cell (PIC) code. We run two simulations with moderate-to-low plasma β (the ratio of thermal pressure to magnetic pressure). The energy decay rate is found to be similar in both cases. The perpendicular wave-number spectrum of magnetic energy shows a slope between $k_{\perp}^{-1.3}$ and $k_{\perp}^{-1.1}$, where the perpendicular ($\perp$) and parallel ($\|$) directions are defined with respect to the magnetic field. The particle kinetic energy distribution function shows the formation of a non-thermal feature in the case of lower plasma β, with a slope close to E^{-1}. The correlation between thin turbulent current sheets and Ohmic heating by the dot product of electric field ($\mathbf{E}$) and current density ($\mathbf{J}$) is investigated. Heating by the parallel $\mathbf{E}_{\|} \cdot \mathbf{J}_{\|}$ term dominates the perpendicular $\mathbf{E}_{\perp} \cdot \mathbf{J}_{\perp}$ term. Regions of strong $\mathbf{E}_{\|} \cdot \mathbf{J}_{\|}$ are spatially well-correlated with regions of intense current sheets, which also appear correlated with regions of strong $\mathbf{E}_{\|}$ in the low β simulation, suggesting an important role of magnetic reconnection in the dissipation of low β plasma turbulence.

1. Introduction

Plasma turbulence is ubiquitous throughout space and astrophysical systems. Turbulence in the solar corona and solar wind is an important candidate mechanism for explaining their heating [1]. Turbulence can also be a possible source of heat in intracluster medium to balance radiative cooling [2]. Alfvén waves in the solar wind [3] nonlinearly interact with each other and produce turbulence. 3D turbulence is characterized by an energy cascade from large to small scales [4]. At small enough scales energy dissipation takes place. In plasmas which are nearly collision-less, kinetic mechanisms have to operate to dissipate turbulent energy. Recent observations of solar wind show the energy cascade even extending from the ion scales to the electron scales [5]. At the same time, current sheets are also observed in the solar wind [6]. These current sheets can also dissipate energy through the mechanism of magnetic reconnection [7]. Therefore it is necessary to understand the correct energy dissipation mechanisms in such turbulence, and for this numerical simulations are needed.

We have simulated decaying plasma turbulence at moderate plasma β (ratio of thermal to magnetic pressure) using both magnetohydrodynamic (MHD) and PIC simulations previously [8]. In that study we found that the simulations compare remarkably well. The turbulent power spectra were well-resolved and for large enough simulation box sizes, the spectral slope matched very well between MHD and PIC simulations. This showed that PIC simulations can reproduce MHD results at large scales. Both simulations showed formation of thin current sheets, which are also observed in solar wind plasmas [6]. The lengths of the current sheets scaled linearly with the driving scale of turbulence, in both MHD and PIC. On the other hand, we saw important differences at small scales. We measured the dissipation fraction as a function of the volume fraction of current sheets. We found that PIC simulations showed more diffuse current sheets compared to MHD. Explicitly measuring the thickness of current sheets revealed that they are of the order of skin-depth scale in PIC whereas they are primarily set by the grid spacing in MHD. This showed that the PIC simulations were successfully capturing the whole range of physics from MHD to kinetic scales.

These simulations were performed at a moderate plasma $\beta = 0.33$. We also observed particle heating in these simulations as the turbulence decays. Recent PIC simulations of magnetic reconnection at low plasma β have revealed strong signatures of non-thermal particle heating [9]. The plasma β can vary significantly in the solar corona, from $\sim 10^{-3}$ at the base of corona to ~ 10 at the top [10]. At the same time, stochastic acceleration of particles in turbulence is considered an important acceleration mechanism in a variety of astrophysical sources, from solar flares to galaxy clusters [11]. Therefore it is important to understand particle heating and acceleration by turbulence at varying plasma β. We present results of a couple of PIC simulations of decaying turbulence with moderate to low plasma β. Sec. 2 describes the setup of the simulations, their energy dynamics and the turbulent power spectra, Sec. 3 shows the particle energization and an analysis of the energization terms along with the role of accompanying current sheets. We conclude with a discussion in Sec. 4.

2. Setup of simulations and energy spectra

We simulate decaying turbulence by setting up an ensemble of waves as the initial condition in the simulation box. These waves are Alfvénic in nature with the characteristic that the magnetic and velocity perturbations are perpendicular to the background field and are parallel/anti-parallel, with no density or pressure perturbations. The form of these perturbations is given in Eqns. (1) and (2) of Ref. [8]. The ion and electron thermal velocities are $v_{th,i} = v_{th,e} = 0.08c$. The plasma is an electron-positron pair plasma with equal ion-electron masses. Pair plasmas are believed to be emitted in astrophysical compact objects like blackholes [12] and pulsars [13]. There is also a need to explain non-thermal particle acceleration in pair-plasmas in pulsars, active galactic nuclei, and gamma-ray-bursts [14]. Therefore our simulations are relevant for such problems. The particles are initialized with a Maxwellian distribution around a mean velocity defined by the linear combination of the fluid velocity and current density. Since it is a pair-plasma, the electrons and ions both contribute equally to the current density and flow. The ratio of ion plasma frequency to ion cyclotron frequency, $\omega_{p,i}/\omega_{c,i}$, sets the ion β which is given by $\beta_i = 2(\omega_{p,i}/\omega_{c,i})^2(v_{th,i}/c)^2$. For pair plasmas, the electron plasma frequency, $\omega_{p,e}$, and electron cyclotron frequency, $\omega_{c,e}$, are the same as their ion counterparts and therefore the plasma β is simply twice this value. In Ref. [8], all the simulations had $\beta = 0.33$. In this study we have two cases, case 1 with $\beta = 0.092$ and case 2 with $\beta = 0.026$. As we vary the plasma β, we keep the ratio of injected energy (in the waves) to background magnetic field energy at a constant level of around 0.2. The simulations are performed with the state-of-the-art PIC code VPIC [15].

The dimensions of the simulation box are $(120d_i, 120d_i, 480d_i)$, where d_i is the ion skin depth, with a resolution of 576 cells in each direction. The simulations are run for at least $2000\omega_{p,i}^{-1}$ time. The setup of initial waves is balanced i.e., there is equal energy in waves moving in opposite directions. The amplitude of the Alfvén waves, $\delta b_\perp$, is initially set such that they are in a state of critical balance with $k_\perp \delta b_\perp \sim k_\parallel B_0$ [16]. In this section perpendicular ($\perp$) and parallel ($\|$) directions are defined w.r.t. the mean magnetic field. As these waves interact non-linearly, they generate turbulence, the energy cascades forward to smaller scales, there it dissipates and the

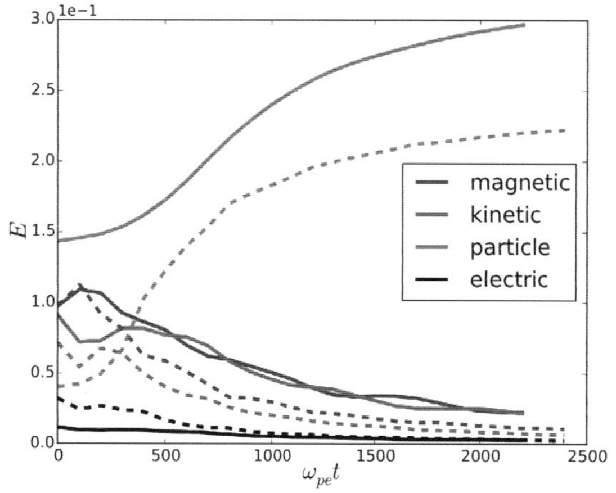

Figure 1. Time evolution of energy. Solid lines are for higher $\beta = 0.092$ case while dashed lines are for the lower $\beta = 0.026$ case. The magnetic energy only includes the perpendicular magnetic field, not the background field. The particle energy here refers to the thermal energy, without the kinetic component. The energy is normalized w.r.t. the background magnetic field energy. That is why the thermal energy in the lower β run is lower.

energy in the waves decays, going into particle energy. The total energy is very well conserved in these simulations, with the finite-grid heating leading to an increase of less than 0.2% in the total energy. We look at the decay of energy in Fig. 1. Since the plasma β changes by changing the background magnetic field, we normalize all energies by the background magnetic field energy. Thus, we can see similar levels of magnetic and kinetic energy in the waves at the initial time for both cases. Naturally the thermal energy is lower for the lower plasma β case. For $\omega_{p,e}t < 200$ we can see an increase of magnetic energy and a decrease of kinetic energy. This is consistent with our earlier simulations [8]. The kinetic energy is smaller than magnetic energy in the lower β case because the relativistic Alfvén velocity becomes smaller compared to the background magnetic field. The decay rate of kinetic and magnetic energy look very similar in both the cases, as also the heating rate of the particles. We see a similar amount of thermal energy being generated in both cases.

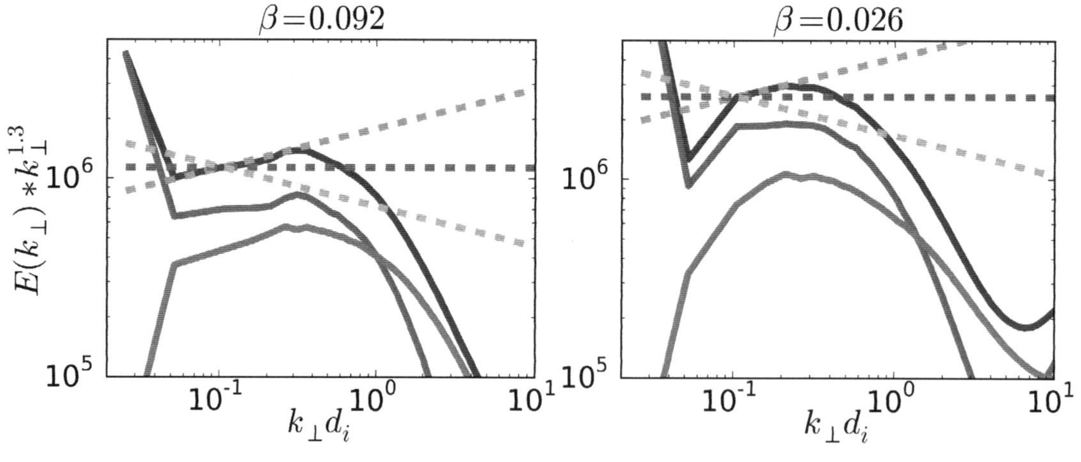

Figure 2. The perpendicular wavenumber spectrum for the two cases. Blue curve is total energy, green is magnetic energy and red is kinetic energy spectrum. The spectrum is normalized by $k_\perp^{1.3}$, so the dashed, horizontal magenta line is a slope of $k_\perp^{-1.3}$, the yellow dashed line is $k_\perp^{-1.5}$, and the blue dashed line is $k_\perp^{-1.1}$ slope.

The power-spectrum of turbulence also decays with energy which we average from $\omega_{p,e}t = 1000$ to $\omega_{p,e}t = 2200$ by compensating for the decay with time. We show this time-averaged perpendicular wave-number spectrum in Fig. 2. The cascade in parallel wave-number is weak. The spectrum is multiplied by $k_\perp^{1.3}$ in order to compare with previous results in Ref. [8]. In the previous simulations at $\beta = 0.33$ at this resolution the spectral slope was closer to $k_\perp^{-1.5}$ (Fig. 3f in Ref. [8]). Also in those simulations the spectral slope tended to $k_\perp^{-1.3}$ when the simulation box was made larger, matching the spectral slope in MHD simulations. In Fig. 2 the wavenumbers $k_\perp d_i \leq 0.16$ are populated with the initially injected waves. We do not get a significant inertial range for wavenumbers larger than $k_\perp d_i = 0.16$. For the $\beta = 0.092$ case the magnetic energy spectrum has a slope between $k_\perp^{-1.3}$ and $k_\perp^{-1.1}$, whereas for the $\beta = 0.026$ case the magnetic energy spectrum has a slope close to $k_\perp^{-1.3}$. The kinetic energy spectrum does not show a clear inertial range. The slope of the total energy spectrum in both the low and moderate β cases is between $k_\perp^{-1.3}$ and $k_\perp^{-1.1}$, but the limited inertial range makes it difficult to identify the spectral slope precisely. We see that the spectrum turns over before reaching the skin-depth scale, at $k_\perp d_i \sim 0.4$. The magnetic energy spectrum dips below the kinetic energy spectrum at $k_\perp d_i \sim 1.0$.

3. Particle energization and E · J analysis

The relativistic particle kinetic energy is given by $(\gamma - 1)$, where γ is the Lorentz factor of the particle and the particle mass and speed of light are normalized to unity. We look at the time evolution of the distribution of particle kinetic energy in these simulations in Fig. 3. The initial distribution is Maxwellian, and the particles are energized to higher energy as the turbulence heats them. By $\omega_{p,e}t = 1000$ the distribution has become steady. For the $\beta = 0.092$ case we see that the distribution function extends to higher energies as the particles are heated, while keeping a Maxwellian shape. For $\beta = 0.026$ case, the distribution develops a flat, non-thermal feature which extends up to $(\gamma - 1) \approx 0.3$, i.e. energies of up to 150 keV for electrons and positrons. The slope of this feature is close to $(\gamma - 1)^{-1}$. This feature begins to appear at around $\omega_{p,e}t = 200$ and reaches a steady level by around $\omega_{p,e}t = 1000$.

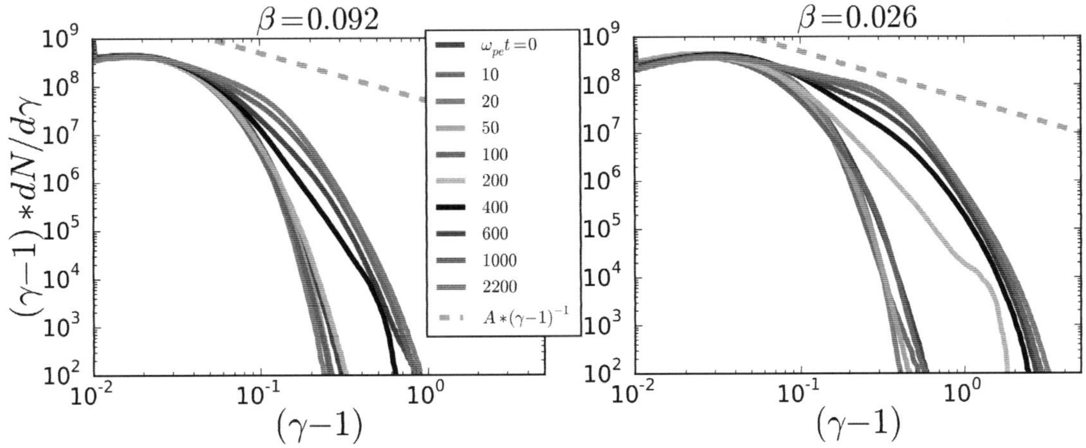

Figure 3. Distribution function of relativistic kinetic energy of particles at various times for the 2 cases. The time is normalized in units of $\omega_{p,e}^{-1}$. The distribution spreads out to higher energies as time increases and the legend applies to both the plots. The dashed line is at a slope of $(\gamma - 1)^{-1}$.

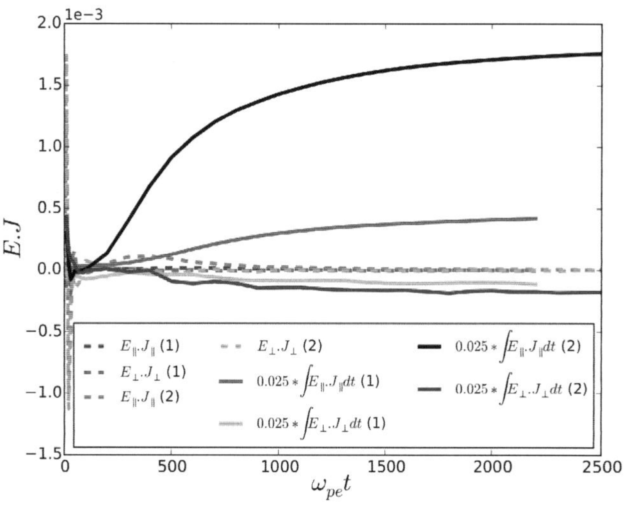

Figure 4. Time evolution of various $\mathbf{E}\cdot\mathbf{J}$ quantities. Parentheses (1) refers to the $\beta = 0.092$ case, while parentheses (2) refers to $\beta = 0.026$ case. The time integrals $\int \mathbf{E}_{\|}\cdot\mathbf{J}_{\|}dt$ and $\int \mathbf{E}_{\perp}\cdot\mathbf{J}_{\perp}dt$ are carried out over arithmetic average of $\mathbf{E}_{\|}\cdot\mathbf{J}_{\|}$ and $\mathbf{E}_{\perp}\cdot\mathbf{J}_{\perp}$ over the entire simulation box. The units of these integrals are arbitrary, multiplied by 0.025 to place them on the same scale. Strong fluctuations are seen in the $\mathbf{E}\cdot\mathbf{J}$ terms initially for $\omega_{p,e}t < 100$, which however do not contribute significantly to the time integral.

We next look at the term $\mathbf{E}\cdot\mathbf{J}$ in Fig. 4 to find the cause of energy dissipation and the corresponding particle heating. We split it up into components parallel to the local magnetic field ($\mathbf{E}_{\|}\cdot\mathbf{J}_{\|}$) and perpendicular to the field ($\mathbf{E}_{\perp}\cdot\mathbf{J}_{\perp}$). In this section perpendicular ($\perp$) and parallel ($\|$) components are defined w.r.t. the local magnetic field. The time evolution of these quantities averaged over the simulation box are shown in Fig. 4. A positive $\mathbf{E}\cdot\mathbf{J}$ value signifies energy dissipation. We find that initially, within the first $\omega_{p,e}t = 100$ time, the $\mathbf{E}\cdot\mathbf{J}$ term is quite noisy, developing large fluctuations. However after $\omega_{p,e}t = 100$ the term settles down and behaves smoothly. We believe the initial perturbations are due to the PIC simulation adjusting to the initial conditions which are not an exact eigen-solution to the PIC dispersion relation. We also plot the time integration of these terms, $\int \mathbf{E}_{\|}\cdot\mathbf{J}_{\|}dt$ and $\int \mathbf{E}_{\perp}\cdot\mathbf{J}_{\perp}dt$ in Fig. 4. It does not show any strong growth during the first $\omega_{p,e}t = 100$ time. Also, if we look at the energy decay in these simulations in Fig. 1, no significant amount of energy is dissipated during this initial time. All this indicates that these initial fluctuations are not affecting energy dissipation in a significant way. We see that $\mathbf{E}_{\|}\cdot\mathbf{J}_{\|}$ is greater than $\mathbf{E}_{\perp}\cdot\mathbf{J}_{\perp}$ by a factor of 3-4 in both cases. The $\mathbf{E}\cdot\mathbf{J}$ terms are much stronger in the lower beta case. The time integrated plot shows this clearly, where $\int \mathbf{E}_{\|}\cdot\mathbf{J}_{\|}dt$ is stronger in the lower beta case. Furthermore, the $\mathbf{E}_{\|}\cdot\mathbf{J}_{\|}$ is positive, whereas $\mathbf{E}_{\perp}\cdot\mathbf{J}_{\perp}$ is negative. This analysis shows that $\mathbf{E}_{\|}\cdot\mathbf{J}_{\|}$ is the important term causing dissipation and heating of particles.

Next we look at the spatial profiles of these heating terms. In Fig. 5 we show the magnitude of current density $|\mathbf{J}|$ and $\mathbf{E}_{\|}\cdot\mathbf{J}_{\|}$ in a cross section of the simulation box perpendicular to the mean field. In case of $\beta = 0.092$ the snapshot is taken at $\omega_{p,e}t = 500$, which corresponds to $0.36\tau_A$, where τ_A is the Alfven crossing time. From Fig. 1 we see strong particle heating is taking place at this time for case 1. In case of $\beta = 0.026$ the snapshot is taken at $\omega_{p,e}t = 300$, which also corresponds to $0.36\tau_A$. Again we see in Fig. 1 strong particle heating at this time for case 2. As a result, we also see similar structure of current sheets in both cases. The current sheets are very thin, of the order of skin depth d_i, in accordance with previous results [8]. The color scale shows that the intensity of current sheets is higher for the lower β case. The $\mathbf{E}_{\|}\cdot\mathbf{J}_{\|}$ term is extremely localized in narrow regions in Fig. 5 for both the cases. These regions are well-correlated with the regions of strong current sheets. This indicates that the dissipation is extremely localized and occurring mostly in the current sheet regions. The dissipation term also looks stronger in the lower β case.

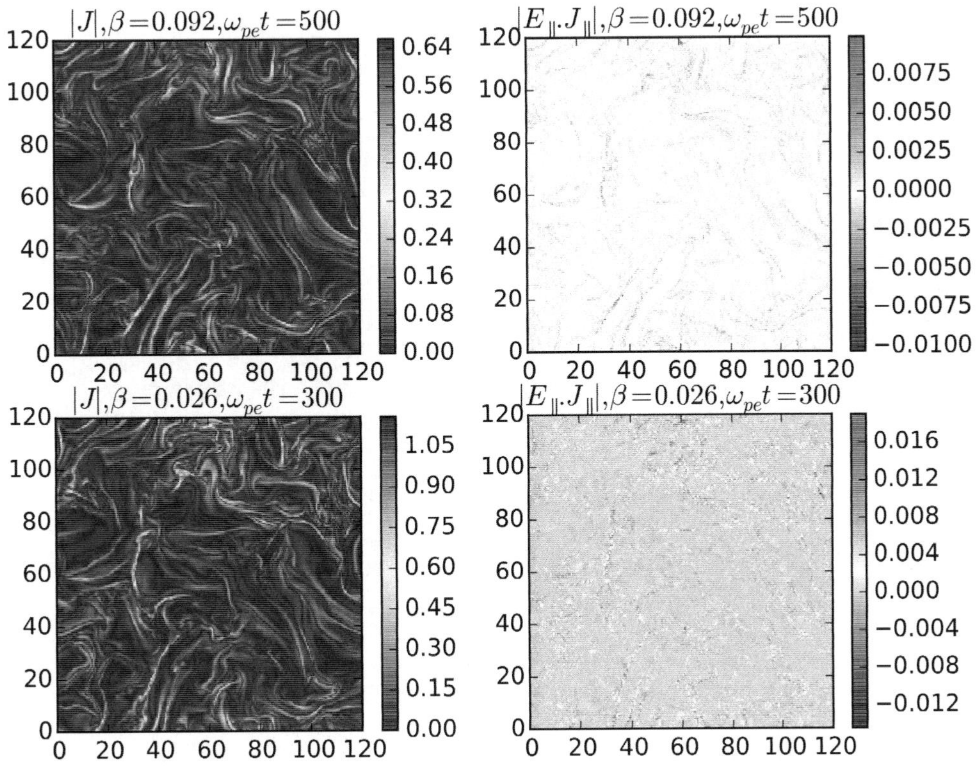

Figure 5. Cross-section color plots of $|\mathbf{J}|$ and $\mathbf{E}_{\|} \cdot \mathbf{J}_{\|}$ for the two cases at the specified times. The cross-section is taken perpendicular to the background magnetic field. The box dimensions are in units of skin-depth.

The parallel electric field $\mathbf{E}_{\|}$ is a necessary condition for magnetic reconnection [17], and often taken as an indicator of reconnection. Is the dissipation we observe in current sheets related to reconnection? We show the $\mathbf{E}_{\|} \cdot \mathbf{J}_{\|}$, $\mathbf{J}_{\|}$, and $\mathbf{E}_{\|}$ cross sections in a plane perpendicular to the background magnetic field in Fig. 6 respectively for the lower β case at $\omega_{p,e}t = 300$. The regions of high $\mathbf{E}_{\|} \cdot \mathbf{J}_{\|}$ are shown by putting circles around them. In Fig. 5 we saw that regions of high dissipation were correlated with regions of high total current density $|\mathbf{J}|$. Here we see that the current density is dominated by the parallel current density $J_{\|}$, and hence regions of strong dissipation are also correlated with strong parallel current. In the third plot of Fig. 6, we see that these circles of strong $\mathbf{E}_{\|} \cdot \mathbf{J}_{\|}$ are also regions of strong $\mathbf{E}_{\|}$. This indicates that reconnection in turbulent current sheets may be the cause of dissipation in this low-β turbulence simulation. It should be noted that there are several regions with high $\mathbf{E}_{\|}$, but they are not corresponding to regions of high current density, thereby not giving strong dissipation.

4. Discussion

We have extended our previous work of kinetic simulations of decaying plasma turbulence at moderate plasma β to low plasma β. We find that the decay rate of turbulence remains similar for this range of plasma β. Consequently, the heating rate of particles is also similar. The perpendicular wavenumber spectrum of magnetic energy shows a slope between $k_{\perp}^{-1.3}$ and $k_{\perp}^{-1.1}$, but the inertial range is very limited in these simulations. The kinetic energy spectrum does not show a clear inertial range and hence it is difficult to accurately identify the spectral slope of

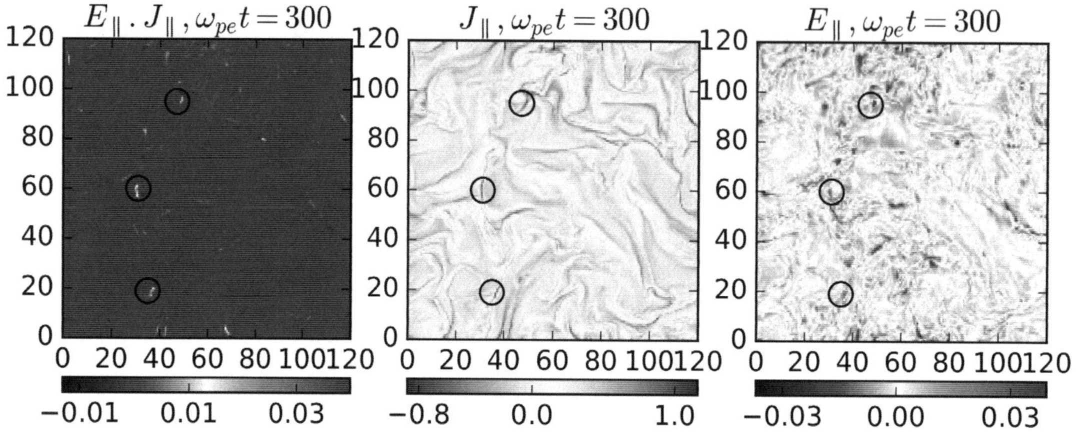

Figure 6. $E_{\|}.J_{\|}$, $J_{\|}$, and $E_{\|}$ (from left to right) in a plane perpendicular to the background magnetic field for the $\beta = 0.026$ case at $\omega_{p,e}t = 300$. This plane is slightly shifted from the plane selected for Fig. 5. The circles are highlighting three regions of high dissipation ($E_{\|}.J_{\|}$).

the total energy in these low β simulations. At low plasma β we also observe strong fluctuations in the electric fields at the beginning of the simulations. At low β the Alfven velocity becomes relativistic, as do the currents and velocities. Also, the waveforms we use to setup waves in the initial conditions are derived from MHD, which are not eigen-functions of the kinetic system. We intend to remove these irregularities from our simulations by utilizing proper wave-dispersion relations in future work.

Nevertheless, it appears that these strong electric fields die down quickly and do not affect the dissipation process significantly. In both moderate and low β simulations, the parallel dissipation term, $\mathbf{E}_{\|} \cdot \mathbf{J}_{\|}$ is stronger than the perpendicular term, $\mathbf{E}_{\perp} \cdot \mathbf{J}_{\perp}$. Regions of strong dissipation are spatially well-correlated with regions of high current density sheets. In the low β case they also correlate well with regions of strong parallel electric field, suggesting that reconnection in turbulent current sheets is playing a major role in this energy dissipation. Also in the low β case, the particle heating results in the development of a non-thermal feature in the particle kinetic energy distribution function, with a slope close to $(\gamma - 1)^{-1}$. However, recent simulations of reconnection also show non-thermal features which can be explained by two Maxwellians at different temperatures, the lower temperature Maxwellian for non-reconnected particles and a higher temperature Maxwellian for particles that have passed through the reconnection region. This aspect will need to be analyzed further in future work.

Acknowledgments

This work was supported in part by the National Science Foundation sponsored Center for Magnetic Self Organization (CMSO) at the University of Chicago, and by the Interuniversity Attraction Poles Programme (initiated by the Belgian Science Policy Office, IAP P7/08 CHARM) and by the KU Leuven GOA/2015-014. This research is part of the Blue Waters sustained-petascale computing project, which is supported by the National Science Foundation (awards OCI-0725070 and ACI-1238993) and the state of Illinois. Blue Waters is a joint effort of the University of Illinois at Urbana-Champaign and its National Center for Supercomputing Applications. This research used resources of the National Energy Research Scientific Computing Center, a DOE Office of Science User Facility supported by the Office of Science of the U.S. Department of Energy under Contract No. DE-AC02-05CH11231.

References

[1] Leamon R J, Matthaeus W H, Smith C W, Zank G P, Mullan D J and Oughton S 2000 *Astrophys. J.* **537** 1054

[2] Zhuravleva I, Churazov E, Schekochihin A A, Allen S W, Arevalo P, Fabian A C, Forman W R, Sanders J S, Simionescu A, Sunyaev R, Vikhlinin A and Werner N 2014 *Nature* **515** 85

[3] Belcher J W, Davis L Jr, *J. Geophys. Res.* **76** 3534

[4] Kolmogorov A N 1941 *Dokl. Akad. Nauk SSSR* **30** 9

[5] Alexandrova O, Saur J, Lacombe C, Mangeney A, Mitchell J, Schwartz S J and Robert P 2009 *Phys. Rev. Lett.* **103** 165003

[6] Li G, 2008 *The Astrophysical Journal Letters* **672** L65

[7] Karimabadi H, Roytershteyn V, Wan M, Matthaeus W H, Daughton W, Wu P, Shay M, Loring B, Borovsky J, Leonardis E, Chapman S C and Nakamura T K M 2013 *Phys. Plasmas* **20** 012303

[8] Makwana K D, Zhdankin V, Li H, Daughton W and Cattaneo F 2015 *Physics of Plasmas* **22** 042902

[9] Li X, Guo F, Li H and Li G 2015 *The Astrophysical Journal Letters* **811** L24

[10] Aschwanden M 2004 *Physics of the Solar Corona An Introduction* (Chichester, UK: Springer) p 29

[11] Petrosian V 2012 *Space Science Reviews* **173** 535

[12] Begelman M C, Blandford D and Rees J 1984 *Rev. Mod. Phys.* **56** 255

[13] Goldreich P and Julian W H 1969 *Astrophys. J.* **157** 869

[14] Hoshino M and Lyubarsky Y 2012 *Space Science Reviews* **173** 521

[15] Bowers K J, Albright B J, Yin L, Bergen B and Kwan T J T 2008 *Phys. Plasmas* **15** 055703

[16] Goldreich P and Sridhar S 1995 *Astrophys. J.* **438** 763

[17] Schindler K and Hesse M 1988 *J. Geophys. Res.* **93** 5547

Cosmic Chandlery with Thermonuclear Supernovae

A C Calder[1,2], B K Krueger[3], A P Jackson[1], D E Willcox[1], B J Miles[4] and D M Townsley[4]

[1] Department of Physics and Astronomy, Stony Brook University, Stony Brook, NY 11794-3800, USA
[2] Institute for Advanced Computational Science, Stony Brook University, Stony Brook, NY 11794-5250, USA
[3] XCP-2, Los Alamos National Laboratory, Los Alamos, NM 87545, USA
[4] Department of Physics and Astronomy, University of Alabama, Tuscaloosa, AL 35487-0324, USA

E-mail: alan.calder@stonybrook.edu

Abstract. Thermonuclear (Type Ia) supernovae are bright stellar explosions, the light curves of which can be calibrated to allow for use as "standard candles" for measuring cosmological distances. Contemporary research investigates how the brightness of an event may be influenced by properties of the progenitor system that follow from properties of the host galaxy such as composition and age. The goals are to better understand systematic effects and to assess the intrinsic scatter in the brightness, thereby reducing uncertainties in cosmological studies. We present the results from ensembles of simulations in the single-degenerate paradigm addressing the influence of age and metallicity on the brightness of an event and compare our results to observed variations of brightness that correlate with properties of the host galaxy. We also present results from "hybrid" progenitor models that incorporate recent advances in stellar evolution.

1. Introduction

Supernovae are bright stellar explosions that signal the violent deaths of stars. These events synthesize heavy elements, may be a site of r-process nucleosynthesis, and create neutron stars and black holes, which are the building blocks of other interesting astrophysical systems such as x-ray binaries. While these events may be divided into multiple classifications observationally, the impetus for the majority of events is understood to be either the release of gravitational binding energy from the collapsing core of a massive star or the release of nuclear binding energy when a thermonuclear runaway incinerates one or more compact stars. Accordingly, these two mechanisms are referred to as "core collapse" and "thermonuclear," respectively, and together these produce and disseminate the majority of heavy elements found in the galaxy and thereby drive galactic chemical evolution. Core collapse supernovae create neutron stars and black holes and may be a site of r-process nucleosynthesis, while thermonuclear supernovae completely disrupt the progenitor, leaving only an unbound remnant. Thermonuclear supernovae are particularly useful as distance indicators for cosmological studies, making understanding these events and properties such as their intrinsic scatter important to fields well beyond stellar astrophysics. This manuscript reports on studies of systematic effects on the brightness of thermonuclear supernovae that follow from the evolutionary history of the progenitor star and

IOP Conf. Series: Journal of Physics: Conf. Series **837** (2017) 012005 doi:10.1088/1742-6596/837/1/012005

host galaxy, with the goal of better constraining uncertainty in the intrinsic scatter of these events.

1.1. Thermonuclear Supernovae

Thermonuclear supernovae are understood to follow from a thermonuclear runaway involving roughly one solar mass of stellar material under degenerate conditions. Discerning the setting(s) of these events, however, is proving to be difficult. Thermonuclear supernovae are unique among supernovae in that they are also very important for cosmological studies because their observational properties (e.g. the light curve) allow calibration of events for use as "standard candles," i.e. objects of known brightness that may be used as distance indicators. Using thermonuclear supernovae in this capacity resulted in the discovery of the accelerating expansion of the Universe and thus the mysterious dark energy driving the acceleration [1, 2, 3].

Observationally, thermonuclear supernovae are classified as "Type Ia" following the classification scheme of Minkowski [4]. The classification is based on observing strong Si lines and a lack of H in the spectrum of an event. Theoretically, these events are understood to follow from the thermonuclear incineration of $\simeq 1\ M_\odot$ of a mixture principally composed of C and O under degenerate conditions. The burning converts much of the material to iron-group elements (IGEs) including $\sim 0.6\ M_\odot$ of radioactive ^{56}Ni, the decay of which powers the light curve [5, 6, 7, 8]. The ability to standardize the light curves is understood to follow from the fact that the source powering the light curve, radioactive ^{56}Ni, is also the principal source of opacity [9]. The result is that the B-band magnitude of the light curves of brighter events decline more slowly from peak than those of dimmer events. This "brighter is broader" result is known as the Phillips relation [10], and it allows the calibration of light curves via a single-parameter "stretch" function [11].

Largely motivated by cosmological studies, many contemporary campaigns are observing Type Ia supernovae. Observations suggest that there may be systematic effects on the peak magnitude of the light curve even after calibration [12], which is a significant source of uncertainty in cosmological studies utilizing these as distance indicators. Spectral diversity raises the possibility of two populations, which may follow from formation channel or from properties of the host galaxy and/or the local environment within the host galaxy. Arguments are made for two populations [13, 14, 15] or for correlation principally with age measured as the delay time from early star formation in a galaxy [16, 11, 17, 18].

One key measure in these studies is the amount of metals, elements heavier than He, that were synthesized by previous generations of stars. The presence of metals influences the evolution of stars and stellar explosions by increasing the number of pathways for nuclear burning [19]. The metallicity of a particular galaxy increases with age, and within a particular galaxy metallicity changes with location, generally increasing toward the densely populated interior. In addition, observations of specific events also can address the issue of progenitors, with some bright observations suggesting more radioactive ^{56}Ni than could be produced by degenerate C and O in a single star [20, 21, 22, 23].

Although many of these events have been observed and these events serve as the premier cosmological distance indicators, a robust theoretical understanding remains elusive. It is widely accepted that compact stars known as white dwarfs (WDs) composed principally of C and O provide the nuclear fuel, but the progenitor systems have not been conclusively identified and questions about the mechanism of the explosion remain. The three most widely accepted proposed progenitor systems are: a single near-Chandrasekhar-mass WD that has gained mass from a companion (the "single degenerate" scenario), the merger of two WDs, and the sub-Chandrasekhar-mass double-detonation scenario in which a detonation in an accreted layer triggers a detonation in the core of the WD [24, and references therein]. For studies of systematic effects, we have adopted the single-degenerate model.

1.2. The Single-Degenerate Explosion Paradigm

The single-degenerate scenario encompasses several explosion mechanisms invoking both deflagrations (subsonic burning fronts) and detonations (supersonic burning fronts). The key to a successful explosion, the production of an unbound remnant with a stratified composition structure similar to observed remnants [25], is the density at which the material of the WD burns. Some of the earliest work addressing the single-degenerate progenitor explored pure detonations and found that the result is the vast majority of the WD material burns to IGEs because of the relatively high density of the WD [26]. In this case, there is no composition stratification in the remnant, and, accordingly, pure detonations have been ruled out as a viable explosion mechanism for some time.

Pure deflagrations have also been explored as the explosion mechanism. In this case, the star reacts to the subsonic burning and expands, which lowers the density of the material prior to it being consumed. The density in the outer regions of the star gets low enough to quench the burning, and the result is an incompletely-burned, bound object that does not release enough energy to explain the majority of observed Type Ia supernovae [27]. Pure deflagrations may explain some unusual events, however [28].

Because the two extremes of pure deflagrations or pure detonations cannot robustly reproduce events with characteristics consistent with the observations, researchers have investigated models evoking a combination of the two. These models may be broadly described as delayed detonation, and a variety of scenarios have been proposed. The classic delayed detonation model was introduced by Khokhlov [29, 30, 31], and variations include pulsational detonations [32, 29, 33, 34, 35], gravitationally confined detonation [36, 37, 38], and deflagration-to-detonation transitions (DDTs) [39, 29, 40, 41, 42, 43]. We adopt a variation of the latter case, a deflagration-to-detonation transition that occurs at the top of a rising plume of hot, burned material when it reaches a threshold density.

The physics by which a DDT may occur are not completely understood, but ideas have been proposed. One idea is that at low densities the flame enters a regime of distributed burning in which it is no longer well-defined but the net burning rate is high enough that it is effectively supersonic [40]. Similarly, others argue that when the flame reaches a certain fractal dimension, it is effectively supersonic [44]. The Zel'dovich mechanism posits that composition and temperature gradients may "prepare" the fuel in just the right way that it will detonate [45, 46, 47]. Our approach is to assume that when the top of a rising, Rayleigh-Taylor unstable plume reaches a particular density, conditions are favorable and the transition occurs [48].

2. Simulation Instrument

The simulations of thermonuclear supernovae we describe here were performed with a modified version of the Flash code, developed at the University of Chicago.[1] Flash is a parallel, adaptive mesh, multi-physics simulation code developed first for nuclear astrophysics applications and subsequently for high-energy-density applications [49, 50, 51, 52, 53]. The modifications to the Flash code for supernova simulations comprise routines to describe thermonuclear burning during both the deflagration and detonation phases, as well as routines to describe the evolution of the dynamic ash.

The significant disparity between radius of a white dwarf ($\sim 10^9$ cm) and the width of laminar nuclear flame at high densities (< 1 cm) requires the use of a model during the deflagration phase because even with many orders of refinement on an adaptive mesh, a macroscopic simulation of the event cannot resolve the actual flame front. Our model propagates an artificially broadened flame with an advection-diffusion-reaction (ADR) scheme [54, 55] via evolution of a reaction

[1] The source code used for these studies is available as a package compatible with the current Flash code from `http://astronomy.ua.edu/townsley/code`.

progress variable ϕ, where $\phi = 0$ indicates unburned fuel and $\phi = 1$ indicates burned ash. The advection-diffusion-reaction equation is

$$\partial_t \phi + \vec{u} \cdot \nabla \phi = \kappa \nabla^2 \phi + \frac{1}{\tau} R(\phi), \quad (1)$$

where $\vec{u}$ is the velocity of the fluid, κ is the diffusion coefficient, τ is the reaction timescale, and $R(\phi)$ is a non-dimensional function. The diffusion and reaction parameters κ, τ, and $R(\phi)$ are tuned to propagate the reaction front at a prescribed speed. We use a modified version of the KPP reaction rate discussed by [55], the "sharpened KPP reaction," with $R \propto (\phi - \epsilon)(1 - \phi + \epsilon)$, where $\epsilon \simeq 10^{-3}$. This scheme has been shown to be acoustically quiet, stable, and to give a unique flame speed [48]. For the input flame speed, we use tabulated flame speeds from direct numerical simulations of thermonuclear burning [56, 57] and boost these to account for enhancements to the burning from unresolved buoyancy and background turbulence [54, 58, 48, 47].

The ADR scheme describes a model flame, but more is needed to adequately describe the burning in the white dwarf. The burning of C and O under degenerate conditions can be described by three stages. First, C is consumed producing reaction products close to Mg. Then O is consumed along with the ashes of C burning, which produces a mixture of silicon group and light elements that is in a statistical quasi-equilibrium [59, 60, 61]. Finally the silicon-group nuclei are converted to IGEs, reaching full nuclear statistical equilibrium (NSE). Each of these stages is described with separate progress variables and separate relaxation times derived from full nuclear network calculations [62, 63].

In both the quasi-equilibrium and full statistical equilibrium, the creation of light elements via photodisintegration balances the creation of heavy elements via fusion, thus maintaining the equilibrium. Energy can continue to be released as the relative abundances change due to hydrodynamic motion, primarily occurring in rising plumes of hot burned material, which changes the local conditions such as density and temperature and thereby adjusts the equilibrium. Electron capture also influences the evolution in three ways: by shifting the binding energy of the material and thereby changing the temperature due to released energy, by changing the Fermi energy and thus the pressure, and by emitting neutrinos that escape and remove energy. Also, electron captures neutronize the material, which produces more neutron rich iron-group material at the expense of ^{56}Ni. As we will see below, this process has a significant impact on the ^{56}Ni yield and thus the brightness of an event. Our flame capturing scheme incorporates these effects via tabulated results from NSE calculations, the details of which may be found in [64]. With all of this included physics, our scheme is able to describe dynamic evolution of the ash in addition to the stages of C-O burning.

In addition to modeling the burning during the deflagration phase, our burning scheme also describes detonations [65, 66] via progress variables. In this case, we use thermally-activated burning with the actual temperature-dependent rate for C consumption to allow a propagating shock to trigger burning, i.e. to propagate a detonation front. In our scheme, the propagating detonation is able to describe the same stages of C burning as the deflagration case, including the relaxation into NSE [63, and references therein].

3. Systematic Effects on the Brightness of an Event

We investigate systematic effects on the brightness of thermonuclear supernovae in the single-degenerate, DDT paradigm (described above). Figure 1 illustrates the progression of thermonuclear burning during a simulation in the DDT paradigm with a massive C-O progenitor WD. The left panel shows the flame during the early deflagration stage with the obvious development of fluid instabilities. The middle panel shows the flame as the first rising bubble reaches the DDT threshold density (marked by the contour). The third panel shows the propagation of two detonations. In all panels, the color scheme shows the composition.

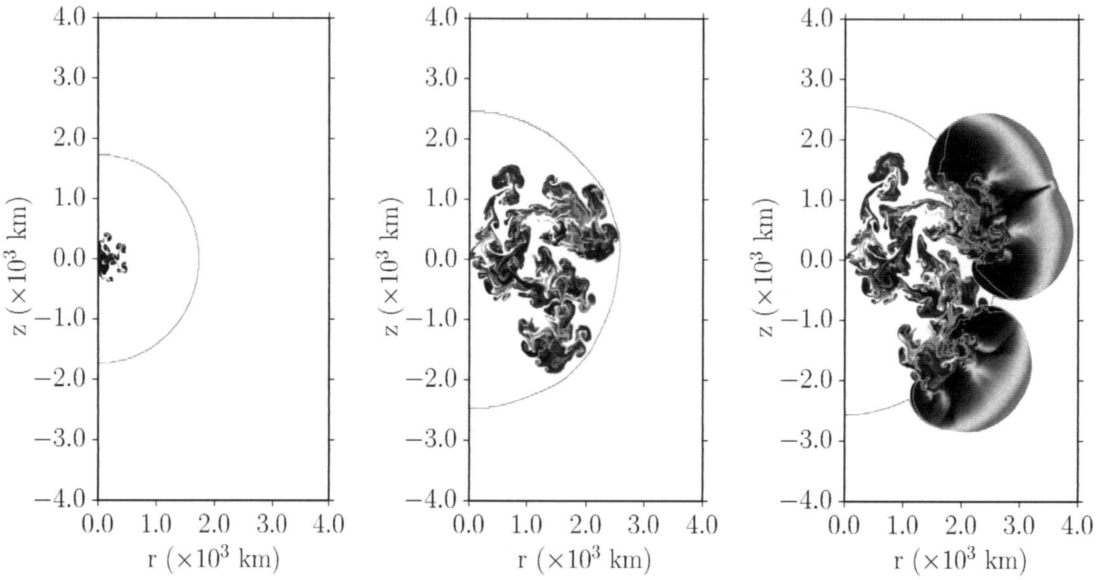

Figure 1. Deflagration to detonation burning progress shown for a C-O WD with $\rho_{\rm DDT} = 10^{7.2}$ g/cm^3 (blue contour). Unburned ^{12}C & ^{16}O fuel is shown in white, and ash from ^{12}C-burning is shown in red. Matter in a quasi-nuclear statistical equilibrium state (primarily intermediate-mass silicon-group elements) is represented in green. Finally, matter in nuclear statistical equilibrium (IGEs and α-particles) is shown in black. From left to right, the burning is shown at 0.75 s, 1.74 s, and 1.84 s after the deflagration is initiated.

We note that an important part of our methodology is to perform suites of simulations in order to investigate statistically-significant trends. In [66] we presented a theoretical framework for these investigations utilizing multiple realizations from randomized initial conditions. Our results have shown that the problem of thermonuclear supernovae in the single-degenerate DDT paradigm is deeply nonlinear, particularly during the deflagration phase, which involves competition between the growth of fluid instabilities and evolution of the flame. Accordingly, these models exhibit a significant degree of variability, necessitating a statistical approach [67]. Finally, we note that all of the supernova explosions presented here are from two-dimensional simulations. We found the large number of simulations needed in ensembles necessitated the use of two-dimensional models to be computationally tractable.

3.1. Effect of Composition

The simplest description of burning in a C-O WD is the alpha chain, a series of alpha capture reactions of the form $X\,(\alpha,\gamma)\,Y$ going through symmetric nuclei ($n = p$) from ^{12}C and ^{16}O all the way to ^{56}Ni. The presence of metals has been known for some time to boost burning rates in WDs by offering additional burning channels beyond just these basic reactions. In addition, metals are neutron rich, and the neutron excess tends to drive the burning away from symmetric nuclei. The result is that the presence of metals affects the brightness of an event in two ways- faster flame speeds lead to different dynamics during the explosion and neutron excess suppresses ^{56}Ni production.

Metal enrichment of the progenitor WD follows from two sources. First, metals present in the massive star that evolved to become the WD will be present in the progenitor WD. Second, convective burning or "simmering" as the progenitor WD approaches the Chandrasekhar limit synthesizes metals. In simulations, we use ^{22}Ne as a proxy for neutron rich metals in the WD progenitor originating from both sources.

Based on essentially a counting argument, [68] found that the presence of ^{22}Ne directly modifies the ^{56}Ni abundance during thermonuclear burning of a WD. Our models allow the investigation of the (nonlinear) effects of metallicity on hydrodynamic models. As noted above, metallicity boosts the burning rate by providing an increased number of reaction channels. Accordingly, the flame speeds in explosion models are increased. The increased flame speed changes the dynamics of the deflagration phase and also will change the DDT density, which follows from a competition between the development of fluid instabilities at the flame front and fire polishing. In [66], we explored the role of ^{22}Ne, incorporating its effect on progenitor structure, laminar flame speed, and energy release into simulations. We found that these influence the duration of the deflagration phase, which in turn influences the amount of expansion prior to the detonation that incinerates the star. The amount of expansion determines the overall density structure, and in particular the amount of mass at densities high enough to burn all the way to IGEs. Thus we found metallicity does influence the yield of ^{56}Ni, but we found that these effects were relatively weak. In [69], we included the effect of the changing flame speed on the DDT density. In this case, we found a more drastic effect on the duration of the deflagration phase and, accordingly, on the yield of ^{56}Ni. Figure 2 shows results from these two studies along with observational results for comparison.

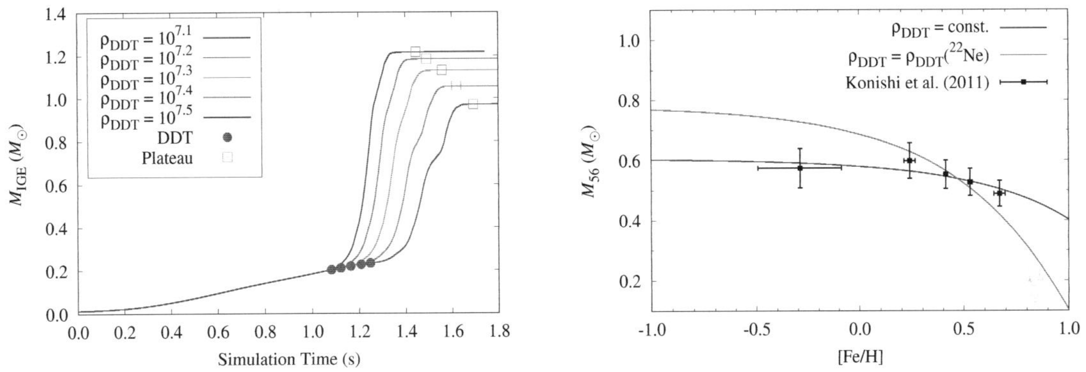

Figure 2. Left panel: Plot of IGE yield vs. time from simulation of the same realization with the DDT density changing to account for increased flame speeds due to increased metallicity. Right panel: Plot of produced mass of ^{56}Ni vs. metallicity comparing simulations to the observational results of [70] (black points). Both panels adapted from [69].

3.2. Effect of Central Density

In [71] and subsequently in [72] we investigated the role of central density on the yield of ^{56}Ni and thus the brightness. Our study constructed models with varying central densities and simulated explosions from these. We found that the overall production of IGEs is statistically independent of progenitor central density, but the mass of stable IGEs is tightly correlated with central density, with a decrease in the production of ^{56}Ni at higher central densities. Thus our results indicate that progenitors with higher central densities produce dimmer events. Our understanding of the source of this trend is the increased neutronization that occurs due to the faster rates of weak reactions at higher densities. By applying the relationship of Lesaffre, et al. [73], which relates the duration of WD cooling prior to the onset of accretion to the central density of the progenitor at ignition, we were able to obtain a relationship between the central density and the cooling time, a significant fraction of the age of the progenitor. With this relationship, we then had relationships between the cooling time and the masses of IGEs and ^{56}Ni.

Results from the central density study are shown in Figure 3. The left panel shows the mass of ^{56}Ni produced in the explosion from progenitors at 5 central densities for 30 realizations each. The right panel shows relationship between age and stretch, a measure of the brightness when considering the Phillips relation [11]. Also shown are the observational results of Neill, et al. [17] for comparison. An obvious result to note is that the trends agree, but our stretches or yields are consistently higher. While evidence suggests that explosions in the single-degenerate paradigm tend to be bright [43], we attribute the difference as primarily due to our choice of DDT density for the suite of simulations.

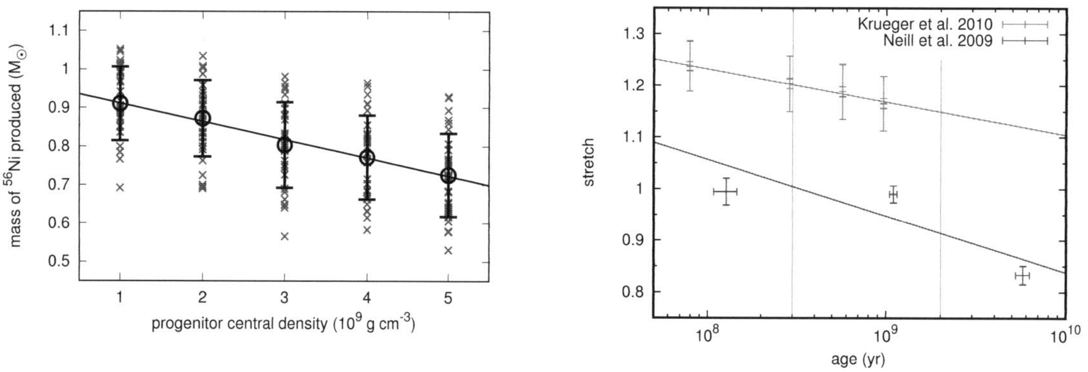

Figure 3. Left panel: Mass of ^{56}Ni as a function of progenitor central density for 30 simulations at each central density. The black lines are the best-fit trend lines with averages and standard deviations marked by the circles and vertical error bars. Right panel: Plots of stretch vs. age with the standard deviation, the standard error of the mean, and a best-fit trend line. In blue are results from Figure 5 of [17] along with a best-fit trend line, and the vertical gray lines mark the cuts between bins. The overall offset to larger stretch in our simulations follows primarily from the choice of DDT density. Adapted from [71].

3.3. Explosions from "Hybrid" Progenitors

Recent work in stellar evolution posits the existence of "hybrid" C-O-Ne white dwarfs consisting of a C core surrounded by a mixture of O and Ne. The structure of these white dwarfs follows from convective boundary mixing quenching a C-flame before it can consume all of the C in the late stages of a super asymptotic giant branch star's life [74]. C burning produces ^{20}Ne that, because of the convective boundary, remains at a larger radius than the unburned C. The result is a hybrid C-O-Ne WD with more mass than a C-O WD (and thus less mass is needed from accretion to approach the Chandrasekhar limit) [75]. During accretion, C burning commences and drives convection that mixes and removes this layering. The result of this simmering is a progenitor WD with a C-depleted core that is cooler than the surrounding material due to energy loss from neutrino emissions. It differs from a C-O progenitor in its temperature profile and because it has a significant amount of ^{20}Ne. Simulating thermonuclear supernovae from these hybrid progenitors thus required adapting the burning module (described above) to account for the ^{20}Ne [76], and these adaptations were incorporated into the source code to be distributed. We note that studies of these hybrid models continue, and very recent work [77] suggests that mixing during the cooling phase of the WD before the onset of accretion plays an important role and may produce progenitors different in composition and structure from the hybrid models we employed in our study.

We investigated the viability of these hybrid progenitors with a series of explosion simulations in [76]. Figure 4 shows the progression of thermonuclear burning during a simulation similarly to the C-O case illustrated in Figure 1. The panels show the early deflagration phase (left), the moment of ignition of the detonation (center) and the propagation of the detonation (right). An obvious difference between the hybrid progenitor and the C-O case is the cool, C-depleted core, which shows little burning (the "hole" in the center of the star seen in each panel) because the deflagration is born in the high-temperature shell above the core. By the end of the simulation, however, these models eventually consume the core during the detonation phase. A noticeable difference from the C-O models is this delay in burning the inner core.

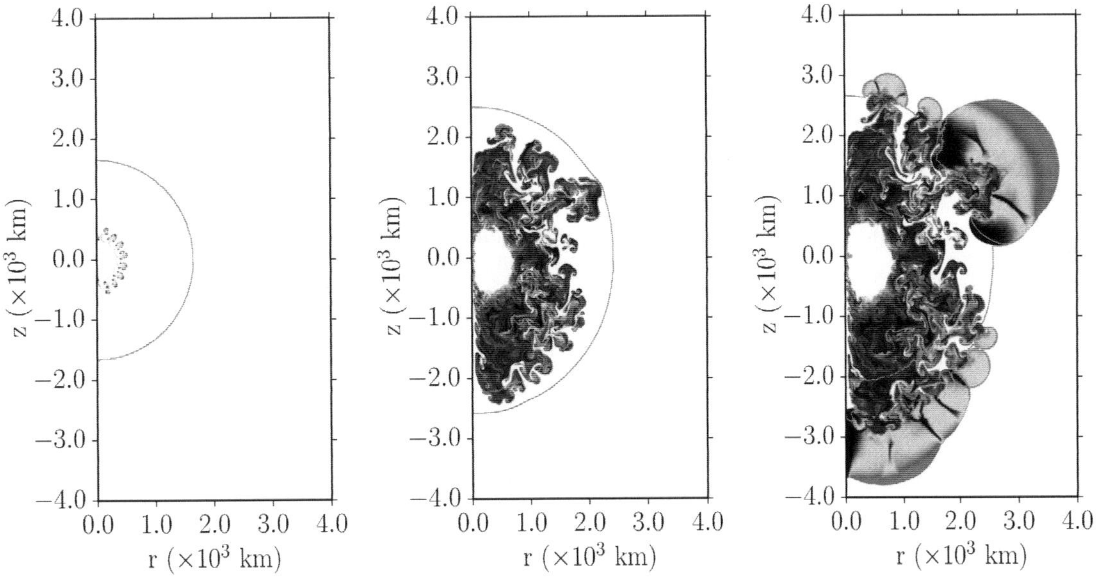

Figure 4. Deflagration to detonation burning progress shown for a C-O-Ne hybrid WD with $\rho_{\rm DDT} = 10^{7.2}$ g/cm^3. The color scheme is the same as that of Figure 1. From left to right, the burning is shown at 0.50 s, 1.48 s, and 1.60 s after the deflagration is initiated.

We found that while there can be considerable variability in the outcome, there are statistically significant trends, with hybrid C-O-Ne models having a lower ^{56}Ni yield and ejecta with lower kinetic energy than similar C-O models. Figure 5 illustrates these trends, with the left panel presenting estimated ^{56}Ni yields and the right panel presenting gravitational binding energy. We concluded that these hybrid progenitors are viable for supernova explosions.

3.4. Analysis via Synthetic Spectra and Light Curves

The flame capturing and energetics scheme described above is sufficient to capture the bulk energetics of the explosion and to estimate the yield of ^{56}Ni and IGEs. Making accurate calculations of the abundances synthesized in an explosion, however, requires much higher fidelity calculations of the thermonuclear burning. We address this issue by postprocessing fluid element histories (temperature and density) from Lagrangian tracer particles embedded in the hydrodynamic flow during a supernova simulation. In [63] we presented a new method for this process with particular attention given to reconstructing the fluid element history within the artificially-thickened model flame. We provided details of the method and verification tests demonstrating that for the problem of deflagration and detonation fronts propagating in a uniform medium, we find good agreement between the post-processing method and a direct simulation of the burning.

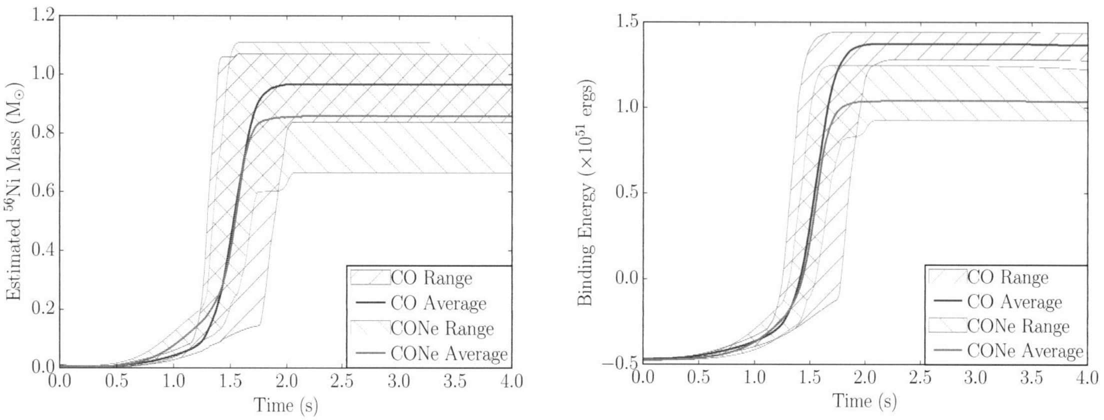

Figure 5. ^{56}Ni yield and gravitational binding energy from suites of simulations of explosions from hybrid C-O-Ne progenitors. The left panel shows the yield of ^{56}Ni from suites of simulations from C-O progenitors (very similar to those of [72]) and from C-O-Ne hybrid progenitors. The right panel shows the binding energy from the two suites. In both panels, the range of results are marked by the hashed regions. Adapted from [76].

With these accurate, detailed abundances, we were able to generate synthetic spectra from our post-processed explosion models to compare with observations in order to explore how signatures of metallicity in the spectral features may be used to constrain the metallicity of the progenitor WD [78]. We performed explosion simulations from progenitors for a range of metallicities, and from the spectra were able to find correlations between the progenitor metallicity and the strength of certain spectral features. Specifically, a Ti feature near 4200 Å and an Fe feature near 5200 Å show a trend of increasing width with increasing metallicity. These results suggest the ability to observationally differentiate between progenitor metallicities.

Figure 6 shows synthetic spectra from post-processed remnants derived from simulations at different metallicities. The top panel shows spectra at 10, 20, 30, and 40 days after the explosion. The middle and bottom panels show spectra from post-processed remnants at 30 days after the explosion that were derived from simulations at low and high metallicities, respectively. The spectra at the two metallicities are "knock out" spectra calculated with and without the contribution to the opacity from an individual transition line [79]. The two knock out spectra are from the same DDT realization simulated at the two metallicities. Changes in the differences between the model spectra and "knock out" spectra serve as a diagnostic for changes following from varying metallicity.

4. Conclusions

We have developed a detailed flame-capturing and energetics scheme to model thermonuclear supernovae and coupled it to post-processing techniques to calculate detailed abundances and synthetic spectra and light curves. With these tools, we have investigated systematic effects on the brightness of thermonuclear supernovae (in the single-degenerate DDT paradigm) that are thought to follow from the evolutionary history of the progenitor star and host galaxy. We investigated the effects of age and composition and our results are in rough agreement with observed trends. We also were able to quantify variations within the suites of simulations and thereby evaluate the variability of these models. While our results are prior to calibration by the Phillips relation, understanding these first-order effects on the brightness is a significant step toward assessing the intrinsic scatter of these events.

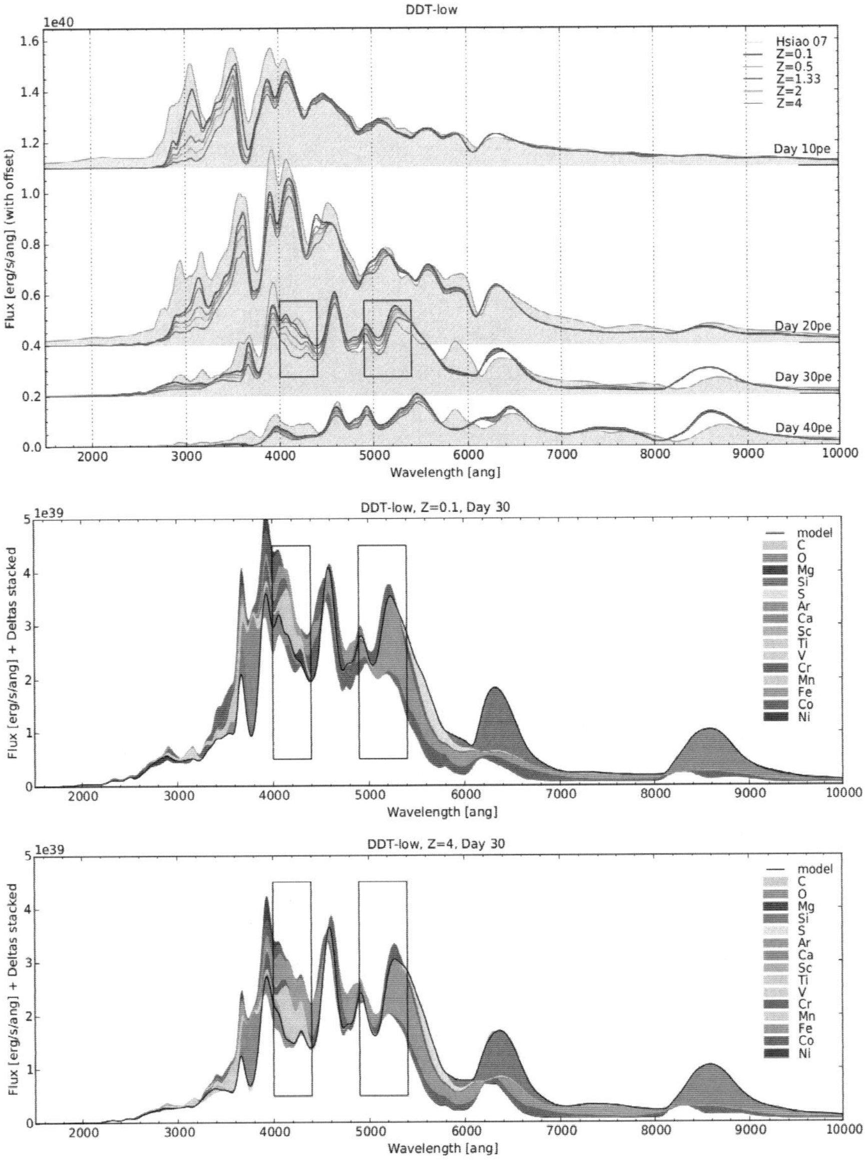

Figure 6. Spectra from a suite of simulations at varying metallicity at 10, 20, 30, and 40 days after the explosion illustrating evolution (top panel). Also shown are "knockout" spectra from one DDT realization simulated with metallicity $Z/Z_{\odot} = 0.1$ (middle panel) and $Z/Z_{\odot} = 4.0$ (bottom panel) at 30 days after the explosion. Contributions from line opacity of individual elements to the full spectrum (black solid lines) are represented by colored patches. These are calculated by removing the line opacities from a single element and recalculating the spectra while keeping the temperatures and population numbers fixed. Emission contributions are represented by colored patches below the full spectra, and absorption contributions are represented by patches above the full spectra. The spectral features highlighted with black boxes indicate two potential spectral indicators for progenitor metallicity. Adapted from [78].

We also investigated explosions from hybrid C-O-Ne progenitors and found that these progenitors produce explosions similar to explosions from traditional C-O progenitors, but with a slightly lower yield of ^{56}Ni and ejecta with slightly lower kinetic energy. The viability of these models is significant in that the hybrid progenitors require less accreted mass and thereby ameliorate a concern with the single-degenerate paradigm.

In the future, we will use these tools to explore simulations from more realistic progenitors as these become available. We are presently performing a study of the convective Urca process with fully three-dimensional simulations. Introduced by Paczyński [80], the convective Urca process occurs when neutrino losses alter the dynamics of convection, in this case in the simmering progenitor WD [81, 82, 83, 84]. The process occurs in a region of the star known as the "Urca shell." Above the density of the shell, electron capture is favorable due to the high Fermi energy, while below this density β-decay occurs. Generally an Urca shell is specific to an Urca pair of nuclides, and a pair that is important for thermonuclear supernova progenitors is ^{23}Ne/^{23}Na. The effect of the Urca shell is twofold—energy is lost to neutrinos and the shell forms a barrier to buoyancy because of the different nucleon/electron ratio across the shell. Through these two effects the Urca shell can influence both the energy content of the star and the flow structure in the convection zone. The results of our study will inform progenitor models for future explosion simulations that will address the effects of these results on the brightness of events.

Acknowledgments

The work described in this paper is a distillation of parts of a large effort exploring thermonuclear supernovae. The authors gratefully acknowledge contributions from Frank Timmes, Ed Brown, David Chamulak, Daan van Rossum, Shimon Asida, Ivo Seitenzahl, Fang Peng, Natalia Vladimirova, Don Lamb, and Jim Truran. This work was supported in part by the US Department of Energy under grants DE-FG02-07ER41516 and DE-FG02-87ER40317. This work was supported in part by the US National Science Foundation under grant AST-0507456. This work was supported in part by the US National Aeronautics and Space Administration under grant NNX09AD19G. D.M.T. received support from the Bart J. Bok fellowship at the University of Arizona for part of this work. A.P.J. received support from a National Research Council Research Associateship for part of this work. The authors acknowledge the hospitality of the Kavli Institute for Theoretical Physics, which is supported by the NSF under grant PHY05- 51164, during the programs "Accretion and Explosion: the Astrophysics of Degenerate Stars" and "Stellar Death and Supernovae." The software used in this work was in part developed by the DOE-supported ASC/Alliances Center for Astrophysical Thermonuclear Flashes at the University of Chicago. Results in this paper were obtained using the high-performance computing system at the Institute for Advanced Computational Science at Stony Brook University and via US National Science Foundation TeraGrid and US Department of Energy INCITE awards. The authors also thank Rachel Losacco and Ishmam Yousuf for previewing the manuscript.

References

[1] Riess A G, Filippenko A V, Challis P, Clocchiatti A, Diercks A, Garnavich P M, Gilliland R L, Hogan C J, Jha S, Kirshner R P, Leibundgut B, Phillips M M, Reiss D, Schmidt B P, Schommer R A, Smith R C, Spyromilio J, Stubbs C, Suntzeff N B and Tonry J 1998 *AJ* **116** 1009 (*Preprint* `astro-ph/9805201`)

[2] Perlmutter S, Aldering G, Goldhaber G, Knop R A, Nugent P, Castro P G, Deustua S, Fabbro S, Goobar A, Groom D E, Hook I M, Kim A G, Kim M Y, Lee J C, Nunes N J, Pain R, Pennypacker C R, Quimby R, Lidman C, Ellis R S, Irwin M, McMahon R G, Ruiz-Lapuente P, Walton N, Schaefer B, Boyle B J, Filippenko A V, Matheson T, Fruchter A S, Panagia N, Newberg H J M, Couch W J and The Supernova Cosmology Project 1999 *ApJ* **517** 565 (*Preprint* `astro-ph/9812133`)

[3] Leibundgut B 2001 *Annu. Rev. Astron. Astrophys* **39** 67–98

[4] Minkowski R 1941 *PASP* **53** 224

[5] Pankey Jr T 1962 *Possible Thermonuclear Activities in Natural Terrestrial Minerals.* Ph.D. thesis HOWARD UNIVERSITY.

[6] Colgate S A and McKee C 1969 *ApJ* **157** 623
[7] Colgate S A, Petschek A G and Kriese J T 1980 *ApJ* **237** L81–L85
[8] Kuchner M J, Kirshner R P, Pinto P A and Leibundgut B 1994 *ApJ* **426** L89
[9] Pinto P A and Eastman R G 2001 *New Astronomy* **6** 307–319
[10] Phillips M M 1993 *ApJ* **413** L105
[11] Howell D A, Sullivan M, Brown E F, Conley A, LeBorgne D, Hsiao E Y, Astier P, Balam D, Balland C, Basa S, Carlberg R G, Fouchez D, Guy J, Hardin D, Hook I M, Pain R, Perrett K, Pritchet C J, Regnault N, Baumont S, LeDu J, Lidman C, Perlmutter S, Suzuki N, Walker E S and Wheeler J C 2009 *ApJ* **691** 661–671 (*Preprint* 0810.0031)
[12] Kirshner R P 2010 *Foundations of supernova cosmology* p 151
[13] Mannucci F, Della Valle M and Panagia N 2006 *MNRAS* **370** 773–783 (*Preprint* arXiv:astro-ph/0510315)
[14] Raskin C, Scannapieco E, Rhoads J and Della Valle M 2009 *ApJ* **707** 74–78 (*Preprint* 0909.4293)
[15] Wang X, Wang L, Filippenko A V, Zhang T and Zhao X 2013 *Science* **340** 170–173 (*Preprint* 1303.2601)
[16] Gallagher J S, Garnavich P M, Caldwell N, Kirshner R P, Jha S W, Li W, Ganeshalingam M and Filippenko A V 2008 *ApJ* **685** 752–766 (*Preprint* 0805.4360)
[17] Neill J D, Sullivan M, Howell D A, Conley A, Seibert M, Martin D C, Barlow T A, Foster K, Friedman P G, Morrissey P, Neff S G, Schiminovich D, Wyder T K, Bianchi L, Donas J, Heckman T M, Lee Y, Madore B F, Milliard B, Rich R M and Szalay A S 2009 *ApJ* **707** 1449–1465 (*Preprint* 0911.0690)
[18] Brandt T D, Tojeiro R, Aubourg É, Heavens A, Jimenez R and Strauss M A 2010 *AJ* **140** 804–816 (*Preprint* 1002.0848)
[19] Chamulak D A, Brown E F and Timmes F X 2006 *Proc. Sci.* **submitted**
[20] Howell D A, Sullivan M, Nugent P E, Ellis R S, Conley A J, Le Borgne D, Carlberg R G, Guy J, Balam D, Basa S, Fouchez D, Hook I M, Hsiao E Y, Neill J D, Pain R, Perrett K M and Pritchet C J 2006 *Nature* **443** 308–311 (*Preprint* arXiv:astro-ph/0609616)
[21] Scalzo R A, Aldering G, Antilogus P, Aragon C, Bailey S, Baltay C, Bongard S, Buton C, Childress M, Chotard N, Copin Y, Fakhouri H K, Gal-Yam A, Gangler E, Hoyer S, Kasliwal M, Loken S, Nugent P, Pain R, Pécontal E, Pereira R, Perlmutter S, Rabinowitz D, Rau A, Rigaudier G, Runge K, Smadja G, Tao C, Thomas R C, Weaver B and Wu C 2010 *ApJ* **713** 1073–1094 (*Preprint* 1003.2217)
[22] Yuan F, Quimby R M, Wheeler J C, Vinkó J, Chatzopoulos E, Akerlof C W, Kulkarni S, Miller J M, McKay T A and Aharonian F 2010 *ApJ* **715** 1338–1343 (*Preprint* 1004.3329)
[23] Tanaka M, Kawabata K S, Yamanaka M, Maeda K, Hattori T, Aoki K, Nomoto K, Iye M, Sasaki T, Mazzali P A and Pian E 2010 *ApJ* **714** 1209–1216 (*Preprint* 0908.2057)
[24] Calder A C, Krueger B K, Jackson A P and Townsley D M 2013 *Frontiers of Physics* **8** 168–188 (*Preprint* 1303.2207)
[25] Mazzali P A, Sauer D N, Pastorello A, Benetti S and Hillebrandt W 2008 *MNRAS* **386** 1897–1906 (*Preprint* 0803.1383)
[26] Arnett W D, Truran J W and Woosley S E 1971 *ApJ* **165** 87
[27] Röpke F K, Hillebrandt W, Schmidt W, Niemeyer J C, Blinnikov S I and Mazzali P A 2007 *ApJ* **668** 1132–1139 (*Preprint* 0707.1024)
[28] Kromer M, Ohlmann S T, Pakmor R, Ruiter A J, Hillebrandt W, Marquardt K S, Röpke F K, Seitenzahl I R, Sim S A and Taubenberger S 2015 *MNRAS* **450** 3045–3053 (*Preprint* 1503.04292)
[29] Khokhlov A M 1991 *A&A* **245** 114–128
[30] Höflich P, Khokhlov A M and Wheeler J C 1995 *ApJ* **444** 831–847
[31] Gamezo V N, Khokhlov A M and Oran E S 2005 *ApJ* **623** 337–346 (*Preprint* astro-ph/0409598)
[32] Ivanova L N, Imshennik V S and Chechetkin V M 1974 *Ap&SS* **31** 497–514
[33] Arnett D and Livne E 1994 *ApJ* **427** 315–329
[34] Arnett D and Livne E 1994 *ApJ* **427** 330–341
[35] Bravo E and García-Senz D 2006 *ApJ* **642** L157–L160 (*Preprint* astro-ph/0604025)
[36] Plewa T, Calder A C and Lamb D Q 2004 *ApJ* **612** L37–L40
[37] Jordan IV G C, Fisher R T, Townsley D M, Calder A C, Graziani C, Asida S, Lamb D Q and Truran J W 2008 *ApJ* **681** 1448–1457
[38] Jordan IV G C, Graziani C, Fisher R T, Townsley D M, Meakin C, Weide K, Reid L B, Norris J, Hudson R and Lamb D Q 2012 *ApJ* **759** 53 (*Preprint* 1202.3997)
[39] Blinnikov S I and Khokhlov A M 1986 *Soviet Astronomy Letters* **12** 131–+
[40] Niemeyer J C and Woosley S E 1997 *ApJ* **475** 740 (*Preprint* astro-ph/9607032)
[41] Niemeyer J C 1999 *ApJ* **523** L57–L60 (*Preprint* arXiv:astro-ph/9906142)
[42] Bell J B, Day M S, Rendleman C A, Woosley S E and Zingale M 2004 *ApJ* **608** 883–906 (*Preprint* astro-ph/0401247)
[43] Fisher R and Jumper K 2015 *ApJ* **805** 150 (*Preprint* 1504.00014)

[44] Woosley S E 1990 *Supernovae* ed A G Petschek pp 182–212
[45] Zel'Dovich Y B, Librovich V B, Makhviladze G M and Sivashinskil G I 1970 *Journal of Applied Mechanics and Technical Physics* **11** 264–270
[46] Khokhlov A M, Oran E S and Wheeler J C 1997 *ApJ* **478** 678–+ (*Preprint* arXiv:astro-ph/9612226)
[47] Jackson A P, Townsley D M and Calder A C 2014 *ApJ* **784** 174 (*Preprint* 1402.4527)
[48] Townsley D M, Calder A C, Asida S M, Seitenzahl I R, Peng F, Vladimirova N, Lamb D Q and Truran J W 2007 *ApJ* **668** 1118–1131 (*Preprint* arXiv:0706.1094)
[49] Fryxell B, Olson K, Ricker P, Timmes F X, Zingale M, Lamb D Q, MacNeice P, Rosner R, Truran J W and Tufo H 2000 *ApJS* **131** 273–334
[50] Calder A C, Curtis B C, Dursi L J, Fryxell B, Henry G, MacNeice P, Olson K, Ricker P, Rosner R, Timmes F X, Tufo H M, Truran J W and Zingale M 2000 *Proceedings of Supercomputing 2000* p http://sc2000.org
[51] Calder A C, Fryxell B, Plewa T, Rosner R, Dursi L J, Weirs V G, Dupont T, Robey H F, Kane J O, Remington B A, Drake R P, Dimonte G, Zingale M, Timmes F X, Olson K, Ricker P, MacNeice P and Tufo H M 2002 *ApJS* **143** 201–229
[52] Dubey A, Calder A, Fisher R, Graziani C, Jordan G, Lamb D, Reid L, Townsley D and Weide K 2013 *International Journal of High Performance Computing Applications* **27** 360–373 ISSN 1094-3420
[53] Dubey A, Antypas K, Calder A, Daley C, Fryxell B, Gallagher J, Lamb D, Lee D, Olson K, Reid L, Rich P, Ricker P, Riley K, Rosner R, Siegel A, Taylor N, Weide K, Timmes F, Vladimirova N and Zuhone J 2014 *International Journal of High Performance Computing Applications* **28** 225–237 ISSN 1094-3420
[54] Khokhlov A M 1995 *ApJ* **449** 695
[55] Vladimirova N, Weirs G and Ryzhik L 2006 *Combust. Theory Modelling* **10** 727–747
[56] Timmes F X and Woosley S E 1992 *ApJ* **396** 649
[57] Chamulak D A, Brown E F and Timmes F X 2007 *ApJ* **655** L93 (*Preprint* astro-ph/0612507)
[58] Gamezo V N, Khokhlov A M, Oran E S, Chtchelkanova A Y and Rosenberg R O 2003 *Science* **299** 77
[59] Imshennik V S, Filippov S S and Khokhlov A M 1981 *Soviet Astronomy Letters* **7** 121
[60] Khokhlov A M 1981 *Soviet Astronomy Letters* **7** 410–413
[61] Khokhlov A M 1983 *Soviet Astronomy Letters* **9** 160
[62] Calder A C, Townsley D M, Seitenzahl I R, Peng F, Messer O E B, Vladimirova N, Brown E F, Truran J W and Lamb D Q 2007 *ApJ* **656** 313–332 (*Preprint* arXiv:astro-ph/0611009)
[63] Townsley D M, Miles B J, Timmes F X, Calder A C and Brown E F 2016 *ApJS* **225** 3 (*Preprint* 1605.04878)
[64] Seitenzahl I R, Townsley D M, Peng F and Truran J W 2009 *Atomic Data and Nuclear Data Tables* **95** 96–114
[65] Meakin C A, Seitenzahl I, Townsley D, Jordan G C, Truran J and Lamb D 2009 *ApJ* **693** 1188–1208 (*Preprint* 0806.4972)
[66] Townsley D M, Jackson A P, Calder A C, Chamulak D A, Brown E F and Timmes F X 2009 *ApJ* **701** 1582–1604 (*Preprint* 0906.4384)
[67] Calder A C, Krueger B K, Jackson A P, Townsley D M, Timmes F X, Brown E F and Chamulak D A 2011 *Proceedings of SciDAC 2010* (Oak Ridge) URL http://computing.ornl.gov/workshops/scidac2010/2010_SciDAC_Proceedings.pdf
[68] Timmes F X, Brown E F and Truran J W 2003 *ApJ* **590** L83 (*Preprint* astro-ph/0305114)
[69] Jackson A P, Calder A C, Townsley D M, Chamulak D A, Brown E F and Timmes F X 2010 *ApJ* **720** 99–113 (*Preprint* 1007.1138)
[70] Konishi K, Cinabro D, Garnavich P M, Ihara Y, Kessler R, Marriner J, Schneider D P, Smith M, Spinka H, Wheeler J C and Yasuda N 2011 *ArXiv e-prints* (*Preprint* 1101.4269)
[71] Krueger B K, Jackson A P, Townsley D M, Calder A C, Brown E F and Timmes F X 2010 *ApJ* **719** L5–L9 (*Preprint* 1007.0910)
[72] Krueger B K, Jackson A P, Calder A C, Townsley D M, Brown E F and Timmes F X 2012 *ApJ* **757** 175 (*Preprint* 1208.1986)
[73] Lesaffre P, Han Z, Tout C A, Podsiadlowski P and Martin R G 2006 *MNRAS* **368** 187–195 (*Preprint* arXiv:astro-ph/0601443)
[74] Denissenkov P A, Herwig F, Truran J W and Paxton B 2013 *ApJ* **772** 37 (*Preprint* 1305.2649)
[75] Denissenkov P A, Truran J W, Herwig F, Jones S, Paxton B, Nomoto K, Suzuki T and Toki H 2015 *MNRAS* **447** 2696–2705 (*Preprint* 1407.0248)
[76] Willcox D E, Townsley D M, Calder A C, Denissenkov P A and Herwig F 2016 *ApJ* **832** 13 (*Preprint* 1602.06356)
[77] Brooks J, Schwab J, Bildsten L, Quataert E and Paxton B 2016 *ArXiv e-prints* (*Preprint* 1611.03061)
[78] Miles B J, van Rossum D R, Townsley D M, Timmes F X, Jackson A P, Calder A C and Brown E F 2016 *ApJ* **824** 59 (*Preprint* 1508.05961)
[79] van Rossum D R 2012 *ArXiv e-prints* (*Preprint* 1208.3781)

[80] Paczyński B 1972 *Astrophys. Lett.* **11** 53–+
[81] Iben Jr I 1978 *ApJ* **219** 213–225
[82] Iben Jr I 1978 *ApJ* **226** 996–1033
[83] Iben Jr I 1982 *ApJ* **253** 248–259
[84] Stein J and Wheeler J C 2006 *ApJ* **643** 1190–1197 (*Preprint* astro-ph/0512580)

Fluid Instabilities of Magnetar-Powered Supernovae

Ke-Jung Chen
Division of Theoretical Astronomy, National Astronomical Observatory of Japan, Tokyo 181-8588, Japan
E-mail: ken.chen@nao.ac.jp

Abstract. Magnetar-powered supernova explosions are competitive models for explaining very luminous optical transits. Until recently, these explosion models were mainly calculated in 1D. Radiation emitted from the magnetar snowplows into the previous supernovae ejecta and causes a nonphysical dense shell (spike) found in previous 1D studies. This suggests that strong fluid instabilities may have developed within the magnetar-powered supernovae. Such fluid instabilities emerge at the region where luminous transits later occur, so they can affect the consequent observational signatures. We examine the magnetar-powered supernovae with 2D hydrodynamics simulations and find that the 1D dense shell transforms into the development of Rayleigh-Taylor and thin shell instabilities in 2D. The resulting mixing is able to fragment the entire shell and break the spherical symmetry of supernovae ejecta.

1. Introduction

Magnetars are neutrons with a strong magnetic field of $10^{14} - 10^{15}$ Gauss (G) and a rapid rotation of period about a few to tens of milliseconds. This strong magnetic field may be generated from the collapse of a rapidly rotating iron core [1, 2, 3, 4]. The magnetic field acts as a brake for the rotating neutron stars by extracting its rotational energy through a dipole radiation. During the early evolution of a magnetar, the energetic radiation appears in the form of x-rays and soft gamma-rays, which have been detected [5, 6, 7, 8, 9, 10]. In addition, recent studies by Woosley and Kasen[11, 12] suggested that magnetars might be able to power supernovae (SNe) with luminous optical transits of $10^{44} - 10^{45}$erg s^{-1}, which is about 10 - 100 times more luminous than that of a generic core-collapse SN. Woosley[11] modeled the magnetar in one dimension (1D) with the `KEPLER` code [13, 14]. In his simulations, a large density spike was found when the magnetar started to emit the dipole radiation. The spike was also seen in [12]. The emerging density spike is originated from the growth of fluid instabilities, and 1D simulations cannot model fluid instabilities, which are intrinsically multidimensional phenomena. It is not clear how mixing inside the magnetar would change the observational signatures of the magnetars. However, the spectrum must be affected by the mixing of different elements within shells.

3D radiation-hydrodynamical simulations are required to study the mixing of magnetar-powered SNe and to obtain the light curves and spectra from the first principles. However, such simulations are still beyond the capability of state-of-the-art numerical codes and computational resources. As a first step to this goal, we have performed the first 2D hydrodynamics simulations based on a realistic magnetar progenitor, but we neglect the full radiation transport. This setup is still effective because the density spike emerges at the early phase of the magnetar when the radiation is still strongly coupled with the gas flow.

The structure of the paper is as follows; in Section 2, we describe the progenitor model and the setup for 2D simulations. In Sections 3 and 4, we present the results of simulations and discuss their astrophysical implications. We conclude in Section 5.

2. Magnetar Model and Numerical Method

We start with the pre-supernova model of a 6 $M_\odot$ carbon-oxygen star that has an initial mass fraction of ^{12}C =0.144 and ^{16}O = 0.856. The ratio is determined by abundances from the full-star stellar evolution models at the end of central helium burning. This particular model approximates the core of a regular star of $20 - 24\,M_\odot$ with a solar metallicity based on the mass loss rate from [15]. The relation between the CO core mass and the full star mass is still unclear due to the uncertainties of mass loss rate during stellar evolution. Mass loss may be driven by stellar wind, rotation, etc. Using a CO star allows us to directly study the evolution of pre-supernova by skipping its main sequence phase. This model is evolved in `KEPLER`, a 1D stellar evolution code, including hydrodynamics, nuclear burning and convection physics [13]. The evolution is followed until an iron core of mass about 1.44 $M_\odot$ forms when the core collapse is about to occur. The radius of the iron core is about 1,400 km, and it promptly collapses into a proto-neutron star about 12 km in size. The gravitational binding energy released from the iron-core collapse powers a supernova. The explosion mechanics behind the core-collapse SNe are still uncertain [16, 17, 18, 19, 20]. Hence, we use the piston explosion model by inserting kinetic energy of 1.2×10^{51} erg at the outer edge of the iron core to blow up the star. The explosion synthesizes about 0.22 $M_\odot$ ^{56}Ni and creates a strong shock of velocity about 10^9 cm s^{-1} shortly after the explosion; we assume that a magnetar shortly forms with a rotation period of 1 millisecond and a magnetic field stress of 4×10^{14} G. This model produces a light curve that fits well with the observational data of a superluminous SN PTF10cwr [21]. However, a prominent density spike forms again in `KEPLER` and its amplitude density grows very rapidly. As shown in Figure 1, the giant spike spanning over four orders of magnitude forms within 1,000 seconds. 1D Density spikes usually originate in the fluid instabilities, which 1D simulations are not capable of modeling. To address this issue, we now map the resulting 1D profiles into `CASTRO` code right before the emergent density spike and follow the simulation in 2D.

2.1. 2D CASTRO Setup

We run 2D simulations with `CASTRO`, a multidimensional adaptive mesh refinement (AMR) hydrodynamics code [22, 23]. It uses an unsplit piecewise parabolic method (PPM) hydro scheme [24] with multispecies advection. We use the helmoltz equation of state from [25], which considers the relativistic electron and positron pairs of arbitrary degeneracy, ions, which are treated as an ideal gas, and photons. 1D `KEPLER` profiles of densities, velocities, temperatures, and different isotope abundances are mapped onto a 2D cylindrical grid of `CASTRO` before the formation of the density spike. The 1D-to-2D mapping is done with a scheme from Chen[26], which conservatively maps the physical quantities such as mass and energy from 1D profiles onto multidimensional grids.

We simulate only an octant part of the star in 2D. The physical size of the domain in r and z is $2\times\ 10^{12}$ cm, which is about fifteen times larger than the radius of the progenitor star. The circumstellar medium (CSM) is filled with an ambient gas of density profile $\rho \propto r^{-3.1}$ starting from the surface of the star. This density profile can prevent any artificial mixing caused by the reverse shock when the forward shock runs into the CSM. The base grid has 256×256 zones, with six levels of AMR for an additional factor of up to 64 (2^6) in spacial resolution. The grid refinement criteria are based on gradients of density, velocity, and pressure. The hierarchy nested grids are also constructed in such a way that the energy-deposited region is highly resolved. Reflecting and outflow boundary conditions are set on the inner and outer boundaries in both r and z, respectively. We use the monopole approximation for self-gravity,

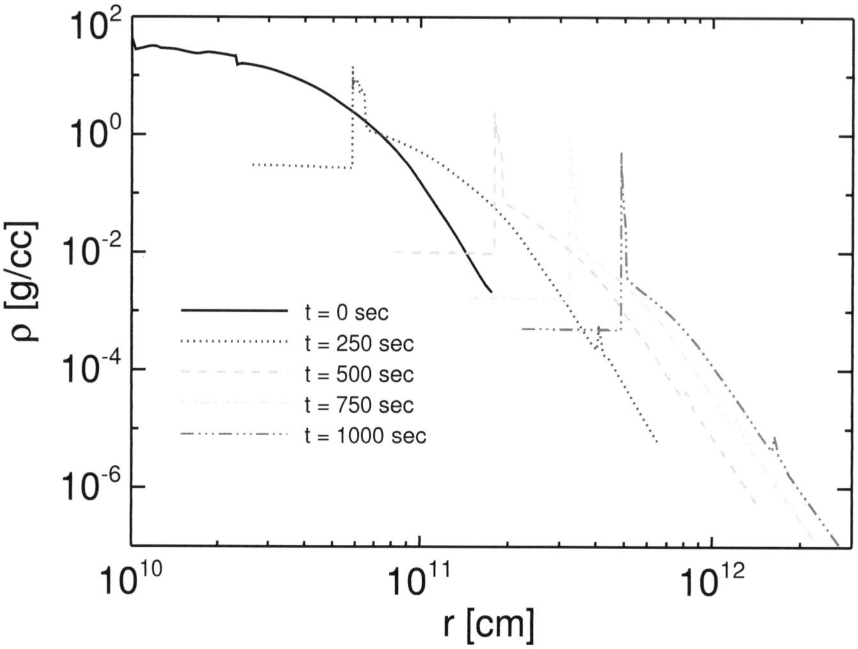

Figure 1. Density evolution of a magnetar-powered SN in 1D. The curves show density at different times after the magnetar formation. A prominent density spike emerges when the radiation-pressure-dominated gas snowplows into previous supernovae ejecta. The amplitude of the spike grows very rapidly and exceeds a density contrast of 10^3 within a thousand seconds.

in which a 1D profile of gravitational force is constructed from the radial average of the density, then the g-field stress of each grid is calculated by the linear interpolation of the 1D profile. This approximation is efficient and well-suited in supernova simulations. A point source gravity of the magnetar is also taken into account.

We assume that the magnetar releases its rotational energy through a dipole radiation. We use the moment of inertia for a typical neutron star, $I \approx 10^{45}$ g cm^2 with a millisecond period, P_{ms}. Then, its rotational energy can be estimated as

$$E = \frac{1}{2} I\omega^2 \approx 2 \times 10^{52} P_{\rm ms}^{-2} \quad \text{erg}; \tag{1}$$

E is dissipated through a dipole radiation described by the Larmor formula [27]. We assume the radius of a neutron star, $R_n \approx 10^6$ cm, and the inclination angle between the magnetic and rotational axes, $\alpha = 30°$. The energy dissipation rate has the form

$$\begin{aligned} \frac{dE}{dt} &= -\frac{32\pi^4}{3c^2}(BR_n^3 \sin\alpha)^2 P^{-4} \\ &\approx -10^{49} B_{15}^2 P_{\rm ms}^{-4} \quad \text{erg}. \end{aligned} \tag{2}$$

Solving Equations (1) and (2) by assuming the constant magnetic field, we can obtain

$$P_{\rm ms}(t) \approx \sqrt{P_0^2 + \frac{B_{15}^2 t}{2000}}, \tag{3}$$

where $B_{15} = B/10^{15}$ G, $P_0 = P_{\rm ms}(0)$. We show the evolution of E with different initial B-field and rotational rates in Figure 2. Briefly speaking, the initial amplitude of energy-release rates, $\epsilon = \dot{E}$, is determined by the initial rotational rate $\propto P_0^{-1}$, its decay rate, $\dot{\epsilon}$ is determined by B_0. Strong B_0 acts like a strong brake that quickly extracts the rotational energy of the magnetar. In 2D CASTRO simulations, the energy released from the dipole radiation is uniformly dumping within a sphere of $r \approx 5 \times 10^9$ cm, which is about the location of the emergent spike seen in the 1D KEPLER model. We deposited the magnetar energy in the small volume near the center of the star by injecting the thermal energy to the gas. At the same time, we added a very small amount of mass of $2.5 \times 10^{-6}\ M_{\odot}{\rm s}^{-1}$ along with the energy injection. Therefore, it can produce a high-velocity wind ($\sim 0.5c$) and does not violate the mass conservation of the simulation. Strictly speaking, the outflow comes from the magnetar should be highly magnetized and relativistic. But we do not know the nature of the magnetar wind. In our current study, we focus on the dynamics of ejecta and fluid instabilities. It is reasonable that we assume the gas and radiation tightly couple at this early phase of magnetar evolution. So we directly dump the magnetar energy to tiny additional mass to form a wind. Once the ejecta becomes very optically thin, this assumption becomes invalid. We then need the radiation transportation simulations to follow the decoupled radiation and gas dynamics.

The CASTRO simulation is evolved until the radiation-pressure-dominated ejecta driven by the dipole radiation has expanded to $r \approx 2 \times 10^{12}$ cm, about 3,000 seconds after the magnetar forms.

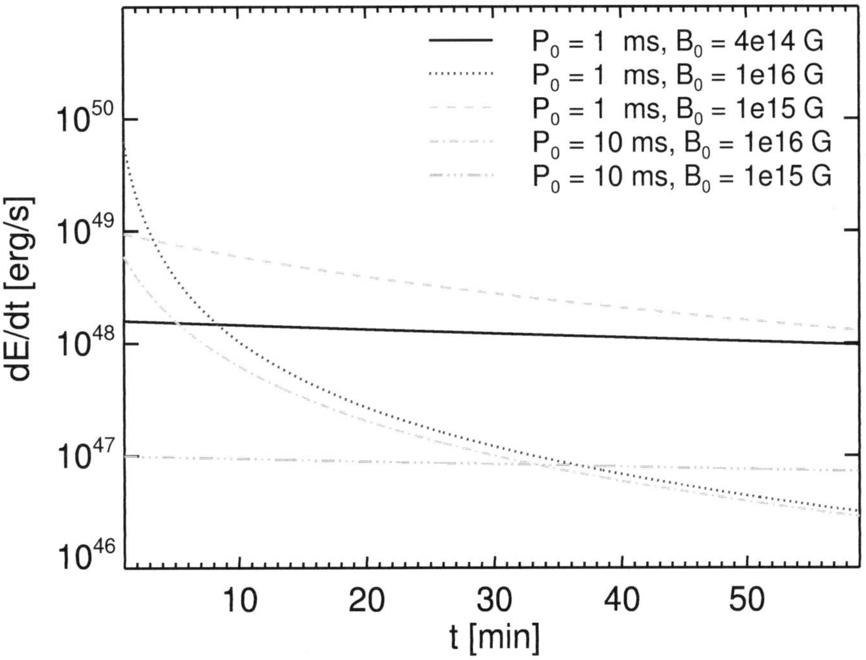

Figure 2. Energy generation rates of a magnetar. Curves show the energy generation rates of magnetars with different initial magnetic fields, B_0, and rotational periods, P_0. A Magnetar of $P = 1$ millisecond and $B_o = 4 \times 10^{14}$ G has a roughly constant rate of about 10^{48}erg s^{-1} within its first hour.

3. Results

The energy injection of the magnetar quickly heats up the surrounding gas to a temperature about $T \approx 5 \times 10^8$ K, apparently too low to ignite any interesting nuclear burning. Instead, the thermalized gas forms an outflow and runs into the previous SN ejecta. The gas driven by the outflow starts to pile up and form into a shell, and the region underneath the shell becomes radiation dominated. We define the entity of within the shell as a "radiative bubble". The bubble quickly expands at a rate of $2 - 5 \times 10^8$ cm/s. It takes about 600 seconds for the bubble to expand to the size of the progenitor star, $r_* \approx 2 \times 10^{11}$ cm. The hot gas flow collides the shell, resulting in the fluid instabilities. In 1D `KEPLER` models, there is no additional dimension to relax such collisions, and gas simply piles up to form a big density spike. The emergent fluid instabilities first form tiny little fingers that grow and fragment as the bubble expands. These fluid instabilities drill the spherical dense shell ahead of them and slightly break down the spherical symmetry of the bubble. The fluid instabilities continue evolving as the bubble expands. We show the evolving fluid instabilities in Figure 3. It is visible that the structure continue evolving as the bubble expands. Significant mixing has occurred inside the bubble and breaks down its spherical symmetry. In Panel (d) of Figure 3, mixing layers start to fulfill the bubble and mix up the isotopes when the bubble has reached to about eight times larger than r_*. In Figure 4, we show the angle-averaged profiles of density. The spikes seen in the 1D `KEPLER` models disappear in 2D and result in a noisy bump, which is the site of mixing in action. Because 2D fluid is able to follow the mixing instead of piling up in a very thin shell, it transforms the density spike into mixing. The density constraint, $\delta\rho = \rho - \langle\rho\rangle/\rho$, inside the mixing region is about 10 – 100 instead of 10^3 (spike) found in 1D models. In 1D models, most radiation is emitted from the density spike, which suggests the radiation may come from the mixing region in the multidimensional models.

To examine the mixing, we show the mixing of ^{56}Ni at end of the simulation in Figure 5. Some fraction of ^{56}Ni appears at edge of the fragmented shell. If such dredging up of ^{56}Ni indeed happens at an early phase of magnetar evolution, there arises the possibility of gamma-ray emission from the ^{56}Ni decay.

4. Discussion

The dipole radiation heats the gas around the magnetar and drives a gas outflow. It creates a pressure gradient between the heated and unheated gas. However, the density gradient of gas is in a reverse direction of the pressure gradient. It fulfills the Rayleigh-Taylor (RT) instability criterion [28]. This creates the initial mixing under the shell, as we have seen. Once the shell expands, the shell propagates away from the magnetar, and it starts to experience nonlinear thin shell instability (NTSI) [29]. NTSI is the strong driver of mixing and eventually breaks down the shells' spherical symmetry. Previous work[30, 31, 32] have studied fluid instabilities in the context of pulsar wind nebula, which is related to our study.

Dipole radiation may directly break out the thin dense layer of the radiative bubble, and escaping radiation can be detected in the form of hard x-ray emission. In addition, the mixing driven by the fluid instabilities has altered the dynamics and chemical compositions of SN ejecta. Since mixing is strongest in the region of the flow from which most of the radiation originates, it likely affects the SN light curves and spectra. There is also a possibility that some freshly synthesized ^{56}Ni about 0.03 $M_\odot$ can appear in the outskirts of the SN ejecta and can be examined in the supernova remnant.

In our simulations, we do not consider the full radiation transport that later may become important in the magnetar-powered SNe because the radiative cooling of ejecta may affect its dynamics as well as its observational signatures. The earlier fragmentation of ejecta in our simulations may seed the large-scale inhomogeneity at an early time. We also assume the constant B field stress in this calculation; however provide B field stress may decay in a spin

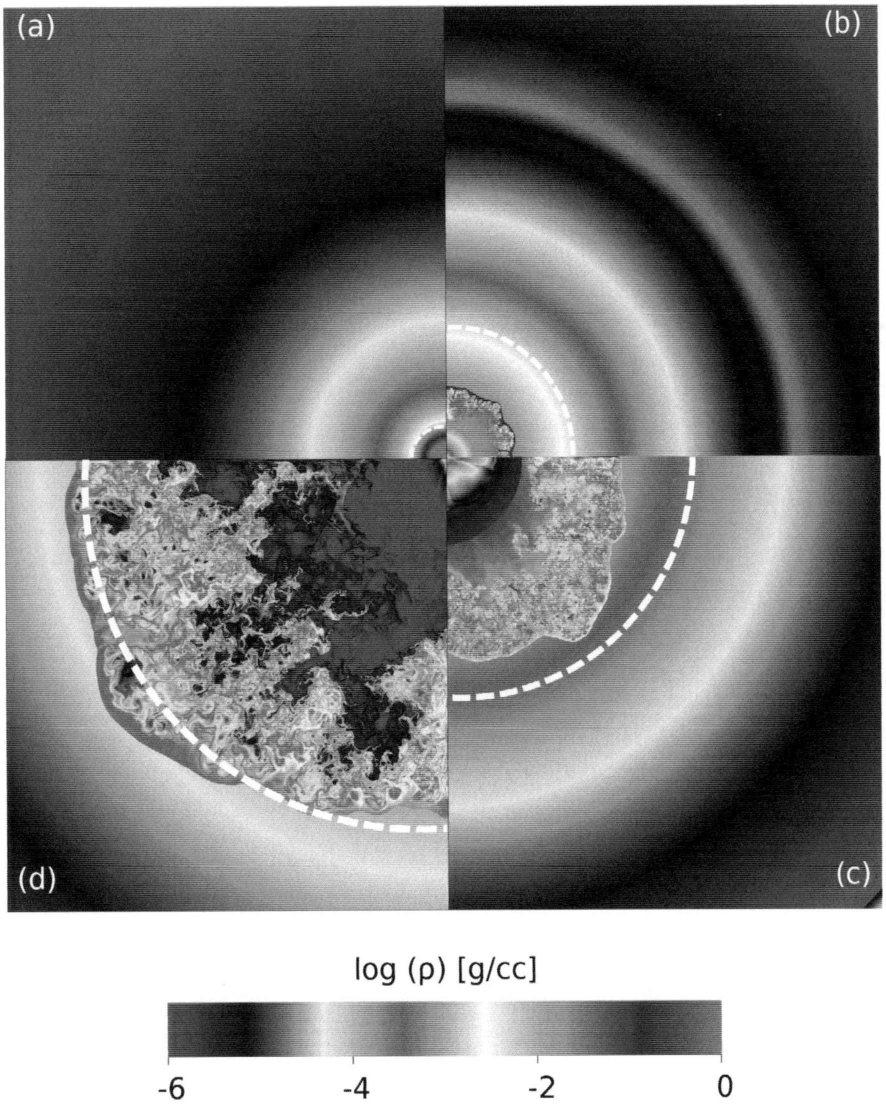

Figure 3. Evolution of fluid instabilities in the early phase of 1 ms model. The white dashed-line arcs indicate the outer boundary of expanding ejecta. Panels (a) – (d) show densities at 0, 800, 1,600, and 2,400 seconds, respectively. In Panel (a), the initial density shell shows a spherical symmetry. After the magnetar forms, fluid instabilities grow from tiny fingers to a large scale mixing as shown in Panel (d). The shell of matter close to the boundary of Panel (b) is caused by the shock wave from the original supernova explosion.

down magnetar. The different parameters of B and P provided different energy injection rates that may affect the growth rates of fluid instabilities and change the amount of escaping dipole radiation of magnetars.

5. Conclusion

We present realistic 2D hydrodynamical simulations of magnetar-powered supernovae. Due to the limits of dimensionality, previous 1D models cannot model the fluid instabilities and

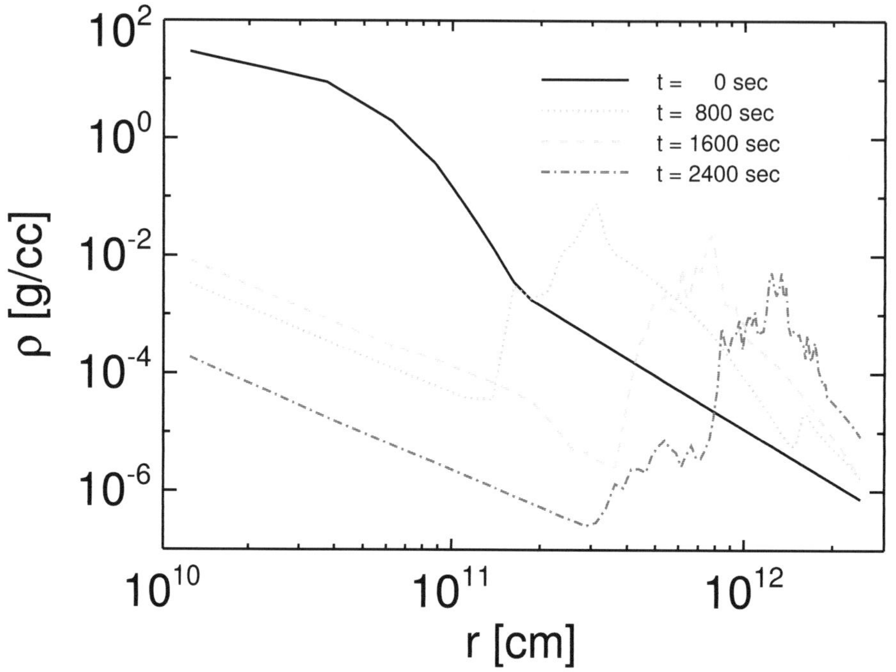

Figure 4. Angle-averaged density profiles of the 2D magnetar. Curves represent angle-averaged profiles of snapshots shown in Figure 3. The fluid instabilities in 2D truncate the 1D density spike into mixing.

mixing from the first principles, thus it produces an unphysical density spike. Instead, our 2D simulations show that strong fluid instabilities and the resulting mixing dampens the formation of the density spike found in 1D. The results suggest that strong fluid instabilities occur in magnetar-powered supernovae. Similar results are expected to appear in 3D. These fluid instabilities are mainly driven by dipole radiation of magnetar, which causes Rayleigh-Taylor instabilities and nonlinear thin shell instabilities. The resulting mixing transforms the supernova ejecta into filamentary structures. The morphology of mixing looks similar to that seen in the Crab Nebula. It may suggest the Crab Nebula might already have formed filamentary structures very early on. The growth of fluid instabilities depends on the stellar structure and the physical properties of the magnetars, such as magnetic field stress and rotation rate. In this paper, we assume a constant magnetic field, which is a crude approximation. We neglect radiative cooling by metals, and dust, which rapidly cool the dense clumps when the ejecta become optically thin. This, in turn, can affect the growth rate of Rayleigh-Taylor instabilities.

Magnetar-powered supernovae are competitive models for the superluminous supernovae, that possibly serve as promising probes for the early universe. It is crucial to obtain more realistic observational signatures of magnetar-powered supernovae, which will become more frequently observed in the coming supernova surveys. But they do demonstrate that multidimensional radiation transport will be required to model how photons are emitted from the complex structures caused by fluid instabilities. In future work, we will use the newly commissioned radiation hydrodynamics version of `CASTRO` [33] to better examine these explosions and to offer realistic observational diagnostics.

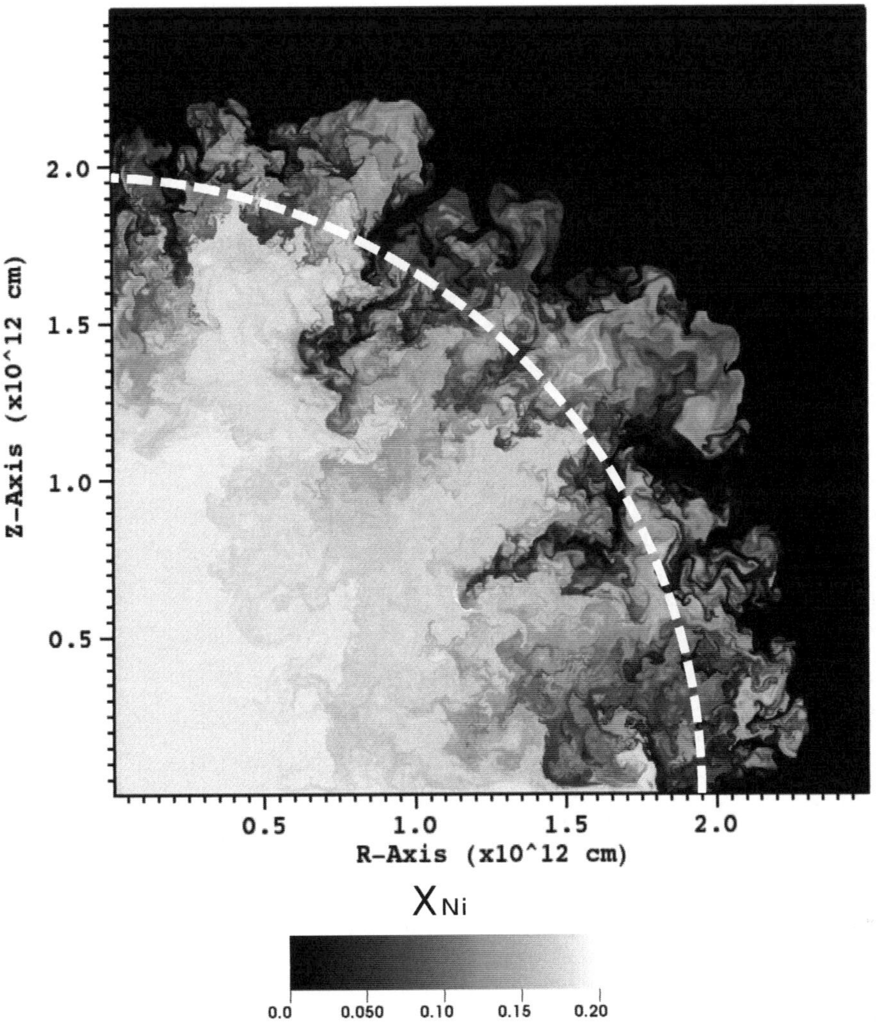

Figure 5. 2D ^{56}Ni distribution at its final snapshot. The white dashed-line arc indicates the outer boundary of the expanding SN ejecta.^{56}Ni is dredged up by the mixing, and some fraction of it starts to expand into the optical thin region. Energetic radiation directly from ^{56}Ni decay may be visible.

Acknowledgments
I thank Stan Woosley and Tuguldur Sukhbold for many useful discussions. I acknowledge the support of the EACOA Fellowship from the East Asian Core Observatories Association. Work at UCSC has been supported by an IAU-Gruber Fellowship, the DOE HEP Program (DE-SC0010676), and the NASA Theory Program (NNX14AH34G). Numerical simulations are supported by the Minnesota Supercomputing Institute (MSI), the National Energy Research Scientific Computing Center (NERSC), and the Center for Computational Astrophysics (CfCA) at National Astronomical Observatory of Japan (NAOJ).

References

[1] Duncan R C and Thompson C 1992 *ApJ* **392** L9–L13
[2] Thompson C and Duncan R C 1993 *ApJ* **408** 194–217
[3] Wheeler J C, Yi I, Höflich P and Wang L 2000 *ApJ* **537** 810–823
[4] Thompson T A, Chang P and Quataert E 2004 *ApJ* **611** 380–393
[5] Woosley S E and Bloom J S 2006 *ARA&A* **44** 507–556
[6] Kouveliotou C, Dieters S, Strohmayer T, van Paradijs J, Fishman G J, Meegan C A, Hurley K, Kommers J, Smith I, Frail D and Murakami T 1998 *Nature* **393** 235–237
[7] Gaensler B M, McClure-Griffiths N M, Oey M S, Haverkorn M, Dickey J M and Green A J 2005 *ApJ* **620** L95–L98
[8] Maeda K, Tanaka M, Nomoto K, Tominaga N, Kawabata K, Mazzali P A, Umeda H, Suzuki T and Hattori T 2007 *ApJ* **666** 1069–1082
[9] Mereghetti S 2008 *A&A Rev.* **15** 225–287
[10] Esposito P, Burgay M, Possenti A, Turolla R, Zane S, de Luca A, Tiengo A, Israel G L, Mattana F, Mereghetti S, Bailes M, Romano P, Götz D and Rea N 2009 *MNRAS* **399** L44–L48
[11] Woosley S E 2010 *ApJ* **719** L204–L207
[12] Kasen D and Bildsten L 2010 *ApJ* **717** 245–249
[13] Weaver T A, Zimmerman G B and Woosley S E 1978 *ApJ* **225** 1021–1029
[14] Heger A, Woosley S E, Martínez-Pinedo G and Langanke K 2001 *ApJ* **560** 307–325
[15] Nieuwenhuijzen H and de Jager C 1990 *A&A* **231** 134–136
[16] Burrows A, Hayes J and Fryxell B A 1995 *ApJ* **450** 830
[17] Janka H T and Mueller E 1996 *A&A* **306** 167
[18] Mezzacappa A, Calder A C, Bruenn S W, Blondin J M, Guidry M W, Strayer M R and Umar A S 1998 *ApJ* **495** 911
[19] Murphy J W and Burrows A 2008 *ApJ* **688** 1159–1175
[20] Nordhaus J, Burrows A, Almgren A and Bell J 2010 *ApJ* **720** 694–703
[21] Chen K J, Woosley S E and Sukhbold T 2016 *ArXiv e-prints*
[22] Almgren A S, Beckner V E, Bell J B, Day M S, Howell L H, Joggerst C C, Lijewski M J, Nonaka A, Singer M and Zingale M 2010 *ApJ* **715** 1221–1238
[23] Zhang W, Howell L, Almgren A, Burrows A and Bell J 2011 *ApJS* **196** 20
[24] Colella P and Woodward P R 1984 *Journal of Computational Physics* **54** 174–201
[25] Timmes F X and Swesty F D 2000 *ApJS* **126** 501–516
[26] Chen K J, Heger A and Almgren A S 2013 *Astronomy and Computing* **34** 70 – 78 ISSN 2213-1337 URL `http://www.sciencedirect.com/science/article/pii/S221313371400002X`
[27] Lyne A G and Graham-Smith F 1990 *Pulsar astronomy*
[28] Chandrasekhar S 1961 *Hydrodynamic and hydromagnetic stability*
[29] Vishniac E T 1994 *ApJ* **428** 186–208
[30] Chevalier R A and Fransson C 1992 *ApJ* **395** 540–552
[31] Jun B I 1998 *ApJ* **499** 282
[32] Blondin J M, Chevalier R A and Frierson D M 2001 *ApJ* **563** 806–815
[33] Zhang W, Howell L, Almgren A, Burrows A, Dolence J and Bell J 2013 *ApJS* **204** 7

Converging on the Initial Mass Function of Stars

Christoph Federrath[1], Mark Krumholz[1], Philip F. Hopkins[2]
[1]Research School of Astronomy and Astrophysics, The Australian National University, Canberra, ACT 2611, Australia
[2]TAPIR, Mailcode 350-17, California Institute of Technology, Pasadena, CA 91125, USA

E-mail: christoph.federrath@anu.edu.au

Abstract. Understanding the origin of stellar masses—the initial mass function (IMF)—remains one of the most challenging problems in astrophysics. The IMF is a key ingredient for simulations of galaxy formation and evolution, and is used to calibrate star formation relations in extra-galactic observations. Modeling the IMF directly in hydrodynamical simulations has been attempted in several previous studies, but the most important processes that control the IMF remain poorly understood. This is because predicting the IMF from direct hydrodynamical simulations involves complex physics such as turbulence, magnetic fields, radiation feedback and mechanical feedback, all of which are difficult to model and the methods used have limitations in terms of accuracy and computational efficiency. Moreover, a physical interpretation of the simulated IMFs requires a numerically converged solution at high resolution, which has so far not been convincingly demonstrated. Here we present a resolution study of star cluster formation aimed at producing a converged IMF. We compare a set of magnetohydrodynamical (MHD) adaptive-mesh-refinement simulations with three different implementations of the thermodynamics of the gas: 1) with an isothermal equation of state (EOS), 2) with a polytropic EOS, and 3) with a simple stellar heating feedback model. We show that in the simulations with an isothermal or polytropic EOS, the number of stars and their mass distributions depend on the numerical resolution. By contrast, the simulations that employ the simple radiative feedback module demonstrate convergence in the number of stars formed and in their IMFs.

1. Introduction

The IMF is the distribution of stellar masses when they are born. We know from observational surveys that most stars have masses of about half the mass of our Sun ($M_\odot$). Stars with smaller masses are rarer and there is a natural cutoff at at the brown dwarf mass of $0.08\,M_\odot$, because gaseous, self-gravitating objects with masses $M \lesssim 0.08\,M_\odot$ cannot fuse hydrogen to helium. Stars more massive than the Sun also become rarer with increasing mass. The high-mass tail of the IMF is indeed a steeply decreasing power-law function with the number of stars $N(M) \propto M^{-1.35}$ [1, 2, 3, 4, 5, 6, 7, 8]. Understanding this seemingly universal power-law tail and the turnover at around $0.5\,M_\odot$ is one of the most challenging open problems in astrophysics. The IMF has far-reaching consequences and applications, e.g., for calibrating extra-galactic star formation relations used to understand galaxy formation and evolution.

A number of numerical simulations exist in the literature that aim to model and understand the IMF from direct hydrodynamical simulations [9, 10, 11, 12, 13, 14, 15, 16]. A critical problem in modeling the IMF is that one must resolve the fragmentation of magnetized, turbulent gas, including radiation feedback [17, 18, 19, 20, 21, 22, 23] and mechanical feedback by jets and outflows [24, 25, 26, 27]. Modeling all of the required physical mechanisms is difficult,

IOP Conf. Series: Journal of Physics: Conf. Series **837** (2017) 012007 doi:10.1088/1742-6596/837/1/012007

and achieving sufficient resolution at the same time is even harder. Furthermore, forming a sufficiently large number of stars in a simulation requires one to either run many small-scale clouds with different realizations of the turbulence or one large-scale, high-mass cloud that spawns many stars to obtain a statistically representative and converged mass function. Currently, there is no set of simulations that fulfills all of these requirements and produces a converged IMF.

Here we present a simple model to include stellar heating feedback in MHD simulations of star cluster formation. We compare this stellar heating model to simulations that use an isothermal or polytropic EOS. Running low-resolution, mid-resolution and high-resolution simulations for all of these models, we show that only the simulations that include the stellar heating feedback converge on the number of stars formed and their cumulative mass distributions.

2. Numerical methods

2.1. Equations of magnetohydrodynamics

We use a modified version of the adaptive mesh refinement (AMR) [28] code FLASH [29, 30] (v4) to solve the three-dimensional, ideal MHD equations including gravity,

$$
\begin{aligned}
&\frac{\partial \rho}{\partial t} + \nabla \cdot (\rho \mathbf{v}) = 0\,, \\
&\rho \left(\frac{\partial}{\partial t} + \mathbf{v} \cdot \nabla \right) \mathbf{v} = \frac{(\mathbf{B} \cdot \nabla)\mathbf{B}}{4\pi} - \nabla P_{\mathrm{tot}} + \rho \mathbf{g}\,, \\
&\frac{\partial E}{\partial t} + \nabla \cdot \left[(E + P_{\mathrm{tot}})\,\mathbf{v} - \frac{(\mathbf{B} \cdot \mathbf{v})\,\mathbf{B}}{4\pi} \right] = \rho \mathbf{v} \cdot \mathbf{g}\,, \\
&\frac{\partial \mathbf{B}}{\partial t} = \nabla \times (\mathbf{v} \times \mathbf{B})\,, \quad \nabla \cdot \mathbf{B} = 0\,,
\end{aligned}
\tag{1}
$$

where the gravitational acceleration of the gas $\mathbf{g}$, is the sum of the self-gravity of the gas and the contribution of sink particles (see Section 2.3 below):

$$
\begin{aligned}
\mathbf{g} &= -\nabla \Phi_{\mathrm{gas}} + \mathbf{g}_{\mathrm{sinks}}\,, \\
\nabla^2 \Phi_{\mathrm{gas}} &= 4\pi G \rho\,.
\end{aligned}
\tag{2}
$$

In these equations, ρ, $\mathbf{v}$, $P_{\mathrm{tot}} = P + 1/(8\pi)\,|\mathbf{B}|^2$, $\mathbf{B}$, and $E = \rho\epsilon_{\mathrm{int}} + (\rho/2)\,|\mathbf{v}|^2 + 1/(8\pi)\,|\mathbf{B}|^2$ denote the gas density, velocity, pressure (thermal+magnetic), magnetic field, and total energy density (internal+kinetic+magnetic), respectively.

We use the positive-definite HLL3R approximate Riemann scheme [31] to solve the MHD equations (1). The HLL3R solver guarantees stability and accuracy of the numerical solution of the hydrodynamical equations [32, 33, 34, 31].

2.2. Turbulence driving

All our simulations include the turbulence driving module in the FLASH code [35]. Turbulence in the interstellar medium is likely driven by a combination of supernova explosions, stellar feedback in the form of jets, outflows and winds, gravitational contraction and accretion, as well as magneto-rotational instability and galactic spiral-arm/disk dynamics [36, 37, 38].

Turbulence is indeed a critical ingredient for star formation [39, 36, 40, 41, 42, 43, 44, 45, 46, 47, 48, 49]. Since turbulence decays quickly [50, 51], but is observed on virtually all spatial scales in the ISM, the turbulence needs to be driven. Here we use a turbulence driving module that produces turbulence similar to the observed turbulence in real molecular clouds, i.e., driving on the largest scales [52, 53] and with a velocity power spectrum $\sim k^{-2}$, consistent with supersonic, compressible turbulence [54, 55, 56]. This type of velocity spectrum is further consistent with simulations of supersonic turbulence [57, 58, 35, 59, 60]. Our turbulence driving module [35] applies a stochastic Ornstein-Uhlenbeck process [61, 62] to construct an acceleration field $\mathbf{F}_{\rm stir}$, which serves as a momentum and energy source term in the MHD equations. Following observations, $\mathbf{F}_{\rm stir}$ only drives large-scale modes, $1 < |\mathbf{k}|\,L/2\pi < 3$, where most of the power is injected at the $k_{\rm inj} = 2$ mode in Fourier space, i.e., on half of the size of the computational domain. The turbulence on smaller scales is not directly affected by the driving and develops self-consistently from the cascade of energy starting on the large scales. In this study we use mixed driving of the turbulence with a turbulence driving parameter $b = 0.5$, typical for clouds in the Milky Way disk [63]. However, cloud-to-cloud variations in the turbulence driving parameter b, which is related to the mixture of driving modes [64, 35, 65, 66, 67, 68], suggest variations from $b \sim 1/3$ (purely solenoidal driving, $\nabla\cdot\mathbf{F}_{\rm stir} = 0$) to $b \sim 1$ (purely compressive driving, $\nabla\times\mathbf{F}_{\rm stir} = 0$) in observed clouds in the Milky Way [69, 70, 71, 72, 73, 63]. Investigating potential variations of the IMF between purely solenoidal driving and purely compressive driving will be the subject of a follow-up study.

2.3. Sink particles and AMR

In order to model star formation and accretion of gas, we use the sink particle method in the FLASH code [74]. Sink particles form dynamically in our simulations when a local region in the simulation undergoes gravitational collapse and forms stars. This is technically achieved by first flagging each computational cell that exceeds the Jeans resolution density,

$$\rho_{\rm sink} = \frac{\pi c_{\rm s}^2}{G\lambda_{\rm J}^2} = \frac{\pi c_{\rm s}^2}{4Gr_{\rm sink}^2}, \qquad (3)$$

where $c_{\rm s}$ is the sound speed, G is the gravitational constant, and $\lambda_{\rm J} = [\pi c_{\rm s}^2/(G\rho)]^{1/2}$ is the local Jeans length. This criterion defines a sink particle radius,

$$r_{\rm sink} = \lambda_{\rm J}/2, \qquad (4)$$

which we set to 2.5 grid cell lengths at the highest level of AMR, to avoid artificial fragmentation [75]. If the gas density in a cell exceeds $\rho_{\rm sink}$, a spherical control volume with radius $r_{\rm sink}$ is gathered around the cell and it is checked that all the gas within this control volume is Jeans-unstable, gravitationally bound and converging towards the central cell of the control volume. A sink particle is formed there, only if all of these checks are passed. This avoids spurious creation of sink particles and guarantees that only bound and collapsing gas forms stars in our simulations [74].

On all AMR levels lower than the maximum AMR level (where sink particles form), we use an AMR criterion based on the local Jeans length, such that $\lambda_{\rm J}$ is always resolved with at least 16 grid cell lengths. This resolution criterion is conservative and computationally costly, but allows us to resolve turbulence on the Jeans scale [76], potential dynamo amplification of the magnetic field in dense cores [77], and allows us to capture the basic structure of accretion discs forming on the smallest scales resolved in the simulations [27]. Accretion of gas occurs if a cell within the accretion radius of an existing sink particle exceeds $\rho_{\rm sink}$, is gravitationally bound to the sink particle and converging towards it. If these accretion criteria are met, the excess mass above $\rho_{\rm sink}$ is accreted onto the sink particle, with conservation of mass, center-of-mass, momentum, and angular momentum enforced by the algorithm. Self-gravity of the gas is computed with a multigrid solver [78], while all gravitational interactions of the sink particles with the gas and with one another are computed by direct summation over all sink particles and grid cells. This procedure avoids inaccuracies in the particle accelerations if particle-mesh and mesh-particle mapping methods were used instead. A second-order leapfrog integrator is used to advance the sink particles with a timestep that allows us to resolve highly eccentric orbits of stars [74].

2.4. Equation of state (EOS)

To model the thermal evolution of the gas, it is common practice in star formation simulations to use an isothermal or polytropic equation of state for the gas pressure,[1]

$$P_{\mathrm{EOS}} = K\rho^{\Gamma}, \tag{5}$$

which, using the ideal gas EOS, defines the temperature due to the EOS,

$$T_{\mathrm{EOS}} = \frac{\mu m_{\mathrm{H}}}{k_{\mathrm{B}}\rho} P_{\mathrm{EOS}} = \frac{\mu m_{\mathrm{H}}}{k_{\mathrm{B}}} K\rho^{\Gamma-1}, \tag{6}$$

where the polytropic exponent Γ may be set to

$$\Gamma = \begin{cases} 1 & \text{for} \quad \rho \le \rho_1 \equiv 2.50 \times 10^{-16}\,\mathrm{g\,cm^{-3}}, \\ 1.1 & \text{for} \quad \rho_1 < \rho \le \rho_2 \equiv 3.84 \times 10^{-13}\,\mathrm{g\,cm^{-3}}, \\ 1.4 & \text{for} \quad \rho_2 < \rho \le \rho_3 \equiv 3.84 \times 10^{-8}\,\mathrm{g\,cm^{-3}}, \\ 1.1 & \text{for} \quad \rho_3 < \rho \le \rho_4 \equiv 3.84 \times 10^{-3}\,\mathrm{g\,cm^{-3}}, \\ 5/3 & \text{for} \quad \rho > \rho_4 . \end{cases} \tag{7}$$

Equations (5–7) approximate previous detailed radiation-hydrodynamic simulations of dense star-forming gas [82]. The polytropic constant K in Equation (5) is adjusted such that $K = c_{\mathrm{s}}^2$, i.e, the square of the sound speed. When the polytropic exponent Γ changes according to the density regimes given by Equation (7), the temperature and sound speed change, and K is adjusted, such that the pressure and temperature are continuous functions of density. In the isothermal regime ($\Gamma = 1$), which serves as the normalization, the sound speed $c_{\mathrm{s}} = 0.2\,\mathrm{km\,s^{-1}}$ and the temperature $T = 11\,\mathrm{K}$ for gas with a typical molecular weight of $2.3\,m_{\mathrm{H}}$ (with m_{H} being the mass of a hydrogen atom). The initial isothermal evolution for $\rho \le 2.5 \times 10^{-16}\,\mathrm{g\,cm^{-3}}$ is a reasonable approximation for dense, molecular gas of solar metallicity, over a wide range of densities [83, 84, 85, 86, 87, 80, 88, 89, 90]. We note that for the numerical resolutions and respective sink particle density thresholds used in the simulations presented below, only the first two density regimes in Equation (7) are relevant.

Equations (5–7) are an approximation to hydrodynamic calculations that take radiative transfer into account [91, 92, 82, 17, 93, 18, 94, 20, 21, 14], covering the phases of isothermal contraction, adiabatic heating during the formation of the first and second core and the effects of H_2 dissociation in the second collapse. However, it does not include the radiative heating feedback from the accreting protostar once it has formed. This additional stellar heating feedback, however, is crucial, because it dominates the heating in the direct vicinity of the protostar, out to distances of the order of $\sim$ 1000–10000 AU from the star. Our main goal here is to account for this additional heating by the protostars.

3. A simple model for stellar heating

We have implemented a new stellar feedback module into the FLASH MHD code, which accounts for the effect of gas heating around newborn protostars until they reach the main sequence. Stellar heating is a direct feedback mechanism (with a spatial correlation based on the position of each star) and must be distinguished from the EOS (which relates pressure, density and

[1] Another common alternative is to use an ideal gas equation of state and to include heating and cooling terms, either by using a cooling curve [79], following the chemical evolution including heating and cooling terms [80, 81], or by directly solving the radiative transfer equation.

temperature of the gas); see §2.4. Since there is no spatial information in the EOS, feedback cannot be modeled with a standard EOS. Here we provide a simple algorithm to add stellar heating to the simulations, which allows us to approximate the direct radiative heading feedback from stars.

3.1. Protostellar and stellar luminosities

The first step in accounting for the direct heating feedback from stars and protostars is to calculate the star's luminosity. Here we use the stellar evolution model developed by Offner et al. [93], which is based on one-zone protostellar evolution models [95, 96, 97] calibrated to match detailed numerical calculations of protostellar evolution [98, 99]. This module provides stellar luminosities from the birth of a star, while it is still accreting, to the main sequence, depending on the evolutionary stage. The following evolutionary phases are distinguished: 'pre-collapse', 'no burning', 'Deuterium burning at fixed core temperature', 'Deuterium burning at variable core temperature', 'Deuterium burning in the shell', and 'main sequence'.

The 'pre-collapse' state applies to a newly formed sink particle with $M_{\mathrm{sink}} < 0.01\,M_{\odot}$ and zero luminosity. During the 'no burning' stage, the star accretes, but the luminosity is determined solely based on the energy released by gravitational contraction of the protostar. As the radius and Deuterium mass of the star evolve further, Deuterium burning first starts in the core and eventually in the shell. Finally, the star reaches the zero-age main sequence, and after that stellar evolution continues on the main sequence.

3.2. Calculating the stellar heating temperature

Based on the stellar luminosity ($L_{\star}$) from the stellar evolution module discussed in the previous subsection, which includes accretion luminosity, we can now estimate the heating of the gas around each sink particle. An accurate calculation of the temperature around each stellar radiation source would require solving the radiative transfer equation [93]. Here we use a simple model for the gas heating that assumes spherical symmetry and a homogenous gas distribution around the star, which allows us to approximate the heating of the gas without solving the radiative transfer equation. We estimate the radial dependence of the temperature due to stellar heating (T_{heat}) by

$$T_{\mathrm{heat}}^4(r) = \frac{L_{\star}}{4\pi\sigma_{\mathrm{SB}}r^2}, \tag{8}$$

where σ_{SB} is the Stefan-Boltzmann constant and r is the distance from the star. This allows us to calculate $T_{\mathrm{heat}}(r)$ based on $L_{\star}$ for each star in the computational domain. For practical reasons and to speed up the computation, we only calculate T_{heat} for $r \leq 10^4\,\mathrm{AU}$, because the contribution of stellar heating at greater distances from the star is essentially negligible [93]. We further apply an inner heating radius of 100 AU, below which $T_{\mathrm{heat}} = T_{\mathrm{heat}}(100\,\mathrm{AU}) = \mathrm{const}$, to approximate the radial dependence of T_{heat} obtained in full radiation hydrodynamical simulations [93]. If the heating radii of multiple stars overlap, we add $T_{\mathrm{heat}}^4(r)$ from each star locally in each cell where the overlap occurs.

3.3. Coupling to the gas EOS

Finally, we add the stellar heating $T_{\mathrm{heat}}(r)$ from Equation (8) to the existing gas temperature (from the EOS, Eq. 6) at every grid point r and for every stellar source. This is similar to the procedure in recent semi-analytic models of the IMF [100] and gives the final gas temperature,

$$T = \left(T_{\mathrm{EOS}}^4 + T_{\mathrm{heat}}^4\right)^{1/4}, \tag{9}$$

Table 1. Key simulation parameters and results.

Simulation	EOS	N_{eff}^3	Δx [AU]	r_{sink} [AU]	ρ_{sink} [g cm^{-3}]	N_{sinks}	Mean M_{sink}	Median M_{sink}
(1)	(2)	(3)	(4)	(5)	(6)	(7)	(8)	(9)
1. LowResIso	Isothermal	2048^3	200	500	8.3×10^{-17}	35	$2.2\,M_\odot$	$1.1\,M_\odot$
2. LowResPoly	Polytropic	2048^3	200	500	8.3×10^{-17}	31	$2.5\,M_\odot$	$2.0\,M_\odot$
3. LowResHeat	Heating FB	2048^3	200	500	8.3×10^{-17}	29	$2.7\,M_\odot$	$1.4\,M_\odot$
4. MidResIso	Isothermal	4096^3	100	250	3.8×10^{-16}	44	$1.7\,M_\odot$	$1.1\,M_\odot$
5. MidResPoly	Polytropic	4096^3	100	250	3.8×10^{-16}	47	$1.7\,M_\odot$	$0.9\,M_\odot$
6. MidResHeat	Heating FB	4096^3	100	250	3.8×10^{-16}	29	$2.6\,M_\odot$	$1.4\,M_\odot$
7. HighResIso	Isothermal	8192^3	50	125	1.8×10^{-15}	88	$0.9\,M_\odot$	$0.4\,M_\odot$
8. HighResPoly	Polytropic	8192^3	50	125	1.8×10^{-15}	62	$1.3\,M_\odot$	$0.4\,M_\odot$
9. HighResHeat	Heating FB	8192^3	50	125	1.8×10^{-15}	30	$2.6\,M_\odot$	$1.8\,M_\odot$

Notes. Column 1: simulation model. Column 2: equation of state: isothermal, polytropic, or polytropic plus stellar heating feedback (FB). Column 3: maximum effective grid resolution (refinement based on Jeans length, guaranteeing a minimum of 16 cells and a maximum of 32 cells per Jeans length). The base grid is always resolved with at least 256^3 cells. Columns 4–6: minimum cell size, sink particle radius, and sink density threshold (see Eq. 3 and 4). Columns 7–9: number of sink particles formed, mean, and median sink particle mass when SFE = 10%, i.e., when a fraction of 10% of the original cloud mass has formed stars.

or in terms of the total gas pressure (directly relevant for integrating the hydrodynamical equations; see Eq. 1),

$$P = \left[P_{\mathrm{EOS}}^4 + \left(\frac{k_{\mathrm{B}}\rho}{\mu m_{\mathrm{H}}}\right)^4 T_{\mathrm{heat}}^4\right]^{1/4}. \tag{10}$$

We note that adding the fourth power of the EOS and heating temperatures is a simplification that may lead to overestimating the heating in regions where $T_{\mathrm{EOS}} \sim T_{\mathrm{heat}}$. This will be addressed in a forthcoming paper. Here we use the simple summation given by Equation (9), suggested in previous work [100].

4. Simulation parameters and initial conditions

All our simulations were carried out in a three-dimensional triple-periodic computational domain with side length $L = 2\,\mathrm{pc}$. The initial gas density is uniform with $\rho = 6.56\times10^{-21}\,\mathrm{g\,cm^{-3}}$, giving a total cloud mass of $M = 775\,M_\odot$ and a mean freefall time of $t_{\mathrm{ff}} = 0.82\,\mathrm{Myr}$. The initial velocities are zero and the turbulence driving (see §2.2) excites turbulent motions. A fully-developed turbulent state is reached after 2 crossing times, $L/(\mathcal{M}c_{\mathrm{s}})$ [101, 35]. The amplitude of the driving was adjusted to drive turbulence with a steady-state sonic Mach number of $\mathcal{M} = \sigma_v/c_{\mathrm{s}} = 5.0$, where the velocity dispersion $\sigma_v = 1.0\,\mathrm{km\,s^{-1}}$ and the initially isothermal sound speed $c_{\mathrm{s}} = 0.2\,\mathrm{km\,s^{-1}}$, appropriate for the initial density in the molecular phase of the ISM. The mass and velocity dispersion define the initial virial parameter, the ratio of twice the kinetic energy to the gravitational energy, $\alpha_{\mathrm{vir}} = 2\,E_{\mathrm{kin}}/E_{\mathrm{grav}} = 0.5$, which is in the range of observed virial parameters [102, 103, 104]. We add an initially uniform magnetic field along the z-axis of the computational domain with $B = 10^{-5}\,\mathrm{G}$, which is subsequently compressed, tangled and twisted by the turbulence, approximating the magnetic field structure in real molecular clouds [105]. This results in an initial plasma beta (ratio of thermal to magnetic pressure)

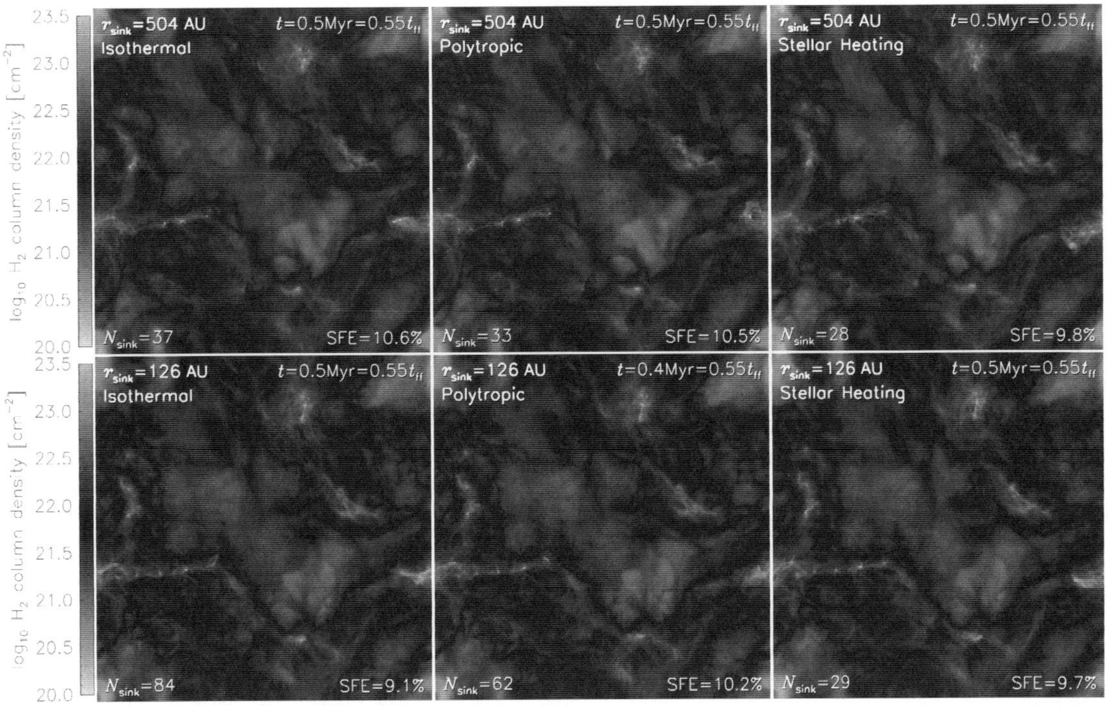

Figure 1. Each image shows a $(2\,\text{pc})^2$ projection of gas density along the z-axis. Each column shows simulations with an isothermal EOS, Eq. (5) with $\Gamma = 1$ (left-hand panels), a polytropic EOS with Γ given by Eq. (7) (middle panels), and the polytropic EOS combined with our new stellar heating feedback introduced in §3 (right-hand panels). The top panels show the low-resolution runs with an effective maximum resolution of 2048^3 cells and the bottom panels show the high-resolution runs with 8192^3 effective resolution (see Tab. 1 for details). Only the simulations that include stellar heating feedback show convergence in the number of sink particles formed.

of $\beta = 0.66$ and an Alfvén Mach number (ratio of velocity dispersion to Alfvén speed) of $\mathcal{M}_A = 2.9$. The magnetic field strengths in the simulations are consistent with observed values [102, 106, 107, 108].

We run a total of 9 simulations (see Table 1): 3 different numerical resolutions (low-resolution, mid-resolution, and high-resolution) times 3 different methods of treating the thermodynamics (isothermal EOS, polytropic EOS, and polytropic EOS + stellar heating feedback). A detailed list of the different simulation parameters (EOS and numerical resolution) and main outcomes (number of sink particles formed and their mean and median mass) is provided in Table 1.

5. Results

We now compare our 3 physical models with different EOS (isothermal, polytropic, and stellar heating) for different numerical resolutions. The main aim is to determine which model may achieve convergence with increasing numerical resolution.

5.1. Spatial distribution of gas and stars

Figure 1 shows the column density along the z-axis (the gas density integrated along the line-of-sight) of simulations 1, 2, and 3 (top panels) and simulations 7, 8, and 9 (bottom panels) from Table 1. All panels show each simulation at the same time, 0.5 Myr (or $0.55\,t_{ff}$) after the turbulence is fully developed, and self-gravity and sink particle formation had been switched on. At this time, all simulations have reached a star formation efficiency (SFE) of about 10% (i.e., a fraction of 10% of the original cloud mass has formed stars). However, the number of sink particles formed varies significantly, depending on the EOS and the numerical resolution. The isothermal simulations form 37 and 84 sink particles in the low- and high-resolution runs, respectively, and the polytopic EOS runs from 33 and 62, respectively. This means that these simulations clearly lack convergence in the number of stars formed. Only the simulations that include stellar heating feedback show convergence, with 28 and 29 sink particles formed in the low- and high-resolution simulations, respectively.

5.2. Time evolution of the number sink particles and cumulative mass distribution functions

We now investigate the numerical convergence in the number of sink particles as a function of time. The left-hand panels of Figure 2 show the number of sink particles in all our simulations as a function of SFE. We see excellent convergence in our stellar heating runs.

Finally, we show the cumulative mass functions in all our simulations in the right-hand panels of Figure 2. We see that in the isothermal and polytropic simulations, the mass functions depend on the numerical resolution. By contrast, the stellar mass functions are well converged for the simulations with stellar heating feedback.

6. Summary and conclusions

We implemented a new simple stellar heating module in the FLASH MHD code (§3). This heating feedback module takes the luminosity of a protostar or evolved star and distributes the radiated energy in a spherically symmetric approximation around each star. This allows us to approximate the effects of full radiation transfer of the stellar radiation field in a computationally fast and simple implementation.

In a series of simulations that employ an isothermal EOS (Eq. 5 with $\Gamma = 1$), a polytropic EOS (with Γ given by Eq. 7), and our new stellar heating module, we investigate the spatial distribution of formed stars, the time evolution of the number of stars, and the cumulative stellar mass functions. We compare low-resolution, mid-resolution, and high-resolution simulations and demonstrate that only the simulations with stellar heating feedback achieve numerical convergence, while simulations with a purely isothermal or polytropic EOS fail to converge on the number of stars and stellar masses formed. We conclude that the direct heating feedback from stars is essential to understanding the IMF of stars and needs to be included in simulations to achieve numerical convergence.

Finally, we caution that given our current resolutions of $\sim 100\,\mathrm{AU}$, these simulations may not be able to accurately capture physical fragmentation on the disk scale. Moreover, the radiation feedback model presented here assumes spherical symmetry, which breaks down on disk scales. Future improvements of the model should take the asymmetry of the accretion disks and the resulting shielding of radiation into account.

Acknowledgements

We thank the anonymous referee for their useful comments on the manuscript. C.F. gratefully acknowledges funding provided by the Australian Research Council's Discovery Projects (grants DP150104329 and DP170100603). The simulations presented in this work used high performance computing resources provided by the Leibniz Rechenzentrum and the Gauss Centre for Supercomputing (grants pr32lo, pr48pi and GCS Large-scale project 10391), the Partnership for Advanced Computing in Europe (PRACE grant pr89mu), the Australian National Computational Infrastructure (grant ek9), and the Pawsey Supercomputing Centre with funding from the Australian Government and the Government of Western Australia, in the

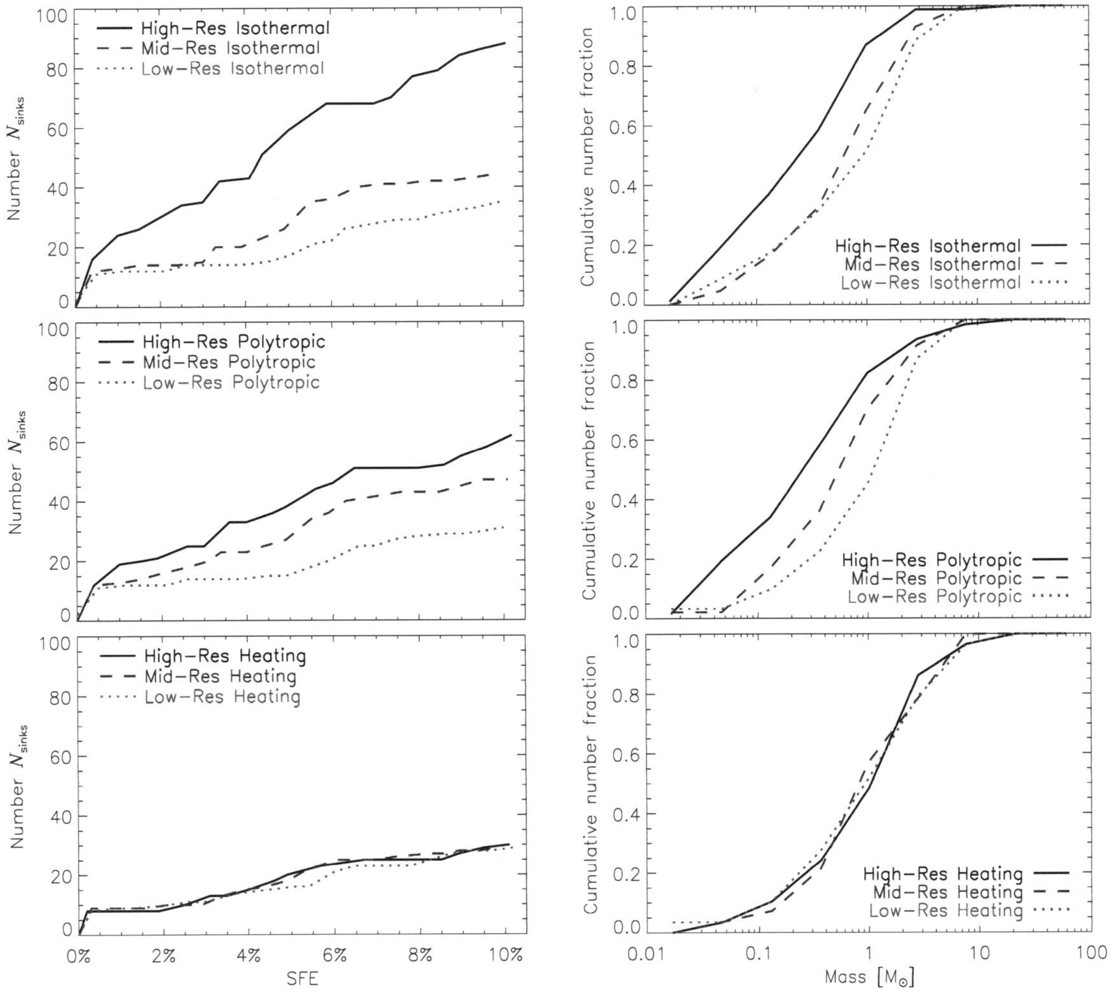

Figure 2. Left-hand panels: the number of sink particles as a function of SFE in runs with the isothermal EOS (top panels), the polytropic EOS (middle panels), and with stellar heating feedback (bottom panels). Right-hand panels: same as left-hand panels, but showing the cumulative stellar mass functions for all our simulations at SFE=10%. Each panel includes runs with three different numerical resolutions: high-resolution (solid), mid-resolution (dashed), and low-resolution (dotted). Only the simulations with stellar heating show numerical convergence in the number of sink particles and their stellar mass functions.

framework of the National Computational Merit Allocation Scheme and the ANU Allocation Scheme. The simulation software FLASH was in part developed by the DOE-supported Flash Center for Computational Science at the University of Chicago.

References

[1] Salpeter E E 1955 *ApJ* **121** 161
[2] Kroupa P 2001 *MNRAS* **322** 231–246 (*Preprint* `arXiv:astro-ph/0009005`)
[3] Muench A A, Lada E A, Lada C J and Alves J 2002 *ApJ* **573** 366–393 (*Preprint* `arXiv:astro-ph/0203122`)
[4] Chabrier G 2003 *Publ. Astron. Soc. Pacific* **115** 763–795 (*Preprint* `arXiv:astro-ph/0304382`)
[5] Alves J, Lombardi M and Lada C J 2007 *A&A* **462** L17–L21 (*Preprint* `arXiv:astro-ph/0612126`)

[6] Smith R J, Clark P C and Bonnell I A 2009 *MNRAS* **396** 830–841 (*Preprint* 0903.3240)
[7] André P, Men'shchikov A, Bontemps S, Könyves V, Motte F, Schneider N, Didelon P, Minier V, Saraceno P, Ward-Thompson D, di Francesco J, White G, Molinari S, Testi L, Abergel A, Griffin M, Henning T, Royer P, Merín B, Vavrek R, Attard M, Arzoumanian D, Wilson C D, Ade P, Aussel H, Baluteau J P, Benedettini M, Bernard J P, Blommaert J A D L, Cambrésy L, Cox P, di Giorgio A, Hargrave P, Hennemann M, Huang M, Kirk J, Krause O, Launhardt R, Leeks S, Le Pennec J, Li J Z, Martin P G, Maury A, Olofsson G, Omont A, Peretto N, Pezzuto S, Prusti T, Roussel H, Russeil D, Sauvage M, Sibthorpe B, Sicilia-Aguilar A, Spinoglio L, Waelkens C, Woodcraft A and Zavagno A 2010 *A&A* **518** L102 (*Preprint* 1005.2618)
[8] Offner S S R, Clark P C, Hennebelle P, Bastian N, Bate M R, Hopkins P F, Moraux E and Whitworth A P 2014 *Protostars and Planets VI* 53–75 (*Preprint* 1312.5326)
[9] Bate M R, Bonnell I A and Bromm V 2003 *MNRAS* **339** 577–599 (*Preprint* arXiv:astro-ph/0212380)
[10] Clark P C, Bonnell I A and Klessen R S 2008 *MNRAS* **386** 3–10 (*Preprint* 0803.4053)
[11] Price D J and Bate M R 2008 *MNRAS* **385** 1820–1834 (*Preprint* 0801.3293)
[12] Girichidis P, Federrath C, Banerjee R and Klessen R S 2011 *MNRAS* **413** 2741–2759 (*Preprint* 1008.5255)
[13] Krumholz M R, Klein R I and McKee C F 2012 *ApJ* **754** 71 (*Preprint* 1203.2620)
[14] Myers A T, McKee C F, Cunningham A J, Klein R I and Krumholz M R 2013 *ApJ* **766** 97 (*Preprint* 1211.3467)
[15] Bertelli Motta C, Clark P C, Glover S C O, Klessen R S and Pasquali A 2016 *MNRAS* **462** 4171–4182 (*Preprint* 1608.01306)
[16] Liptai D, Price D J, Wurster J and Bate M R 2016 *MNRAS* Accepted (arXiv:1610.07619) (*Preprint* 1610.07619)
[17] Krumholz M R, Klein R I and McKee C F 2007 *ApJ* **656** 959–979 (*Preprint* arXiv:astro-ph/0609798)
[18] Bate M R 2009 *MNRAS* **392** 1363–1380 (*Preprint* 0811.1035)
[19] Kuiper R, Klahr H, Beuther H and Henning T 2010 *ApJ* **722** 1556–1576 (*Preprint* 1008.4516)
[20] Commerçon B, Hennebelle P, Audit E, Chabrier G and Teyssier R 2010 *A&A* **510** L3 (*Preprint* 0912.3138)
[21] Commerçon B, Hennebelle P and Henning T 2011 *ApJ* **742** L9 (*Preprint* 1110.2955)
[22] Peters T, Banerjee R, Klessen R S and Mac Low M 2011 *ApJ* **729** 72 (*Preprint* 1010.5905)
[23] Myers A T, Klein R I, Krumholz M R and McKee C F 2014 *MNRAS* **439** 3420–3438 (*Preprint* 1401.6096)
[24] Nakamura F and Li Z Y 2007 *ApJ* **662** 395–412 (*Preprint* astro-ph/0703152)
[25] Nakamura F and Li Z Y 2011 *ApJ* **740** 36 (*Preprint* 1107.3616)
[26] Cunningham A J, Klein R I, Krumholz M R and McKee C F 2011 *ApJ* **740** 107 (*Preprint* 1104.1218)
[27] Federrath C, Schrön M, Banerjee R and Klessen R S 2014 *ApJ* **790** 128 (*Preprint* 1406.3625)
[28] Berger M J and Colella P 1989 *Journal of Computational Physics* **82** 64–84
[29] Fryxell B, Olson K, Ricker P, Timmes F X, Zingale M, Lamb D Q, MacNeice P, Rosner R, Truran J W and Tufo H 2000 *ApJ* **131** 273–334
[30] Dubey A, Fisher R, Graziani C, Jordan IV G C, Lamb D Q, Reid L B, Rich P, Sheeler D, Townsley D and Weide K 2008 *Numerical Modeling of Space Plasma Flows* (*Astronomical Society of the Pacific Conference Series* vol 385) ed Pogorelov N V, Audit E and Zank G P p 145
[31] Waagan K, Federrath C and Klingenberg C 2011 *Journal of Computational Physics* **230** 3331–3351
[32] Bouchut F, Klingenberg C and Waagan K 2007 *Numerische Mathematik* **108** 7–42
[33] Bouchut F, Klingenberg C and Waagan K 2010 *Numerische Mathematik* **115** 647–679
[34] Waagan K 2009 *Journal of Computational Physics* **228** 8609–8626
[35] Federrath C, Roman-Duval J, Klessen R S, Schmidt W and Mac Low M 2010 *A&A* **512** A81
[36] Mac Low M M and Klessen R S 2004 *RvMP* **76** 125–194
[37] Elmegreen B G 2009 *IAU Symposium* (*IAU Symposium* vol 254) ed J Andersen, J Bland-Hawthorn, & B Nordström p 289
[38] Federrath C, Rathborne J M, Longmore S N, Kruijssen J M D, Bally J, Contreras Y, Crocker R M, Garay G, Jackson J M, Testi L and Walsh A J 2016 *IAU Symposium 322* In press (arXiv:1609.08726) (*Preprint* 1609.08726)
[39] Padoan P and Nordlund Å 2002 *ApJ* **576** 870–879
[40] Elmegreen B G and Scalo J 2004 *ARAA* **42** 211–273 (*Preprint* astro-ph/0404451)
[41] Krumholz M R and McKee C F 2005 *ApJ* **630** 250–268 (*Preprint* arXiv:astro-ph/0505177)
[42] McKee C F and Ostriker E C 2007 *ARAA* **45** 565–687 (*Preprint* 0707.3514)
[43] Hennebelle P and Chabrier G 2008 *ApJ* **684** 395–410 (*Preprint* 0805.0691)
[44] Hennebelle P and Chabrier G 2009 *ApJ* **702** 1428–1442 (*Preprint* 0907.2765)
[45] Hennebelle P and Falgarone E 2012 *A&A Rev.* **20** 55 (*Preprint* 1211.0637)
[46] Federrath C and Klessen R S 2012 *ApJ* **761** 156 (*Preprint* 1209.2856)
[47] Hopkins P F 2013 *MNRAS* **430** 1653–1693 (*Preprint* 1210.0903)

[48] Krumholz M R 2014 *Physics Reports* **539** 49 (*Preprint* 1402.0867)
[49] Padoan P, Federrath C, Chabrier G, Evans II N J, Johnstone D, Jørgensen J K, McKee C F and Nordlund Å 2014 *Protostars and Planets VI* 77–100 (*Preprint* 1312.5365)
[50] Mac Low M M, Klessen R S, Burkert A and Smith M D 1998 *PhRvL* **80** 2754–2757 (*Preprint* arXiv:astro-ph/9712013)
[51] Stone J M, Ostriker E C and Gammie C F 1998 *ApJ* **508** L99–L102 (*Preprint* arXiv:astro-ph/9809357)
[52] Heyer M H, Williams J P and Brunt C M 2006 *ApJ* **643** 956–964 (*Preprint* arXiv:astro-ph/0511441)
[53] Brunt C M, Heyer M H and Mac Low M 2009 *A&A* **504** 883–890 (*Preprint* 0910.0398)
[54] Larson R B 1981 *MNRAS* **194** 809–826
[55] Heyer M H and Brunt C M 2004 *ApJ* **615** L45–L48 (*Preprint* arXiv:astro-ph/0409420)
[56] Roman-Duval J, Federrath C, Brunt C, Heyer M, Jackson J and Klessen R S 2011 *ApJ* **740** 120 (*Preprint* 1107.4569)
[57] Kritsuk A G, Norman M L, Padoan P and Wagner R 2007 *ApJ* **665** 416–431 (*Preprint* astro-ph/07043851)
[58] Schmidt W, Federrath C, Hupp M, Kern S and Niemeyer J C 2009 *A&A* **494** 127
[59] Federrath C and Klessen R S 2013 *ApJ* **763** 51 (*Preprint* 1211.6433)
[60] Federrath C 2013 *MNRAS* **436** 1245–1257 (*Preprint* 1306.3989)
[61] Eswaran V and Pope S B 1988 *Computers and Fluids* **16** 257–278
[62] Schmidt W, Hillebrandt W and Niemeyer J C 2006 *Computers and Fluids* **35** 353–371
[63] Federrath C, Rathborne J M, Longmore S N, Kruijssen J M D, Bally J, Contreras Y, Crocker R M, Garay G, Jackson J M, Testi L and Walsh A J 2016 *ApJ* In press (arXiv:1609.05911) (*Preprint* 1609.05911)
[64] Federrath C, Klessen R S and Schmidt W 2008 *ApJ* **688** L79–L82
[65] Molina F Z, Glover S C O, Federrath C and Klessen R S 2012 *MNRAS* **423** 2680–2689
[66] Konstandin L, Federrath C, Klessen R S and Schmidt W 2012 *JFM* **692** 183–206 (*Preprint* 1111.2748)
[67] Nolan C A, Federrath C and Sutherland R S 2015 *MNRAS* **451** 1380–1389 (*Preprint* 1504.04370)
[68] Federrath C and Banerjee S 2015 *MNRAS* **448** 3297–3313 (*Preprint* 1412.2756)
[69] Padoan P, Jones B J T and Nordlund A P 1997 *ApJ* **474** 730 (*Preprint* arXiv:astro-ph/9603061)
[70] Brunt C M 2010 *A&A* **513** A67 (*Preprint* 1002.1239)
[71] Price D J, Federrath C and Brunt C M 2011 *ApJ* **727** L21 (*Preprint* 1010.3754)
[72] Ginsburg A, Federrath C and Darling J 2013 *ApJ* **779** 50 (*Preprint* 1310.0809)
[73] Kainulainen J, Federrath C and Henning T 2013 *A&A* **553** L8
[74] Federrath C, Banerjee R, Clark P C and Klessen R S 2010 *ApJ* **713** 269–290 (*Preprint* 1001.4456)
[75] Truelove J K, Klein R I, McKee C F, Holliman II J H, Howell L H and Greenough J A 1997 *ApJ* **489** L179
[76] Federrath C, Sur S, Schleicher D R G, Banerjee R and Klessen R S 2011 *ApJ* **731** 62
[77] Sur S, Schleicher D R G, Banerjee R, Federrath C and Klessen R S 2010 *ApJ* **721** L134–L138
[78] Ricker P M 2008 *ApJ* **176** 293–300 (*Preprint* 0710.4397)
[79] Sutherland R S and Dopita M A 1993 *ApJ* **88** 253–327
[80] Glover S C O, Federrath C, Mac Low M and Klessen R S 2010 *MNRAS* **404** 2–29 (*Preprint* 0907.4081)
[81] Grassi T, Bovino S, Schleicher D R G, Prieto J, Seifried D, Simoncini E and Gianturco F A 2014 *MNRAS* **439** 2386–2419 (*Preprint* 1311.1070)
[82] Masunaga H and Inutsuka S i 2000 *ApJ* **531** 350–365
[83] Wolfire M G, Hollenbach D, McKee C F, Tielens A G G M and Bakes E L O 1995 *ApJ* **443** 152–168
[84] Omukai K, Tsuribe T, Schneider R and Ferrara A 2005 *ApJ* **626** 627–643 (*Preprint* arXiv:astro-ph/0503010)
[85] Pavlovski G, Smith M D and Mac Low M M 2006 *MNRAS* **368** 943–958 (*Preprint* arXiv:astro-ph/0504504)
[86] Glover S C O and Mac Low M M 2007 *ApJ* **169** 239–268 (*Preprint* arXiv:astro-ph/0605120)
[87] Glover S C O and Mac Low M M 2007 *ApJ* **659** 1317–1337 (*Preprint* arXiv:astro-ph/0605121)
[88] Hill T, Motte F, Didelon P, Bontemps S, Minier V, Hennemann M, Schneider N, André P, Men'shchikov A, Anderson L D, Arzoumanian D, Bernard J P, di Francesco J, Elia D, Giannini T, Griffin M J, Könyves V, Kirk J, Marston A P, Martin P G, Molinari S, Nguyen Luong Q, Peretto N, Pezzuto S, Roussel H, Sauvage M, Sousbie T, Testi L, Ward-Thompson D, White G J, Wilson C D and Zavagno A 2011 *A&A* **533** A94 (*Preprint* 1108.0941)
[89] Hennemann M, Motte F, Schneider N, Didelon P, Hill T, Arzoumanian D, Bontemps S, Csengeri T, André P, Konyves V, Louvet F, Marston A, Men'shchikov A, Minier V, Nguyen Luong Q, Palmeirim P, Peretto N, Sauvage M, Zavagno A, Anderson L D, Bernard J P, Di Francesco J, Elia D, Li J Z, Martin P G, Molinari S, Pezzuto S, Russeil D, Rygl K L J, Schisano E, Spinoglio L, Sousbie T, Ward-Thompson D and White G J 2012 *A&A* **543** L3 (*Preprint* 1206.1243)
[90] Glover S C O and Clark P C 2012 *MNRAS* **426** 377–388 (*Preprint* 1203.4251)
[91] Larson R B 1969 *MNRAS* **145** 271

[92] Yorke H W, Bodenheimer P and Laughlin G 1993 *ApJ* **411** 274–284
[93] Offner S S R, Klein R I, McKee C F and Krumholz M R 2009 *ApJ* **703** 131–149 (*Preprint* 0904.2004)
[94] Peters T, Banerjee R, Klessen R S, Mac Low M, Galván-Madrid R and Keto E R 2010 *ApJ* **711** 1017–1028 (*Preprint* 1001.2470)
[95] Nakano T, Hasegawa T and Norman C 1995 *ApJ* **450** 183
[96] Nakano T, Hasegawa T, Morino J I and Yamashita T 2000 *ApJ* **534** 976–983
[97] Tan J C and McKee C F 2004 *ApJ* **603** 383–400 (*Preprint* astro-ph/0307414)
[98] Palla F and Stahler S W 1991 *ApJ* **375** 288–299
[99] Hosokawa T and Omukai K 2009 *ApJ* **691** 823–846 (*Preprint* 0806.4122)
[100] Guszejnov D, Krumholz M R and Hopkins P F 2016 *MNRAS* **458** 673–680 (*Preprint* 1510.05040)
[101] Federrath C, Klessen R S and Schmidt W 2009 *ApJ* **692** 364–374
[102] Falgarone E, Puget J L and Perault M 1992 *A&A* **257** 715–730
[103] Kauffmann J, Pillai T and Goldsmith P F 2013 *ApJ* **779** 185 (*Preprint* 1308.5679)
[104] Hernandez A K and Tan J C 2015 *ApJ* **809** 154 (*Preprint* 1502.00183)
[105] Federrath C 2016 *Journal of Plasma Physics* In press (arXiv:1610.08132) (*Preprint* 1610.08132)
[106] Falgarone E, Troland T H, Crutcher R M and Paubert G 2008 *A&A* **487** 247–252
[107] Crutcher R M, Wandelt B, Heiles C, Falgarone E and Troland T H 2010 *ApJ* **725** 466–479
[108] Crutcher R M 2012 *ARAA* **50** 29–63

IOP Conf. Series: Journal of Physics: Conf. Series **837** (2017) 012008 doi:10.1088/1742-6596/837/1/012008

Toward realistic simulations of magneto-thermal winds from weakly-ionized protoplanetary disks

Oliver Gressel

Niels Bohr International Academy, The Niels Bohr Institute, Blegdamsvej 17, DK-2100, Copenhagen Ø, Denmark

E-mail: gressel@nbi.ku.dk

Abstract. Protoplanetary disks (PPDs) accrete onto their central T Tauri star via magnetic stresses. When the effect of ambipolar diffusion (AD) is included, and in the presence of a vertical magnetic field, the disk remains laminar between 1-5 au, and a magnetocentrifugal disk wind forms that provides an important mechanism for removing angular momentum. We present global MHD simulations of PPDs that include Ohmic resistivity and AD, where the time-dependent gas-phase electron and ion fractions are computed under FUV and X-ray ionization with a simplified recombination chemistry. To investigate whether the mass loading of the wind is potentially affected by the limited vertical extent of our existing simulations, we attempt to develop a model of a realistic disk atmosphere. To this end, by accounting for stellar irradiation and diffuse reprocessing of radiation, we aim at improving our models towards more realistic thermodynamic properties.

1. Introduction

Interpreting observational properties of T Tauri systems [1] is closely tied to understanding the complex dynamical evolution of gaseous protoplanetary disks (PPDs), both in terms of their chemistry, and in terms of the microphysics that govern the evolution of the embedded magnetic fields. Moreover, with PPDs being the birth sites for extrasolar planets, a sound physical picture of the dust [2] and planetesimal [3] evolution is needed to ultimately provide the building blocks for a comprehensive theory of planet formation.

The fundamental drivers of disk evolution are mass loss processes and redistribution of angular momentum [4]. In sufficiently ionized parts of the disk, the latter can be achieved by turbulent stresses. When accounting for the ionization structure of a typical PPD, large parts of the disk, however, remain laminar owing to the effect of ambipolar diffusion (AD). In this situation, angular momentum is primarily transported via a magnetocentrifugal disk wind [5, 6]. While this picture is further complicated when accounting for the Hall effect [7, 8], it has nevertheless become clear that the thermal structure of the disk plays an important role in setting the mass loading of the disk wind [9], and hence the timescale on which the system evolves [10, 11].

2. Methods

As in our previous work [6], we are solving the single-fluid MHD equations including Ohmic resistivity and ambipolar diffusion, that is, the electromotive force resulting from the mutual collision of ions and neutrals. The diffusion coefficients, $\eta_{\rm O}$, and $\eta_{\rm AD}$ are specified by means of a look-up table, which has been obtained by solving a simple chemical ionization/recombination network. The resulting electromotive forces stemming from the Ohmic and ambipolar diffusion terms are then given by

$$\mathbf{E}_{\rm O} \equiv -\eta_{\rm O}\,(\nabla\times\mathbf{B})\,, \quad \text{and} \tag{1}$$

$$\mathbf{E}_{\rm AD} \equiv \eta_{\rm AD}\left[(\nabla\times\mathbf{B})\times\hat{\mathbf{b}}\right]\times\hat{\mathbf{b}}\,, \tag{2}$$

with $\hat{\mathbf{b}} \equiv \mathbf{B}/|\mathbf{B}|$ being the unit vector along $\mathbf{B}$, and where the double vector product results in an additional minus sign, such that positive coefficients $\eta_{\rm O}$ and $\eta_{\rm AD}$ signify diffusion of magnetic fields, with the latter only being sensitive to currents perpendicular to the field direction.

For the typical number densities in the context of PPDs, the frictional coupling between the ions and neutrals happens so quickly (compared to dynamical timescales of interest) that we assume the charged species move at their terminal velocity with respect to the neutrals, that is, where the Lorentz force is balanced by the drag. Because of the low degree of ionization, we formulate our continuity equation and momentum conservation in terms of the neutral component, although it does experience the Lorentz force as mediated by particle collisions.

We use a modified version of the NIRVANA-III finite volume Godunov code [12, 13]. The code adopts a total-energy formulation[1] with conserved variables ρ, $\rho\mathbf{v}$, and $e \equiv \epsilon + \rho\mathbf{v}^2/2 + \mathbf{B}^2/2$. Together with the conservation of radiation energy, $E_{\rm R}$, and defining the total pressure, $p^\star$, as the sum of the gas and magnetic pressures, the system of equations we solve reads

$$\begin{aligned}
\partial_t\rho + \nabla\cdot(\rho\mathbf{v}) &= 0\,, && (3)\\
\partial_t(\rho\mathbf{v}) + \nabla\cdot\left[\rho\mathbf{v}\mathbf{v} + p^\star I - \mathbf{B}\mathbf{B}\right] &= -\rho\nabla\Phi + \rho\kappa_{\rm R}/c\,\mathbf{F}_{\rm R}\,, && (4)\\
\partial_t e + \nabla\cdot\left[(e+p^\star)\mathbf{v} - (\mathbf{v}\cdot\mathbf{B})\mathbf{B}\right] &= \nabla\cdot\left[(\mathbf{E}_{\rm O}+\mathbf{E}_{\rm AD})\times\mathbf{B}\right] - \rho(\nabla\Phi)\cdot\mathbf{v} && \\
&\quad + c\,\rho\kappa_{\rm P}\left(E_{\rm R} - a_{\rm R}\,T^4\right) + \rho\kappa_{\rm R}/c\,\mathbf{F}_{\rm R}\cdot\mathbf{v} + \mathcal{Q}^+_{\rm irr}\,, && (5)\\
\partial_t E_{\rm R} + \nabla\cdot(E_{\rm R}\mathbf{v}) &= -\,c\,\rho\kappa_{\rm P}\left(E_{\rm R} - a_{\rm R}\,T^4\right) - \nabla\cdot\mathbf{F}_{\rm R} - \mathcal{P}_{\rm R}:\nabla\mathbf{v}\,, && (6)\\
\partial_t\mathbf{B} - \nabla\times\left[\mathbf{v}\times\mathbf{B} + \mathbf{E}_{\rm O} + \mathbf{E}_{\rm AD}\right] &= 0\,, && (7)
\end{aligned}$$

where $\mathbf{F}_{\rm R}$ is the radiation flux, $\mathcal{P}_{\rm R}$ is the radiation pressure tensor, $\kappa_{\rm R}$ and $\kappa_{\rm P}$ are the Rosseland and Planck mean opacities, respectively, $a_{\rm R} \equiv 4\,\sigma/c$ is the radiation density constant, $T \equiv \bar{\mu}m_{\rm H}/k_{\rm B}\;p/\rho$ is the gas temperature, and the gas pressure is $p = (\gamma-1)\epsilon$, where $\gamma = 7/5$ is chosen as appropriate for an ideal diatomic gas. The gravitational term $\Phi(r) \equiv -G\,M_\odot/r$ represents the point-mass potential of the solar-mass star at the center of our spherical-polar coordinate system, and $\mathcal{Q}^+_{\rm irr}$ represents external irradiation heating due to the star.

2.1. Flux limited diffusion

The above equations can only be solved once the radiation flux and pressure tensor are specified. An attractive method for obtaining the Eddington tensor that relates $\mathcal{P}_{\rm R}$ with $E_{\rm R}$ is to solve for the steady-state but angle-dependent radiation intensity and integrate the first (for $\mathbf{F}_{\rm R}$) and second (for $\mathcal{P}_{\rm R}$) order moments directly [14]. In the interest of maintaining minimum algorithmic complexity and computational expense, we instead adopt the classical *ad hoc* closure

$$\mathbf{F}_{\rm R} = -\lambda(R)\,\frac{c}{\rho\kappa_{\rm R}}\nabla E_{\rm R}\,, \tag{8}$$

[1] Note that we do not include the radiation energy density, $E_{\rm R}$, in the total energy (as it is sub-dominant in the problem we consider). We however retain the radiation flux in the momentum equation for consistency.

that specifies $\mathbf{F}_\mathrm{R}$ in terms of a diffusive flux with diffusion coefficient $D \equiv \lambda(R)c/\rho\kappa_\mathrm{R}$, where

$$R \equiv \frac{|\nabla E_\mathrm{R}|}{\rho\kappa_\mathrm{R}\langle E_\mathrm{R}\rangle} \tag{9}$$

is a dimensionless number specifying how abruptly the radiation energy density varies compared to the length scale defined by the optical extinction coefficient $\rho\kappa_\mathrm{R}$, and where $\lambda(R) \to 1/R$ (for $R \gg 1$) is a limiter function [15] that guarantees $|\mathbf{F}_\mathrm{R}| < cE_\mathrm{R}$ in regions of low optical depth, that is, in regions where the diffusion approximation is not valid. In the optically thick limit, $\lambda(R) \to 1/3$ (for $R \ll 1$), which corresponds to the Eddington approximation.

The described approach has its known shortcomings over characteristics-based methods [14], but in combination with stellar irradiation heating (determined using a simplified ray tracing algorithm) it has been deemed an acceptable compromise [16, 17, 18] in terms of being able to incorporate radiation thermodynamics in fully dynamical 3D simulations, whereas more accurate Monte Carlo methods are comparatively expensive and have to deal with statistical noise [19].

We currently treat both irradiation and diffuse redistribution in the *gray* approximation, but multigroup approaches are straightforward [20], especially for the irradiation component [17, 18], wherein computational expenditure scales with the number of frequency bins, and where increased realism can be achieved for the thermal structure of the outer disk ($R \gtrsim 20\,\mathrm{au}$). For our PPD model, we precompute look-up tables for mean opacities, $\kappa_\mathrm{R}(\rho, T)$ and $\kappa_\mathrm{P}(\rho, T)$, using D. Semenov's `opacity.f` [21], where a fixed dust-to-gas mass ratio of 0.01395 is assumed. To account for depletion of small dust by grain growth, we enable the reduction of the obtained opacities by a scale factor. Since the opacity tables combine contributions from dust grains and gas molecules, this is only valid for temperatures below the dust sublimation threshold.

2.2. Reduced speed of light approximation

The thermodynamic coupling of the gaseous matter with the radiation field is described by the $c\,\rho\kappa_\mathrm{P}\left(E_\mathrm{R} - a_\mathrm{R}T^4\right)$ terms in eqns. (5) and (6), respectively. Subsuming the $-\nabla\cdot\mathbf{F}_\mathrm{R}$ term in eqn. (6) with the definition of the diffusive flux (8) amounts to a diffusion equation for the redistribution of radiation energy. Both effects can be combined into the subsystem

$$\partial_t \epsilon = +c\,\rho\kappa_\mathrm{P}\left(E_\mathrm{R} - a_\mathrm{R}T^4\right), \tag{10}$$

$$\frac{c}{\hat{c}}\,\partial_t E_\mathrm{R} = -c\,\rho\kappa_\mathrm{P}\left(E_\mathrm{R} - a_\mathrm{R}T^4\right) + \nabla\cdot\left[D\,\nabla E_\mathrm{R}\right], \tag{11}$$

which we solve by means of operator splitting. Unlike in ref. [18], we do not include the $\mathcal{P}_\mathrm{R}:\nabla\mathbf{v}$ term in this subsystem but instead treat it as a regular source term when updating the main hyperbolic system of equations. Since, for a large diffusion coefficient D, the parabolic system (11) becomes stiff, the most common approach is to use implicit methods to solve it. Especially in view of applications employing adaptive mesh refinement, we have chosen to avoid an implicit update for $\nabla\cdot(D\,\nabla E_\mathrm{R})$, as it demands costly non-local communication patterns, and a potentially expensive matrix inversion.

Instead, we integrate the diffusion part of (11) in a time-explicit fashion, and use the *reduced speed of light* approach [22] to ameliorate the strict time-step constraints. This method has recently been employed in the context of simulations of the interstellar medium [23]. The approximation is valid as long as the radiative timescales resulting from the adopted artificial value of $\hat{c} = \phi c$ (with $\phi = \mathrm{const.} < 1$) are still short compared to any relevant dynamical timescales. In the context of PPDs, we are mainly interested in the role of radiative effects in setting the consistent temperature structure of the disk, and true radiation hydrodynamic effects are believed to only be of minor importance during the T Tauri phase [24].

The approximation is introduced by amplifying the left-hand-side of (11) by a factor of $c/\hat{c} > 1$. It is crucial that, because all other terms remain unaffected, this implies the modification does not alter the late-time steady-state solution, where $\partial_t \to 0$, but only changes the *timescale* on which this solution is achieved. On practical grounds, we multiply (11) by $\hat{c}/c$, which implies replacing c for $\hat{c}$ in the radiation matter coupling, and attenuating the diffusion term by a factor of ϕ. This illustrates how the method works to weaken the stiffness of the diffusion term. To explicitly integrate (11), we employ the second-order accurate Runge-Kutta-Legendre (RKL2) super-time-stepping scheme [25], that is already used for the updates of the other parabolic terms (such as, viscosity, thermal conduction, resistive and ambipolar diffusion) in the code.

2.3. Radiation matter coupling

Even with the reduced speed, $\hat{c} = \phi\, c$, the radiation-matter-coupling term $\hat{c}\,\rho\kappa_{\rm P}\left(E_{\rm R} - a_{\rm R}\,T^4\right)$ can itself contribute a severe timestep constraint in regions of high opacity, where the coupling becomes stiff. Ignoring, for the moment, the diffusion term in (11), the coupling amounts to an ordinary differential equation, similar to production/destruction equations that are common in other fields of science. While these are typically solved by explicit Runge-Kutta (RK) methods, higher order predictor/corrector schemes do not typically guarantee positivity or conservation of, in our case, the energy $\bar{e} \equiv (\phi\,\epsilon + E_{\rm R})$.

There however exists a class of modified RK methods [26] that employ the so-called Patankar trick – an implicit weighting of the production/destruction terms with the ratio of the evolved quantity before and after the update. It can be shown that for a single-step update, such a weighting precludes that negative values are obtained. Moreover, since the weighting factors are overall symmetric, conservation is guaranteed. Specifically, we use the MPRK scheme given by eqn. (27) in ref. [26]. The resulting update is formally implicit, but can algebraically be manipulated into fully explicit form, that is,

$$\begin{aligned} \Delta e &\equiv \epsilon^{(n+1)} - \epsilon^{(n)} \\ &\equiv \left(E_{\rm R}^{(n)} - E_{\rm R}^{(n+1)}\right)\phi^{-1} = \frac{E_{\rm R}^{(n)} - \left(a_{\rm R}\,T^4\right)^{(n)}}{\phi + \left(a_{\rm R}\,T^4\right)^{(n)}/\epsilon^{(n)} + \left(c\rho\kappa_{\rm P}\Delta t\right)^{-1}}\,, \end{aligned} \tag{12}$$

for the predictor step, and

$$\Delta e = \frac{E_{\rm R}^{\star}\,\epsilon^{\dagger} E_{\rm R}^{(n)} - \left(a_{\rm R}\,T^4\right)^{\star} E_{\rm R}^{\dagger}\epsilon^{(n)}}{\phi E_{\rm R}^{\star}\epsilon^{\dagger} + \left(a_{\rm R}\,T^4\right)^{\star} E_{\rm R}^{\dagger} + E_{\rm R}^{\dagger}\epsilon^{\dagger}\left(c\rho\kappa_{\rm P}\Delta t\right)^{-1}}\,, \tag{13}$$

for the corrector step, where $\epsilon^{\dagger} \equiv \epsilon^{(n)} + \Delta e$ is the forward-Euler predictor value of $\epsilon^{(n+1)}$, and $\epsilon^{\star} \equiv 0.5\,(\epsilon^{(n)} + \epsilon^{\dagger})$ is the time-averaged state.[2] Compared to implicit methods, that demand iterative root-finding, or so-called θ-schemes (see, e.g., section 3.4 in ref. [23]) the method presented here offers a relatively inexpensive, parameter-free non-iterative alternative.

2.4. Irradiation heating

Our existing global disk models [27, 28] have either assumed a locally-isothermal temperature $T = T(R)$, with R being the cylindrical radius from the star, or have used an adiabatic equation of state with a Newtonian cooling term in the energy equation that reinstated the $T(R)$ profile on a specified timescale (typically a short fraction of the local orbital period).

Compared to this, even relatively basic models [29] of dust absorption and re-radiation of star light in the disk surface, obtain a much more complex temperature structure within the PPD

[2] The same conventions, of course, apply for the terms $(a_{\rm R}\,T^4)$, and $E_{\rm R}$.

– with superheated surface layers and cool interiors. To account for such effects, we include a frequency-integrated radially attenuated irradiation flux

$$\mathbf{F}_{\mathrm{R,irr}}(r) = F(r_\star)\left(\frac{r_\star}{r}\right)^2 \exp\left(-\tau_{\mathrm{P}}(r)\right)\hat{\mathbf{r}} \tag{14}$$

from the central star with effective temperature $T_\star = 5780\,\mathrm{K}$, and $r_\star = r_\odot$, and where the optical depth, $\tau_{\mathrm{P}}(r) \equiv \int_{r_\star}^{r} \rho(r')\,\kappa_{\mathrm{P}}(\rho, T_\star)\,\mathrm{d}r'$ is obtained by integrating along radial rays. Following previous work [16, 17, 18], we obtain the irradiation energy source term $\mathcal{Q}^{+}_{\mathrm{irr}}$ by computing the negative divergence of (14) over each grid cell. In regions of low optical depth, we use the integral formulation

$$\mathcal{Q}^{+}_{\mathrm{irr}}(r_i) = \frac{\rho\kappa_{\mathrm{P}}}{\Delta V_r}\int_{r_{i-1/2}}^{r_{i+1/2}} \hat{\mathbf{r}}\cdot\mathbf{F}_{\mathrm{R,irr}}(r')\,r'^2\mathrm{d}r'\,, \tag{15}$$

with $\Delta V_r = {}^1\!/_3\,(r^3_{i+1/2} - r^3_{i-1/2})$ instead, as this formulation has been found to produce a more accurate solution on the discretized mesh, when differences across cells are small [30].

3. Results

3.1. Radiative transfer test problems

To verify our implementation of the radiation-matter-coupling term, we have performed a simple one-zone model [31], where the thermal energy density, ϵ, is initially out of balance with the radiation energy density $E_{\mathrm{R}} = 10^{12}\,\mathrm{erg\,cm^{-3}}$. In Fig. 1, we show three cases with $\epsilon_0 = 10^2$, 10^6, and $10^{10}\,\mathrm{erg\,cm^{-3}}$, which all converge to the same final equilibrium state. For the purpose of plotting the curves, we have limited the timestep artificially to sample timescales shorter than the radiative equilibration timescale. We have, however, tested that even for numerical time steps somewhat larger than the coupling timescale the scheme remains stable, as expected from the implicit-like integration scheme (see sect. 2.3) that we use.

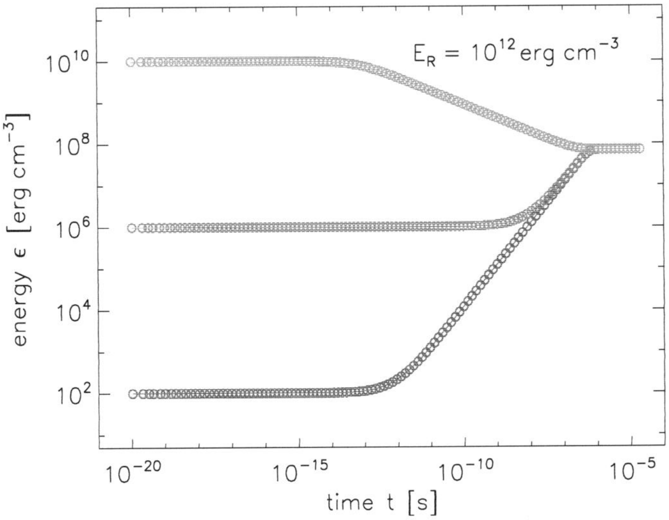

Figure 1. A simple single zone radiation matter coupling test, with $\kappa_{\mathrm{P}} = 0.4\,\mathrm{cm^2\,g^{-1}}$, $\bar{\mu} = 0.6$, $\gamma = 5/3$, and $\rho = 10^{-7}\,\mathrm{g\,cm^{-3}}$.

A standard test case for assessing the interplay of the radiation-matter-coupling with the radiation diffusion are radiative shocks, for which there exist semi-analytic solutions in simplified situations [32]. In Fig. 2, we plot the solution of the Ma = 2 case from ref. [32], using $n = 128$ grid cells in the x direction, as well as two levels of adaptive mesh refinement (triggered by gradients in the thermal energy, ϵ), which are shown as gray shaded areas in the plot. As seen in

IOP Conf. Series: Journal of Physics: Conf. Series **837** (2017) 012008 doi:10.1088/1742-6596/837/1/012008

the lower panel of Fig. 2, apart from the shock location (which has been shifted by $x_{\rm offs} = 22.5\,{\rm cm}$ with respect to the semi-analytic solution), all quantities agree to within ‰ accuracy. We have also successfully performed the harder Ma = 5 test, which we omit here for brevity. An excellent description of the radiative shock test, including the precise values used for initial conditions, can be found in sect. 4.4 of ref. [18].

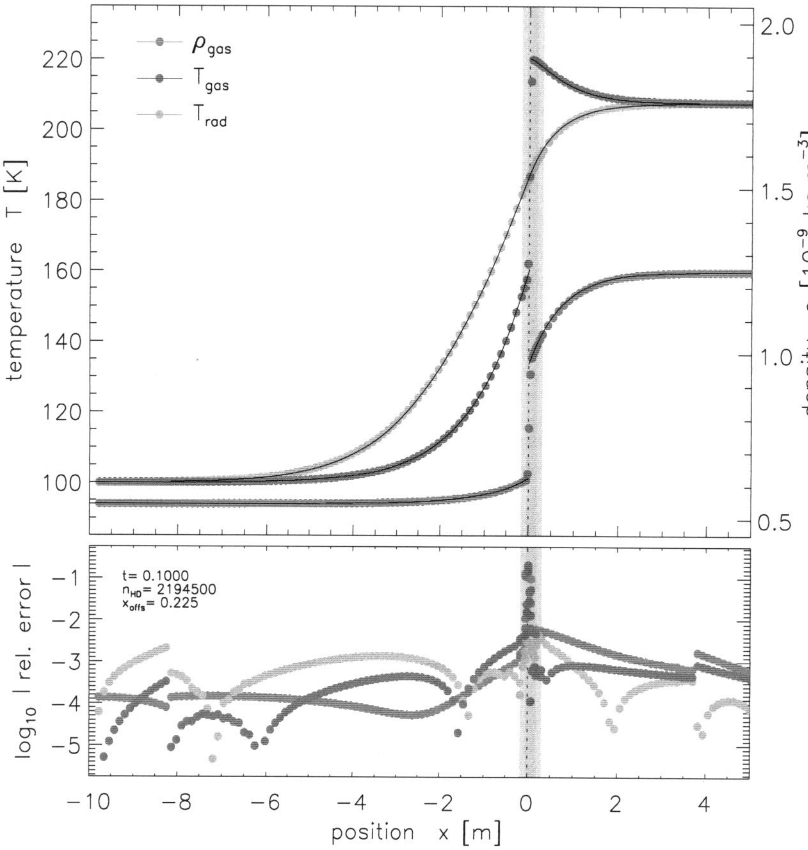

Figure 2. Radiative shock with Ma = 2 and resolution $n = 128$ +2 levels of adaptive mesh refinement (shaded areas). Upper panel: Profiles of $\rho_{\rm gas}$ (blue), $T_{\rm gas}$ (red), and $T_{\rm rad}$ (yellow). Lower panel: deviation from the semi-analytic solution (black).

3.2. Preliminary global MHD simulations with irradiation

Returning to the original motivation for implementing radiative physics in our modified version of the NIRVANA-III code, we conclude by presenting a preliminary snapshot of a global axisymmetric protoplanetary disk simulation including Ohmic resistivity, ambipolar diffusion, radiation diffusion, and stellar irradiation. The ultimate goal of these simulations is to study how the mass-loading of the magnetocentrifugal wind depends on the disk thermal structure.

As a simple proof-of-concept, we present a close-up of the inner disk in a simulation covering seven (initial) pressure scale heights in latitude (see Fig. 3). The basic disk setup is very similar to the one used in ref. [6], and we have additionally used opacities that correspond to a dust-depletion of a factor of ten compared to the typical interstellar abundance. Similar to our previous simulations, the magnetic field lines (white) bend outward, and in the upper disk layers, where the matter is sufficiently coupled to the magnetic field, a magnetocentrifugal disk wind ensues (black vectors).

Iso-contour lines (gray) of the radiation temperature $T_{\rm rad} \equiv E_{\rm R}^{1/4} a_{\rm R}^{-1/4}$ illustrate the disk's thermal structure that deviates noticeably from the constant-on-cylinders radial temperature

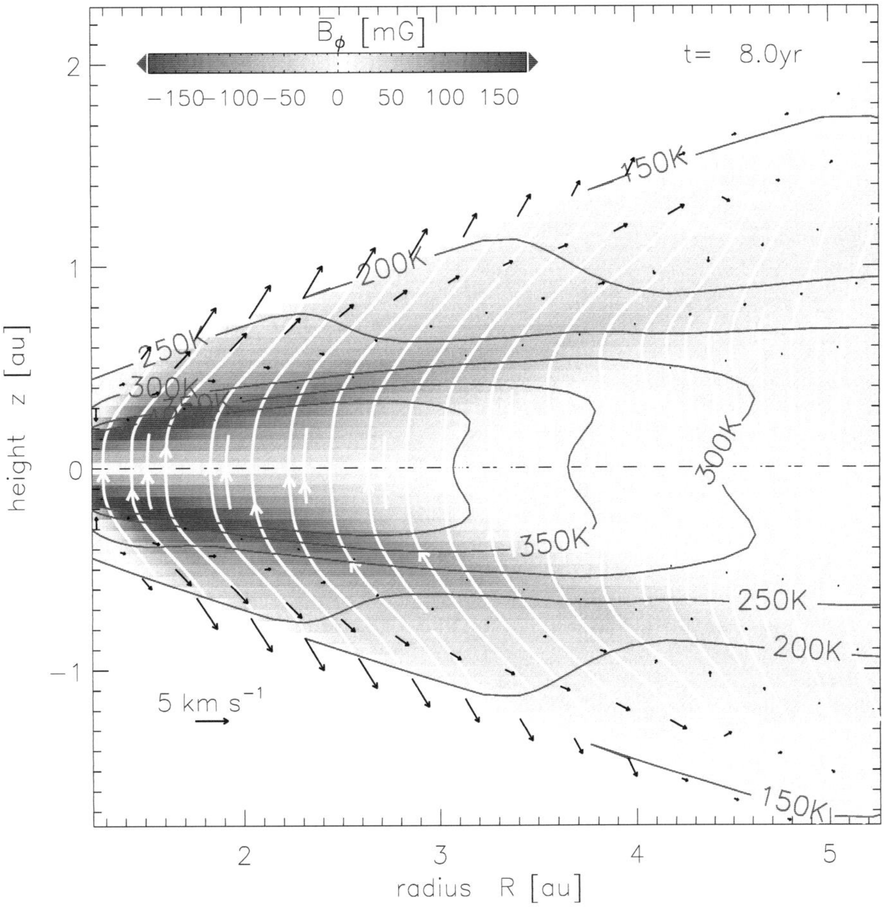

Figure 3. Detail from a proof-of-concept global MHD simulation of a PPD with ambipolar diffusion, radiative transfer and stellar irradiation, showing $\bar{B}_\phi$ (colour), poloidal velocity (black) and magnetic field lines (white), and iso-contours of the radiation temperature (grey).

profile $T = T(R)$, that we have assumed previously. The presented preliminary run used a reduced-speed-of-light factor $\phi = 10^{-4}$. Further tests will show whether the chosen time-explicit framework is powerful and efficient enough to be of use in realistic situations.

Acknowledgments

The author thanks Jon Ramsey for many useful discussions, for carefully reading this manuscript, and for providing his code to compute semi-analytic reference solutions of radiative shocks. Dmitry Semenov is acknowledged for providing helpful clues in connection with his opacity code. The research leading to these results has received funding from the European Research Council (ERC) under the European Union's Horizon 2020 research and innovation programme (grant agreement No 638596).

References

[1] Williams J P and Cieza L A 2011 ARA&A **49** 67–117 (*Preprint* 1103.0556)
[2] Johansen A, Blum J, Tanaka H, Ormel C, Bizzarro M and Rickman H 2014 PPVI 547
[3] Gressel O, Nelson R P and Turner N J 2012 MNRAS **422** 1140–1159 (*Preprint* 1202.0771)
[4] Turner N J, Fromang S, Gammie C, Klahr H, Lesur G, Wardle M and Bai X N 2014 PPVI 411–432

[5] Bai X N and Stone J M 2013 ApJ **769** 76 (*Preprint* 1301.0318)
[6] Gressel O, Turner N J, Nelson R P and McNally C P 2015 ApJ **801** 84 (*Preprint* 1501.05431)
[7] Lesur G, Kunz M W and Fromang S 2014 A&A **566** A56 (*Preprint* 1402.4133)
[8] Simon J B, Lesur G, Kunz M W and Armitage P J 2015 MNRAS **454** 1117–1131 (*Preprint* 1508.00904)
[9] Bai X N, Ye J, Goodman J and Yuan F 2016 ApJ **818** 152 (*Preprint* 1511.06769)
[10] Bai X N 2016 ApJ **821** 80 (*Preprint* 1603.00484)
[11] Suzuki T K, Ogihara M, Morbidelli A, Crida A and Guillot T 2016 A&A (*Preprint* 1609.00437)
[12] Ziegler U 2004 JCoPh **196** 393–416
[13] Ziegler U 2011 JCoPh **230** 1035–1063
[14] Jiang Y F, Stone J M and Davis S W 2012 ApJS **199** 14 (*Preprint* 1201.2223)
[15] Levermore C D and Pomraning G C 1981 ApJ **248** 321–334
[16] Bitsch B, Crida A, Morbidelli A, Kley W and Dobbs-Dixon I 2013 A&A **549** A124 (*Preprint* 1211.6345)
[17] Kuiper R and Klessen R S 2013 A&A **555** A7 (*Preprint* 1305.2197)
[18] Ramsey J P and Dullemond C P 2015 A&A **574** A81 (*Preprint* 1409.3011)
[19] Noebauer U, Sim S, Kromer M, Röpke F and Hillebrandt W 2012 MNRAS **425** 1430 (*Preprint* 1206.6263)
[20] González M, Vaytet N, Commerçon B and Masson J 2015 A&A **578** A12 (*Preprint* 1504.01894)
[21] Semenov D, Henning T, Helling C, Ilgner M and Sedlmayr E 2003 A&A **410** 611–621
[22] Gnedin N Y and Abel T 2001 NewA **6** 437–455 (*Preprint* astro-ph/0106278)
[23] Skinner M A and Ostriker E C 2013 ApJS **206** 21 (*Preprint* 1306.0010)
[24] Hartmann L 1998 Accretion Processes in Star Formation
[25] Meyer C D, Balsara D S and Aslam T D 2012 MNRAS **422** 2102–2115
[26] Burchard H, Deleersnijder E and Meister A 2003 AppliedNumMath **47** 1 – 30 ISSN 0168-9274
[27] Nelson R P, Gressel O and Umurhan O M 2013 MNRAS **435** 2610–2632 (*Preprint* 1209.2753)
[28] Gressel O, Nelson R P, Turner N J and Ziegler U 2013 ApJ **779** 59 (*Preprint* 1309.2871)
[29] Chiang E I and Goldreich P 1997 ApJ **490** 368–376 (*Preprint* astro-ph/9706042)
[30] Bruls J H M J, Vollmöller P and Schüssler M 1999 A&A **348** 233–248
[31] Turner N J and Stone J M 2001 ApJS **135** 95–107 (*Preprint* astro-ph/0102145)
[32] Lowrie R B and Edwards J D 2008 Shock Waves **18** 129–143

Theoretical Models of Protostellar Binary and Multiple Systems with AMR Simulations

Tomoaki Matsumoto[1], Kazuki Tokuda[2], Toshikazu Onishi[2], Shu-ichiro Inutsuka[3], Kazuya, Saigo[2], Shigehisa Takakuwa[4],

[1] Faculty of Sustainability Studies, Hosei University, Fujimi, Chiyoda-ku, Tokyo 102-8160, Japan

[2] Department of Physical Science, Graduate School of Science, Osaka Prefecture University, 1-1 Gakuen-cho, Naka-ku, Sakai, Osaka 599-8531, Japan

[3] Department of Physics, Nagoya University, Chikusa-ku, Nagoya 464-8602, Japan

[4] Department of Physics and Astronomy, Graduate School of Science and Engineering, Kagoshima University, 1-21-35 Korimoto, Kagoshima 890-0065, Japan

E-mail: matsu@hosei.ac.jp

Abstract. We present theoretical models for protostellar binary and multiple systems based on the high-resolution numerical simulation with an adaptive mesh refinement (AMR) code, SFUMATO. The recent ALMA observations have revealed early phases of the binary and multiple star formation with high spatial resolutions. These observations should be compared with theoretical models with high spatial resolutions. We present two theoretical models for (1) a high density molecular cloud core, MC27/L1521F, and (2) a protobinary system, L1551 NE. For the model for MC27, we performed numerical simulations for gravitational collapse of a turbulent cloud core. The cloud core exhibits fragmentation during the collapse, and dynamical interaction between the fragments produces an arc-like structure, which is one of the prominent structures observed by ALMA. For the model for L1551 NE, we performed numerical simulations of gas accretion onto protobinary. The simulations exhibit asymmetry of a circumbinary disk. Such asymmetry has been also observed by ALMA in the circumbinary disk of L1551 NE.

1. Introduction

Most of low-mass young stars are members of multiple systems of two or more stars, and multiple star formation is considered to be a major mode of low-mass star formation [1]. Recent high-resolution observations with Atacama Large Millimeter/Submillimeter Array (ALMA) have revealed early phase of low-mass star formation. In order to compare these detailed observations with numerical simulations, the numerical simulations should be performed also with a high-resolution. An adaptive mesh refinement (AMR) technique allows us to obtain a local high resolution for grid-based numerical simulations [2]. In this paper, we report two cases of numerical models for comparison with the recent ALMA observations.

The first case is for a dense core in the Taurus molecular cloud, MC27 or L1521F [3, 4]. Observations have suggested that MC27/L1521F is in a very early stage of star formation [5]. The recent ALMA Cycle 0 and 1 observations toward MC27/L1521F has been carried out [6, 7], and revealed an arc structure with a length of $\sim$ 2000 au at the center of the cloud core [6]. In order to investigate the origin of the arc structure, we performed numerical simulations for protostellar collapse of cloud cores [8].

The second case is for a protobinary system, L1551 NE, which is a class I object located in the Taurus region. The recent ALMA Cycle 0 observations have resolved fine structures around the protobinary stars: a circumstellar disk around each star and a circumbinary disk around the binary stars [9]. We performed numerical simulations for gas accretion onto binary stars in order to investigate complex structures around the binary stars.

2. A high-density molecular cloud core, MC27/L1521F

2.1. Models and methods

As the initial condition, a filamentary molecular cloud is considered because recent observations have shown that molecular clouds often exhibit filamentary structures [10, 11, 12]. The filament is assumed to be infinitely long and in an equilibrium state where thermal pressure supports an isothermal cloud against its self-gravity. The filamentary cloud has a density distribution of $\rho(R) = \rho_0(1 + R^2/R_0^2)^{-2}$ [14, 13] with $R_c = 0.05\,\mathrm{pc}$ and $\rho_0 = 2c_s^2/(\pi G R_0^2) = 1.45 \times 10^{-19}\,\mathrm{g\,cm^{-3}}$ (the corresponding number density is $n_0 = 3.79 \times 10^4\,\mathrm{cm^{-3}}$), where the isothermal sound speed is assumed as $c_s = 0.190\,\mathrm{km\,s^{-1}}$ and the corresponding gas temperature is $T = 10$ K. The radius of the cloud R_c comes form a typical width of filamentary clouds indicated by the Herschel survey [11]. For comparison, we performed other simulations assuming a spherical cloud, and confirmed that the main results described in this section are insensitive to the initial cloud shape.

Turbulence was imposed at the initial stage, of which velocity field is incompressible with a power spectrum of $P(k) \propto k^{-4}$, generated according to [15], where k is the wavenumber. This power spectrum results in a velocity dispersion of $\sigma(\lambda) \propto \lambda^{1/2}$, in agreement with the Larson scaling relations [16]. The barotropic equation of state was assumed to be $P(\rho) = c_s^2\rho + \kappa\rho^{7/5}$ with $\kappa = c_s^2\rho_{\mathrm{cr}}^{-2/5}$, where the critical density is set at $\rho_{\mathrm{cr}} = 10^{-13}\,\mathrm{g\,cm^{-3}}$ (the corresponding number density is $n_{\mathrm{cr}} = 2.62 \times 10^{10}\,\mathrm{cm^{-3}}$), which was taken from the numerical results of [17]. The gas is approximated to be isothermal for $\rho \lesssim \rho_{\mathrm{cr}}$ and the gas temperature increases adiabatically with the density for $\rho \gtrsim \rho_{\mathrm{cr}}$. The adopted equation of state does not take into account shock heating, because we implicitly assume that the shocked gas is instantaneously cooled by emission of radiation. The magnetic field was ignored for simplicity.

A three-dimensional AMR code, SFUMATO [18], was employed as a simulation code. The hydrodynamic scheme was modified to have a third order of accuracy in space and a second order in time. The MUSCL scheme with the MINMOD limiter is adopted, and it has a first order accuracy only at the discontinuities. The self-gravity of the gas is solved by the multigrid method. The computational domain is resolved on a base grid of $l = 0$ with 256^3 cells. The maximum grid level was set at $l = 11$. The cell width is $\Delta x_{\mathrm{min}} = 0.613$ au on the finest grid of $l = 11$, and $\Delta x_{\mathrm{max}} = 1.26 \times 10^3$ au on the base grid of $l = 0$. The Jeans condition was employed as a refinement criterion [19]; blocks are refined when the Jeans length is shorter than 8 times the cell width, i.e., $\lambda_J < 8\Delta x$, where λ_J is the Jeans length.

The sink particles were employed as a sub-grid model of protostars. The detailed implementations of the sink particles are shown in [20]. The critical density for the sink particle formation is set at $\rho_{\mathrm{sink}} = 1 \times 10^{-11}\,\mathrm{g\,cm^{-3}}$ ($n_{\mathrm{sink}} = 2.62 \times 10^{12}\,\mathrm{cm^{-3}}$), and the sink radius is set at $r_{\mathrm{sink}} = 4\Delta x_{\mathrm{min}} = 2.45$ au.

2.2. Results

Figure 1 shows the column density distribution on four different spatial scales at the stage of $t_p = 1.760 \times 10^4$ yr, where t_p denotes the time after the first sink particle formation. The simulation shows that the initial filamentary cloud fragments into several cloud cores as shown in Figure 1a. A cloud core undergoes gravitational collapse to exhibit a sharply peaked density distribution (Figure 1b) in agreement with molecular line observations toward MC27/L1521F [5]. In the early phase, the cloud core fragments into four gas condensations, which evolve into four protostars (sink particles). Two sink partices merge because of their chaotic orbits,

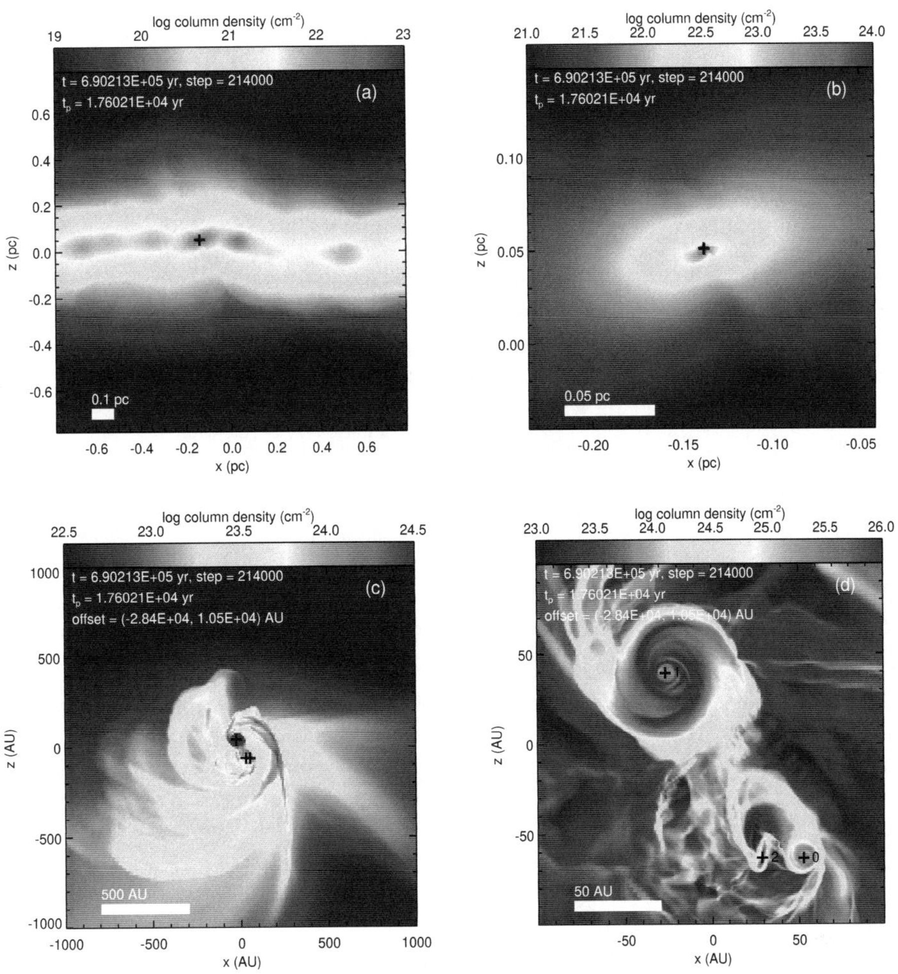

Figure 1. Column density distribution along the y-direction on four different spatial scales at $t = 0.690$ Myr ($t_p = 1.760 \times 10^4$ yr). The color scales depict the column densities. The black crosses indicate the positions of the sink particles. In panels (c) and (d), the coordinates are offset so that the center of each panel coincides with the center of mass of the sink particles. The sink particles are labeled with identification numbers 0, 1, and 2 in panel (d). Taken from [8].

and three sink particles survive to form a triple system as shown in Figures 1c and 1d. Every protostar has a circumstellar disk around it. The most massive protostar, which is labeled 1, has the largest disk. The other protostars have small disks because successive close encounters between protostars 0 and 2 prevent growth of the circumstellar disks around them.

The simulation reproduces formation of a triple system. The ALMA observations have also revealed a similar system: a protostar and two gas condensation at the center of the MC27/L1521F [6].

The simulation exhibits arc structures in the infalling envelope as shown in Figure 1c. The arc structures extend up to a 1000 au scale in agreement with the ALMA observations [6]. These structures are caused by the gravitational interaction between the sink particles and the envelope gas. The gas is accelerated by the gravitational torque from the orbiting protostars.

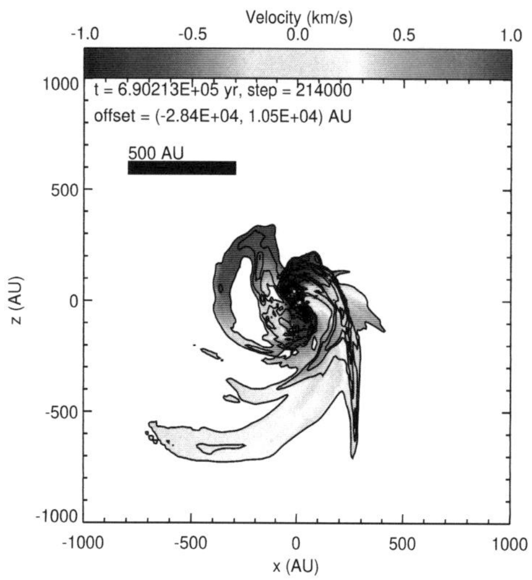

Figure 2. Line-of-sight velocity distribution in the arc structures. The color scale shows the density-weighted velocity along the line of sight. The contour shows the column density on a logarithmic scale with $\log(N/\mathrm{cm}^{-2}) = 23.4, 23.6, \cdots$, where N denotes the column density. The velocity is shown only in the regions where the column density is greater than the lowest contour level. Taken from [8].

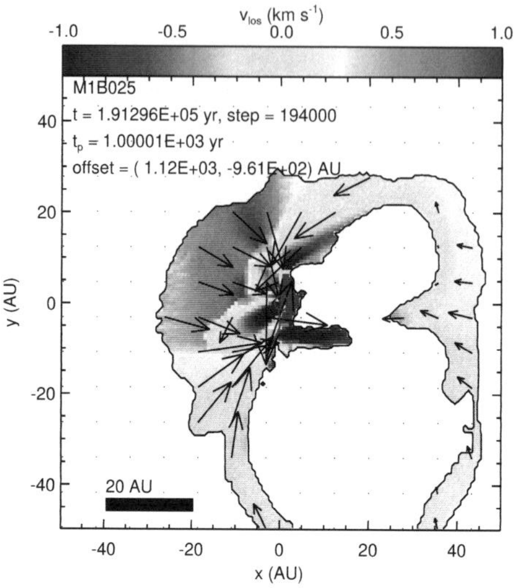

Figure 3. Line-of-sight velocity distribution of the magnetic walls in MHD simulations. The color scale shows the density-weighted velocity distribution along the line of sight (the z-direction). The arrows denote the density-weighted velocity on the $x-y$ plane. The black contour is an outline of a volume of $\rho \geq \rho_{\mathrm{cr}}$.

Figure 2 shows line-of-sight velocity distribution in the arc structures. This figure is comparable to Figure 4 of [6]. The velocity ranges typically from $-0.5\,\mathrm{km\,s^{-1}}$ to $0.5\,\mathrm{km\,s^{-1}}$ in agreement with the observation. This suggests that the observed arc structure in MC27/L1521F is caused by the gravitational interaction as shown in the simulation.

We performed another simulation, taking account of magnetic field [22]. The initial magnetic field is assumed to be uniform, and a relatively strong magnetic field is assumed with a mass-to-flux ratio of 1.12. The MHD simulations also reproduce arc structures as shown in Figure 3. The arc structures are caused by magnetic walls, which have been predicted by the theoretical arguments [23] and the one-dimensional MHD simulations [24]. The MHD simulations however do not reproduce multiple dense gas condensations, which are observed at the center of MC27/L1521F [6]. This is because the magnetic field suppresses fragmentation of cloud cores in typical cases in agreement with the previous MHD simulations [21]. If the observed dense gas condensations are produced by fragmentation of the cloud core, the magnetic field is weak enough to promote fragmentation of the cloud core.

3. A binary protostar system, L1551 NE

3.1. Models and methods

L1551 NE is a protobinary system embedded in the envelope gas, and the outflows are associated with the protobinary stars, suggesting that the protobinary stars actively accrete the gas of the envelope. In order to mimic this protobinary system, we consider the situation where the gas is injected at the computational boundaries, and it falls onto the protobinary stars.

In the three-dimensional computational domain of $[-12D, 12D]^2 \times [-6D, 6D]$, two protostars are considered, of which barycenter is located at the origin, where $D = 145$ au is a binary separation of L1551 NE. The coordinates of the primary and secondary stars are therefore $(x, y, z) = (-M_2D/M_{\rm tot}, 0, 0)$ and $(x, y, z) = (M_1D/M_{\rm tot}, 0, 0)$, respectively, where M_1 and M_2 are masses of the primary and secondary stars, and $M_{\rm tot} = M_1 + M_2$ is a total mass of the binary stars. The total mass and mass ratio are set at $M_{\rm tot} = 0.8M_\odot$ and $M_2/M_1 = 0.19$ according to the observations [25].

Gas with a constant density and a specific angular momentum is injected at the cylindrical boundaries: the cylindrical surface of $R = 12D$, and the top and bottom surfaces of $z = \pm 6D$. On these surfaces, the gas has a radial velocity of $v_r = [2GM_{\rm tot}/r - (j_{\rm inf}/R)^2]^{1/2}$, assuming a freefall from an infinity direction, where $j_{\rm inf}$ is the specific angular momentum of the gas, r is the spherical radius, and R is the cylindrical radius. When the centrifugal force is larger than the gravity ($j_{\rm inf}^2/R^3 > GM_{\rm tot}R/r^3$), no gas is injected there, corresponding the centrifugal barrier. These boundary conditions are similar to those of [26]. The specific angular momentum of the injected gas is given by $j_{\rm inf} = (GM_{\rm tot}R_{\rm cent})^{1/2}$, where $R_{\rm cent}(= 300$ au) is the centrifugal radius as determined from the observations [25, 27]. For simplicity, an isothermal gas is assumed with a sound speed of $c_s = 0.1(GM_{\rm tot}/D)^{1/2} = 0.221\,{\rm km\,s^{-1}}$. The gas temperature is therefore 13.5 K, which is typical for dense molecular cloud cores and infalling envelopes.

A three-dimensional AMR code, SFUMATO [18], was employed. The hierarchical grid was fixed during the calculation, and a so-called fixed mesh refinement (FMR) was adopted. The grid configuration is shown in Figure 4. The block size is set at 16^3 cells, and the computational domain is resolved by $8^2 \times 4$ blocks. The base grid of $l = 0$ therefore has a size of $128^2 \times 64$. The cell width is $\Delta x_{\rm min} = 0.85$ au on the finest grid of $l = 5$, and $\Delta x_{\rm max} = 27$ au on the base grid of $l = 0$.

The binary protostars are represented by two sink particles. The sink particles accrete gas within a given radius of $4\Delta x_{\rm min} = 3.4$ au. The simulation was performed on a rotating frame where the sink particles are at rest, taking into account both the centrifugal and Coriolis forces. The calculation was halted at ~ 250 rotations of the binary orbit.

3.2. Results

The simulation shows that the injected gas falls toward the center, and the angular momentum prevents the gas from falling into the binary stars directly, providing a circumbinary disk with a central gap. The gravitational torque from the orbiting binary stars produces the spiral arms in the circumbinary disk, and it also induces accretion onto the circumstellar disks (Figure 5). The gases in each circumstellar disk rotates in circular orbits around the sink particle. The angular momenta are transported presumably because of numerical viscosity and the gas finally falls onto the sink particle. Similar structures have been also shown in the many previous simulations [28, 26, 29, 30, 31].

Based on the simulation here, observational visualization was performed to compare the simulation with the observation [9]. The comparison indicates that the observed velocity features exhibits qualitative agreement with those associated with the spiral arms in the numerical simulations. L1551 NE has active outflows, implying that the magnetic fields play an important role in the acceleration of the outflow in the regions near the protostars. The hydrodynamical model presented here should be extended to MHD models in order to investigate the effects of the magnetic fields on the circumbinary structure.

The ALMA observations toward L1551 NE suggest that the circumbinary disk on the secondary side is brighter than on the primary side [9]. Such an asymmetry of a circumbinary disk has been also observed toward a class II binary, UY Aur [32]. Asymmetry in the circumbinary disk is reproduced in the numerical simulation here. Figure 5 shows asymmetry

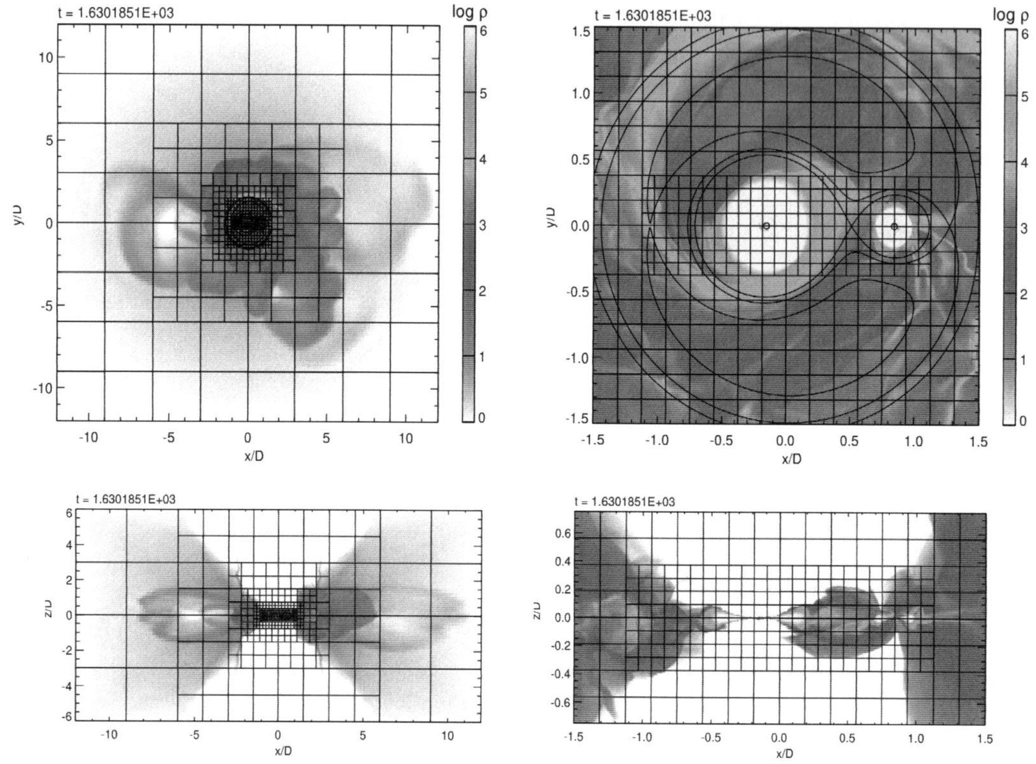

Figure 4. A snapshot of the simulation for binary accretion. The upper panels show the density distributions in logarithmic scale in the $z = 0$ plane, while the lower panels show those in the $y = 0$ plane. The right panels are enlargements of the left panels. The The black horizontal and vertical lines are the boundaries of the AMR blocks. In upper panels, the contours of the Roche potential are shown for comparison.

in the circumbinary disk, where the surface density is higher on the secondary side (the right side) than on the primary side (the left side) in this snapshot. The asymmetric structure rotates with respect to the position angle of the binary stars. Figure 6 shows the surface density as a function of time in a ring region with a radius of $R = (2 \pm 0.1)D$, indicating that the asymmetric structure rotates with an angular velocity of $\Omega = \Omega_\star/4$, where $\Omega_\star$ is an angular velocity of the binary orbit. Such commensurability between the asymmetric circumbinary disk and the binary orbit indicates a resonance caused by the gravitational torque from the binary stars

4. Summary

We performed numerical simulations for modeling of the recent ALMA observations toward a high-density molecular cloud core, MC27/L1521F, and a binary protostar system, L1551 NE. The AMR technique was adopted for the numerical simulations to obtain a high resolution because the corresponding ALMA observations also exhibit a high spatial resolution. The numerical simulations for MC27/L1521F suggest that the observed dense gas condensations are caused by fragmentation of the cloud core, and the arc structure is a consequence of the dynamical interaction between the fragments and the infalling envelope. The numerical simulations for L1551 NE suggest that existence of the spiral arms in the circumbinary disk. The simulations also exhibit asymmetry in the circumbinary disk, which is caused by the resonance

IOP Conf. Series: Journal of Physics: Conf. Series **837** (2017) 012009 doi:10.1088/1742-6596/837/1/012009

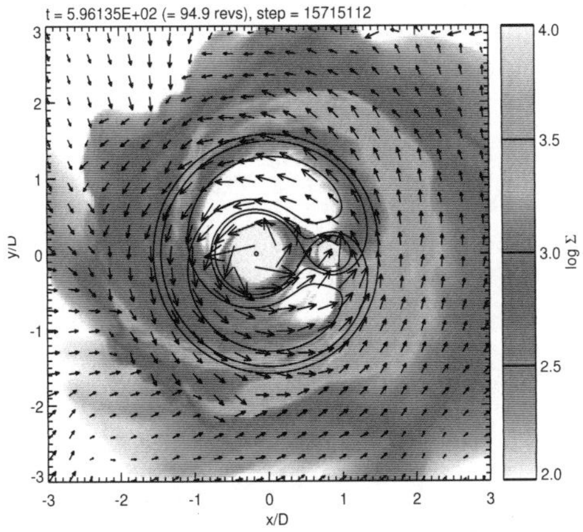

Figure 5. Surface density of the central box of $[-3D, 3D]^2 \times [-1.5D, 1.5D]$. The arrows denote the velocity distribution in the $z = 0$ plane. The velocity in the rest frame is plotted.

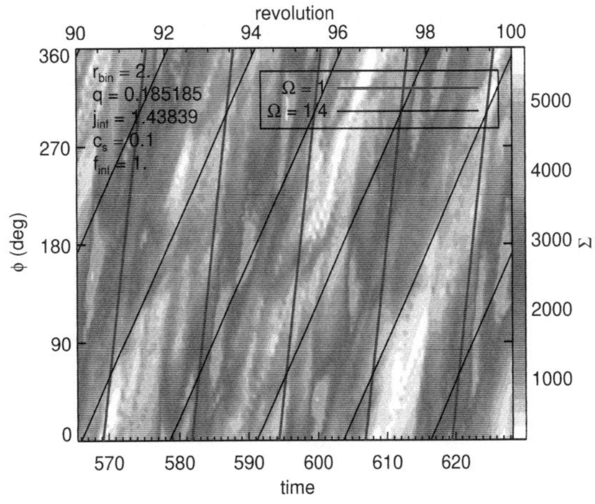

Figure 6. Surface density distribution within a thin ring with a radius of $R = (2 \pm 0.1)D$. The ordinate and abscissa denote the azimuthal angle ϕ and time t, respectively. The stages between 90 and 100 binary periods are shown. Blue and black lines show the relationships of $\phi = \Omega t$ with $\Omega = \Omega_\star$ and $\Omega_\star/4$, respectively.

of the rotation of the circumbinary disk and the binary orbit.

Acknowledgments

Numerical computations were carried out on Cray XC30 at the Center for Computational Astrophysics, National Astronomical Observatory of Japan. This research was supported by JSPS KAKENHI Grant Numbers 26400233, 26287030, 24244017, 23540270.

References

[1] Reipurth, B., Clarke, C. J., Boss, A. P., et al. 2014, Protostars and Planets VI, 267
[2] Berger, M. J., & Colella, P. 1989, Journal of Computational Physics, 82, 64
[3] Mizuno, A., Onishi, T., Hayashi, M., et al. 1994, Nature, 368, 719
[4] Codella, C., Welser, R., Henkel, C., Benson, P. J., & Myers, P. C. 1997, A&A, 324, 203
[5] Onishi, T., Mizuno, A., & Fukui, Y. 1999, PASJ, 51, 257
[6] Tokuda, K., Onishi, T., Saigo, K., et al. 2014, ApJL, 789, L4
[7] Tokuda, K., Onishi, T., Matsumoto, T., et al. 2016, ApJ, 826, 26
[8] Matsumoto, T., Onishi, T., Tokuda, K., & Inutsuka, S.-i. 2015, MNRAS, 449, L123
[9] Takakuwa, S., Saito, M., Saigo, K., et al. 2014, ApJ, 796, 1
[10] André, P., Men'shchikov, A., Bontemps, S., et al. 2010, A&A, 518, L102
[11] Arzoumanian, D., André, P., Didelon, P., et al. 2011, A&A, 529, L6
[12] Hacar, A., Tafalla, M., Kauffmann, J., & Kovács, A. 2013, A&A, 554, A55
[13] Ostriker, J. 1964, ApJ, 140, 1056
[14] Stodólkiewicz, J. S. 1963, Acta Astron., 13, 30
[15] Dubinski, J., Narayan, R., & Phillips, T. G. 1995, ApJ, 448, 226
[16] Larson, R. B. 1981, MNRAS, 194, 809
[17] Masunaga, H., Miyama, S. M., & Inutsuka, S. 1998, ApJ, 495, 346.
[18] Matsumoto, T. 2007, PASJ, 59, 905
[19] Truelove, J. K., Klein, R. I., McKee, C. F., Holliman, J. H., II, Howell, L. H., & Greenough, J. A. 1997, ApJL, 489, L179

[20] Matsumoto, T., Dobashi, K., & Shimoikura, T. 2015, ApJ, 801, 77
[21] Machida, M. N., Matsumoto, T., Hanawa, T., & Tomisaka, K. 2005, MNRAS, 362, 382
[22] Matsumoto, T., Machida, M. N., & Inutsuka, S.-i. 2016 in preparation
[23] Li, Z.-Y., & McKee, C. F. 1996, ApJ, 464, 373
[24] Tassis, K., & Mouschovias, T. C. 2005, ApJ, 618, 783
[25] Takakuwa, S., Saito, M., Lim, J., et al. 2012, ApJ, 754, 52
[26] Bate, M. R., & Bonnell, I. A. 1997, MNRAS, 285, 33
[27] Takakuwa, S., Saito, M., Lim, J., & Saigo, K. 2013, ApJ, 776, 51
[28] Artymowicz, P., & Lubow, S. H. 1996, ApJL, 467, L77
[29] Ochi, Y., Sugimoto, K., & Hanawa, T. 2005, ApJ, 623, 922
[30] Gómez de Castro, A. I., López-Santiago, J., Talavera, A., Sytov, A. Y., & Bisikalo, D. 2013, ApJ, 766, 62
[31] Young, M. D., Baird, J. T., & Clarke, C. J. 2015, MNRAS, 447, 2907
[32] Tang, Y.-W., Dutrey, A., Guilloteau, S., et al. 2014, ApJ, 793, 10

IOP Conf. Series: Journal of Physics: Conf. Series **837** (2017) 012010 doi:10.1088/1742-6596/837/1/012010

The status of general relativistic simulations of compact binary mergers as engines of short gamma-ray bursts

Vasileios Paschalidis*

Department of Physics, Princeton University, Princeton, NJ 08544, USA

E-mail: *vp16@princeton.edu

Abstract. Black hole - neutron star (BHNS) and neutron star - neutron star (NSNS) binaries are perhaps the most popular progenitors for short-hard gamma ray bursts. After about two decades of numerical relativity simulations of binary compact objects we are beginning to understand the necessary ingredients for jets to emerge from these systems following merger. We report on the latest development of this field summarizing the results from state-of-the-art numerical relativity (magnetohydrodynamic) simulations of compact binary mergers as progenitors of short-hard gamma-ray bursts.

1. Introduction

A few months ago, the LIGO and Virgo collaborations marked the beginning of the era of gravitational wave astronomy with the announcement of the detection of two signals that were consistent with the gravitational wave (GW) signal generated by the inspiral and merger of binary black hole systems [1, 2]. Over the next few years, GWs are expected to also be detected from neutron star–neutron star (NSNS) and black hole–neutron star (BHNS) binary mergers. The merger rates for these binaries are uncertain but upper limits have been estimated to 12,600 (3,600) $\mathrm{Gpc}^{-3}\mathrm{yr}^{-1}$ for NSNS (BHNS) [3].

Inspiraling and merging NSNSs and BHNSs are sources not only of GWs, but also of electromagnetic (EM) signals that can be generated prior to [4, 5, 6, 7, 8, 9] and after [10, 11] merger. The detection of both GW and EM signals arising from the same source would provide a wealth of information about the source, test relativistic gravitation (see e.g. [12]), and constrain the nuclear equation of state (EOS) (see e.g. [13]) among other things.

Among the different EM signals NSNSs and BHNSs may have, a short gamma-ray burst (sGRB) is the one these systems are particularly famous for. It is widely accepted that NSNS and BHNS are the most likely progenitors of sGRBs [14, 15, 16, 17, 18, 19, 20]. A simultaneous detection of an sGRB with a GW consistent with the coalescence of a BHNS or NSNS system would solidify the BHNS and NSNS binaries as the progenitors of sGRBs. However, due to the merger rates of these objects such an association may require a lucky nearby event or third-generation GW observatories [19]. Until then, studying these binaries via simulations in full general relativity is the only reliable avenue for gaining a complete theoretical understanding of these systems as sGRB engines. However, this involves a large number of obstacles that have to be overcome. For example one needs to solve the Einstein equations for the spacetime metric [10 partial differential equations (PDEs)] coupled to the energy-momentum and neutrino transport equations (8 PDEs), coupled to Maxwell's equations for the electromagnetic field (8 PDEs), coupled to baryon number conservation (1 PDE) for a total of 27 coupled PDEs in 3+1 dimensions. In fact the neutrino transport equation is 6+1 dimensional problem because in addition to the 3 spatial and one temporal dimension it has two angular and one frequency (energy) dimension. Note that above we even excluded the lepton number conservation equations and that the entire system need to be supplemented with a hot nuclear equation of state, which complicate the problem even further. Modern codes that are able to solve all or a subset of the above equations adopt a finite volume approach (typically third-order accurate or higher) for the (magneto)hydrodynamics, and (fourth-order accurate or higher)finite difference or pseudo-spectral methods for the Einstein equations. Other unique challenges of modeling compact binaries in full general relativity involve the presence of physical singularities – infinities, when evolving BH spacetimes and the choice of appropriate coordinates. However, many of these problems are now under control due to the development of sophisticated theoretical and computational numerical relativity tools (see [21] for a textbook). As far as modeling these objects as sGRB engines is concerned, simulations indicate that general relativity, and magnetic fields are crucial for launching jets (see below). Recent simulations in [22] suggest that neutrinos are unable to power outflows following NSNS mergers, and inefficient at driving jets following BHNS mergers. However, neutrinos can launch winds that make the funnel area above the black hole less baryon loaded. As a result neutrinos may be important for the magnetic fields to be able to ultimately accelerate jets to the Lorentz factors of $O(100)$, which are necessary for explaining sGRB phenomenology.

In recent years a large number of NSNS and BHNS numerical relativity simulations have augmented our understanding of the processes involved in these mergers. Despite the challenge posed (see e.g. [23] for a detailed discussion) due to the very high-resolution requirements necessary to capture the relevant magnetic effects, the large range of length and time scales involved in the problem and because of a very challenging neutrino transport problem that has to be solved, with the current tools and available codes we have taken great strides toward understanding the necessary ingredients for jet launching following BHNS and NSNS mergers. Here, we will provide a status report of numerical relativity (magneto)hydrodynamic simulations of compact binary mergers as sGRB engines, and summarize what has been learned so far.

The article is structured as follows: In Sec. 2 we present results from recent simulations of binary BHNSs and in Sec. 3 results from recent simulations of binary NSNSs. We conclude in Sec. 4 with a brief summary and list of open questions. Unless otherwise specified, below we adopt geometrized units where $G = c = 1$.

2. BHNS mergers

For a review of simulations of BHNS binaries we refer the reader to [24]. Here we focus on simulations of these binaries as sGRB engines. The first numerical relativity simulations testing the viability of BHNSs as progenitors of sGRB engines, addressed the question of whether quasicircular, irrotational BHNSs (the NS is practically non-spinning, but the BH can be spinning) can form a sizable accretion disk outside the remnant BH. Having a disk outside the BH is necessary to form the BH-disk engine that is thought to power the sGRB. Even a small fraction of the accretion power can account for typical sGRB luminosities: $L_\gamma = \epsilon \dot{M} c^2 \sim 10^{51}\mathrm{erg/s}(\epsilon/0.01)(M_{\rm disk}/0.1M_\odot)(t_{\rm acc}/1\mathrm{s})^{-1}$, where ϵ is a conversion efficiency from accretion power to gamma-ray luminosity, $M_{\rm disk}$ the disk mass, and $t_{\rm acc}$ the accretion time scale. But, forming a disk outside the BH is not trivial, because this requires that the NS tidal disruption occur outside the innermost stable circular orbit (ISCO) of the remnant BH, and this is possible only for small BH to NS mass ratios, and for binaries either involving rapidly spinning BHs and/or NSs with small compaction [25, 26].

Multiple BHNS simulations have investigated the BHNS parameter space in which a relatively large amount matter can remain outside the BH following the NS tidal disruption. and the results were summarized in [27], who derived fitting formulae for predicting the disk mass onto the BH. The main finding can be summarized as follows: *at the most likely BH to NS mass ratio* $(7:1)$ *the disk mass will be* $\sim 0.1M_{\odot}$ *only if the initial BH dimensionless spin is* $a = J_{\rm BH}/M_{\rm BH}^2 \gtrsim 0.8$, which can be a very tight constraint.

This result holds for quasicircular, irrotational BHNSs which probably dominate the rates of BHNS mergers in the Universe. However, recent work [28, 29, 30] argues that in galactic nuclei and globular clusters (GCs), compact binaries can merge with sizable eccentricities at rates which may be relevant for GW detection [31, 32]. It is also plausible that mergers in GCs could account for up to $\sim 30\%$ of the observed sGRBs [33] (see also [34, 35, 36]). Thus, the progenitors of a fraction sGRBs may be eccentric compact objects mergers occurring in GCs. An additional aspect of GC NSs is that $\gtrsim 80\%$ of pulsars in GCs in our Galaxy have millisecond spin periods, and hence compact binary mergers in GCs involving neutron stars are likely to occur with rapidly spinning neutron stars [37, 38, 39].

Using the above as motivation, numerical relativity hydrodynamic simulations of the final encounter of hydrodynamic eccentric BHNS mergers in full general relativity with $4:1$ mass ratio have been performed in [31, 32], and in [37] where the effects of the NS spin were considered for the first time. Depending on the value of the pericenter distance, the amount of disk mass found in these simulations ranges from $\sim 1\% - 20\%$ of the NS rest mass even for non-spinning BHs, and for stiffer equations of state it can be even 30%. We note that a $4:1$ mass ratio, quasi-circular BHNS merger with a non-spinning NS results in negligibly small accretion disk [27]. Thus, eccentric BHNS mergers as may occur in GCs can form BH-disk engines more easily than quasi-circular binaries.

However, showing that compact binary mergers involving NSs can form a BH-disk system is one requirement for demonstrating viability of BHNSs as progenitors of sGRBs. The second requirement is to demonstrate that jets can be launched by the resulting BH-disk systems – a necessary ingredient for many models of sGRBs [17]. Next we describe efforts to demonstrate jet launching via numerical relativity magnetohydrodynamic simulations of BHNS mergers.

The first MHD simulations of BHNSs in full general relativity were performed in [41], who did not report jet formation after simulating for ~ 30 ms. The first detailed, numerical relativity studies exploring the parameter space of magnetized BHNSs with MHD simulations in full general relativity were performed in [42, 43]. These studies endowed the initial neutron star with a dipole magnetic field restricted in the interior of the star. The results about jet emergence were negative, even when the maximum initial magnetic field pressure at the center was 5% of the gas pressure! The lack of jets was initially attributed to the resolution being too low to capture the magnetorotational instability (MRI) in the remnant disk. However, even when increasing the resolution to resolve the MRI wavelength by 10 grid points – which has been found empirically to suffice for capturing the basic effects of MRI – again no jet was found. More recent MHD simulations of BHNSs in full general relativity [44] that adopt even higher resolution, and resolve the fastest growing MRI mode, confirmed the results of [42, 43].

These results were very intriguing because for over 15 years, fixed-spacetime GRMHD accretion flows onto BHs showed that jets are a natural outcome [45]. So, what is the problem with accretion disks formed following BHNS mergers?

The answer to this question is simple: for a magnetized torus with magnetic fields initially restricted in its interior, a jet is launched only for special magnetic field geometries, and in particular, a magnetic field with a vertical component having consistent sign is necessary [46]. Thus, if one seeded toroidal magnetic fields in the torus no jets would be launched [46]. What happens in a BHNS merger is that during and following the NS disruption is that the bulk of the magnetic energy is swallowed by the BH during the main merger event, while the initially

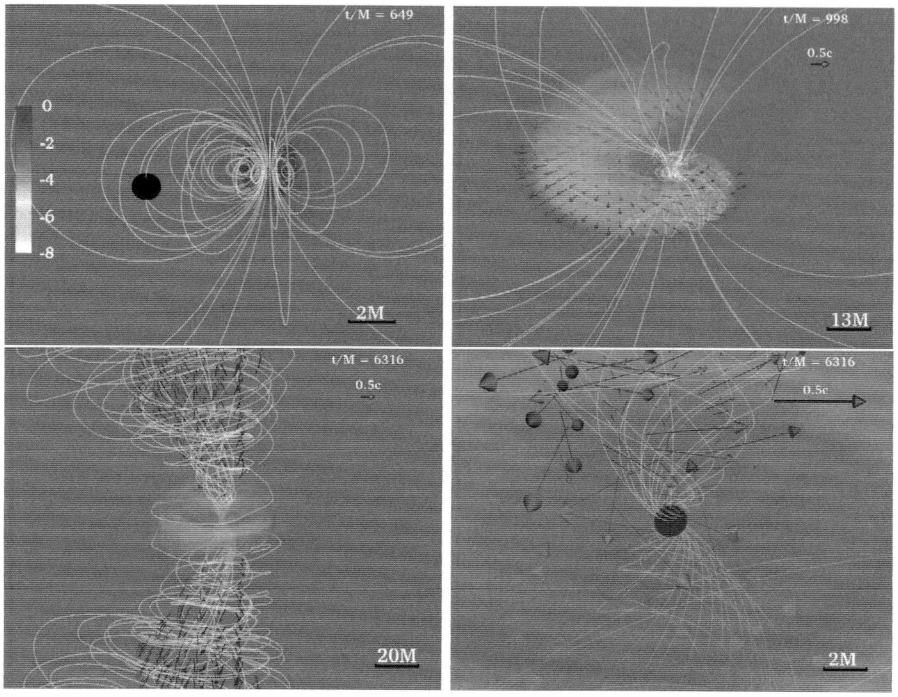

Figure 1. Snapshots of the ratio of the rest-mass density divided by its initial maximum value $\rho_{0,\max} = 8.92 \times 10^{14}(1.4M_\odot/M_{\rm NS})^2$g cm^{-3} (log scale). Arrows designate matter velocities and white curves the magnetic field lines. The bottom panels show the system after an incipient jet has been launched. Here, M $= 2.5 \times 10^{-2}(M_{\rm NS}/1.4M_\odot)ms=$ $7.58(M_{\rm NS}/1.4M_\odot)$km. Figure from [40].

poloidal magnetic field in the remaining matter is stretched and wound into an almost purely toroidal geometry. Therefore, per the results of [46] no jets can be launched.

The missing link was finally found in [40], who made the realization that these early BHNS simulations adopted magnetic fields restricted in the NS interior. However, pulsars suggest that a more realistic magnetic field geometry is a dipole extending from the NS interior well into its exterior [40]. While this sounds like it would be a straightforward change, it is not trivial for ideal MHD codes to evolve magnetic-field dominated regions as would be required in a magnetosphere. For this reason, the authors in [40] designed new types of initial conditions, which satisfy one aspect of magnetic-field dominance, namely the magnetic field dominates the pressure in the exterior atmosphere, but not the energy density. These atmospheric initial conditions allow the evolution of exterior magnetic fields with an ideal GRMHD code, and via a sequence of simulations varying the degree of magnetic-pressure dominance one can test the validity of the results as the force-free regime is approached [40]. This is the procedure developed and followed by [40], who performed MHD simulations of BHNS mergers in full GR and demonstrated that an incipient jet launched about $\sim$ 100 ms following the BHNS merger (see Fig. 1). Interestingly, the disk lifetime was found to be $\sim$0.5 s, and the jet EM luminosity 10^{51} erg/s, i.e., consistent with characteristic sGRB values. The flow was only mildly relativistic $\Gamma_L \sim 1.3$, hence the term "incipient". However, the magnetization in the incipient jet outflow was $B^2/8\pi\rho c^2 \sim 100$, and $B^2/8\pi\rho c^2$ is a crucial ratio because it approximately equals the terminal Lorentz factor of a magnetically powered jet [47]. Thus, the jets found in [40], in principle can be accelerated to the Lorentz factors required to explain sGRB phenomenology. Therefore, BHNSs can indeed be

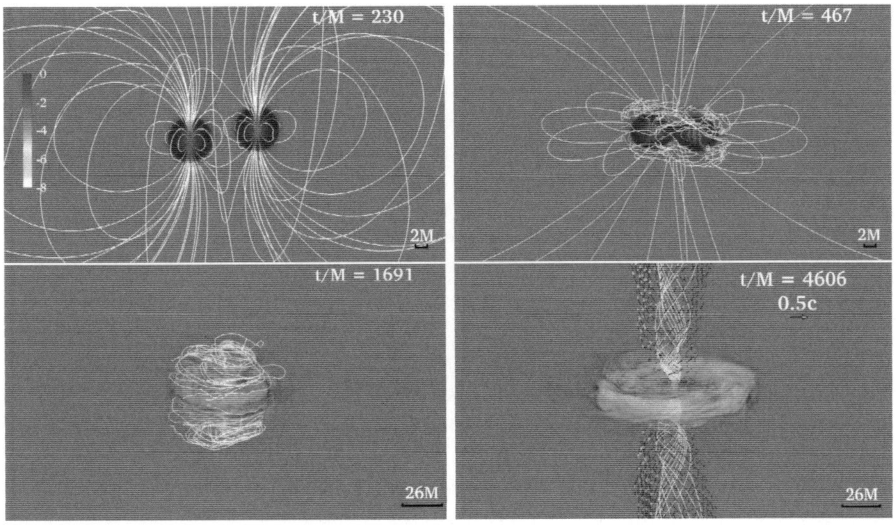

Figure 2. Same as in Fig. 1, but here $\rho_{0,\text{max}} = 5.9 \times 10^{14}(1.625\, M_\odot/M_{\text{NS}})^2$ g cm^{-3} and $M = 1.47 \times 10^{-2}(M_{\text{NS}}/1.625 M_\odot)$ ms $= 4.43(M_{\text{NS}}/1.625 M_\odot)$ km. The appearance of an organized, large scale magnetic field inside the incipient jet is clear in the bottom right panel. Figure from [50].

the progenitors of sGRBs!

3. Simulations of NSNS mergers

For a comprehensive review of simulations of NSNS binaries we refer the reader to [48], as well as [49] for a more recent review. Here we focus on work related to NSNS mergers as sGRB engines and in particular on the latest MHD simulations in full GR.

Only few numerical relativity MHD simulations of NSNSs have been performed. Early MHD calculations of NSNS mergers in full general relativity reported the emergence of "jet-like structures" [51]. This means the formation of a structure resembling a funnel, but no jet outflow was reported. A recent high-resolution MHD study of NSNS mergers in full GR [52] reports that even when the MRI is captured the fallback material has so strong a ram pressure that not only a jet cannot be launched, but also a wind cannot be launched (unlike in BHNSs where winds are possible). More recent MHD simulations of NSNSs in full GR [53, 54], also do not find jets, however the initial magnetic fields were too weak to resolve MRI.

Motivated by the jet emergence in the BHNS calculations of [40], [50] adopted similar methods for the magnetic field initial data and evolution as in [40], but this time in a NSNS setting adopting the same matter and spacetime initial data as those in [51]. The authors report that MRI operates and that about 60 ms after merger, an incipient jet emerges even in this NSNS scenario (see Fig. 2). The accretion timescale in this simulation was found to be ~ 0.2 s, and the jet EM luminosity 10^{51} erg/s – again consistent with characteristic values from sGRBs. The magnetization in the collimated outflow was $B^2/8\pi\rho c^2 \sim 100$, implying that terminal Lorentz factors of $\Gamma_L \sim 100$ could potentially be achieved. Also, consistent with the findings of [52] the authors found that the jet emerged shortly after the ambient density of the fallback matter decreased to levels satisfying $B^2/(8\pi\rho c^2) \gg 1$.

We note here, that unlike the other numerical relativity MHD studies of NSNSs that do not report jet launching, the authors of [50] endowed the neutron stars with magnetar-strength magnetic fields. While such fields are high by pulsar standards, they are still dynamically unimportant initially, thus offering a natural means to equipartition-level magnetic fields in the post-merger hypermassive neutron star. These high magnetic fields in the hypermassive neutron star are expected to naturally arise because of the Kelvin-Helmholtz instability (KHI) and the MRI operating during and after merger, respectively [55, 56]. Thus, [50] argued that if the KHI and MRI have sufficient time to operate and amplify the post-merger magnetic field to equipartition levels, then a jet is possible following BH formation.

4. Conclusions and future challenges

Numerical relativity simulations of NSNSs and BHNSs over the last 15 years have improved our understanding of how these binaries may form the BH-disk engines that can power jets and explain sGRBs. The works of [40, 50] have provided the first "existence proofs" that BHNSs and NSNSs can form jet engines and hence are viable progenitors of sGRBs. However, these works endowed the initial neutron stars with magnetic fields that are larger by a couple of orders of magnitude than typical inferred pulsar magnetic fields. Even though such strong magnetic fields are dynamically unimportant and hence one would expect the results to scale to weaker field strengths, these calculations should be performed with weaker magnetic fields to corroborate the results. But, this would then require much better grid resolutions (almost 10 times better) to be able to resolve all the MHD processes. Using such high resolution would make these simulations at best impractical. However, as more accurate numerical methods are adopted and the scaling is improved, and as the computers become faster such simulations should be possible within the next decade. Furthermore, it is very difficult to accurately evolve magnetically dominated areas with ideal MHD codes. Thus, simulations including magnetic fields extending from the NS interior out to its exterior should also be revisited with more sophisticated methods (perhaps with resistive MHD codes [57, 58]).

Despite the great progress, open questions always remain and here is an incomplete list: Do the incipient jets persist for an accretion timescale? How can incipient jets reach $\Gamma_L \gtrsim 100$? What mechanisms give rise to the prompt gamma ray emission? Do neutrinos play any role? These and many other questions pose the challenge for future simulations modeling compact binaries as sGRB engines.

Acknowledgments

This work was supported by NSF grant PHY-1607449, the Simons Foundation, and NASA grant NNX16AR67G (Fermi).

References

[1] Abbott B P *et al.* (LIGO Scientific Collaboration and Virgo Collaboration) 2016 *Phys. Rev. Lett.* **116**(6) 061102 URL `http://link.aps.org/doi/10.1103/PhysRevLett.116.061102`
[2] Abbott B P *et al.* (LIGO Scientific Collaboration and Virgo Collaboration) 2016 *Phys. Rev. Lett.* **116**(24) 241103 URL `http://link.aps.org/doi/10.1103/PhysRevLett.116.241103`
[3] Abbott B P *et al.* (Virgo, LIGO Scientific) 2016 *Astrophys. J.* **832** L21 (*Preprint* 1607.07456)
[4] Hansen B M and Lyutikov M 2001 *Mon.Not.Roy.Astron.Soc.* **322** 695 (*Preprint* astro-ph/0003218)
[5] McWilliams S T and Levin J 2011 *Astrophys. J.* **742** 90 (*Preprint* 1101.1969)
[6] Paschalidis V, Etienne Z B and Shapiro S L 2013 *Phys. Rev. D* **88** 021504 (*Preprint* 1304.1805)
[7] Palenzuela C, Lehner L, Ponce M, Liebling S L, Anderson M, Neilsen D and Motl P 2013 *Physical Review Letters* **111** 061105 (*Preprint* 1301.7074)
[8] Ponce M, Palenzuela C, Lehner L and Liebling S L 2014 *Phys. Rev. D* **90** 044007 (*Preprint* 1404.0692)
[9] Metzger B D and Zivancev C 2016 *Mon. Not. R. Astron. Soc.* **461** 4435–4440 (*Preprint* 1605.01060)
[10] Metzger B D and Berger E 2012 *Astrophys. J.* **746** 48 (*Preprint* 1108.6056)
[11] Metzger B D, Bauswein A, Goriely S and Kasen D 2015 *Mon. Not. R. Astron. Soc.* **446** 1115–1120 (*Preprint* 1409.0544)
[12] Yunes N, Yagi K and Pretorius F 2016 *ArXiv e-prints* (*Preprint* 1603.08955)
[13] Lackey B D and Wade L 2015 *Phys. Rev. D* **91** 043002 (*Preprint* 1410.8866)

[14] Eichler D, Livio M, Piran T and Schramm D N 1989 *Nature* **340** 126–128
[15] Narayan R, Paczynski B and Piran T 1992 *Astrophys. J. Lett.* **395** L83–L86 (*Preprint* astro-ph/9204001)
[16] Mochkovitch R, Hernanz M, Isern J and Martin X 1993 *Nature* **361** 236–238
[17] Meszaros P 2006 *Rept.Prog.Phys.* **69** 2259–2322
[18] Lee W H and Ramirez-Ruiz E 2007 *New Journal of Physics* **9** 17 (*Preprint* astro-ph/0701874)
[19] Berger E 2014 *Ann. Rev. Astron. Astroph.* **52** 43–105
[20] Paschalidis V, Ruiz M and Shapiro S L 2015 *Astrophys. J. Lett.* **806** L14 (*Preprint* 1410.7392)
[21] Baumgarte T W and Shapiro S L 2010 *Numerical Relativity: Solving Einstein's Equations on the Computer* (Cambridge University Press)
[22] Just O, Obergaulinger M, Janka H T, Bauswein A and Schwarz N 2015 *ArXiv e-prints* (*Preprint* 1510.04288)
[23] Paschalidis V 2016 (*Preprint* 1611.01519)
[24] Shibata M and Taniguchi K 2011 *Living Reviews in Relativity* **14** 6–+
[25] Etienne Z B, Faber J A, Liu Y T, Shapiro S L, Taniguchi K *et al.* 2008 *Phys.Rev.* **D77** 084002
[26] Etienne Z B, Liu Y T, Shapiro S L and Baumgarte T W 2009 *Phys. Rev. D* **79** 044024 (*Preprint* 0812.2245)
[27] Foucart F 2012 *Phys. Rev. D* **86** 124007 (*Preprint* 1207.6304)
[28] Lee W H, Ramirez-Ruiz E and van de Ven G 2010 *Astrophys. J.* **720** 953–975 (*Preprint* 0909.2884)
[29] Kocsis B and Levin J 2012 *Phys. Rev. D* **85** 123005 (*Preprint* 1109.4170)
[30] Samsing J, MacLeod M and Ramirez-Ruiz E 2014 *Astrophys.J.* **784** 71 (*Preprint* 1308.2964)
[31] Stephens B C, East W E and Pretorius F 2011 *Astrophys. J. Lett.* **737** L5 (*Preprint* 1105.3175)
[32] East W E, Pretorius F and Stephens B C 2012 *Phys. Rev. D* **85**(12) 124009 URL http://link.aps.org/doi/10.1103/PhysRevD.85.124009
[33] Grindlay J, Portegies Zwart S and McMillan S 2006 *Nature Physics* **2** 116–119 (*Preprint* astro-ph/0512654)
[34] Narayan R, Paczynski B and Piran T 1992 *Astrophys. J.* **395** L83 astro-ph/9204001 (*Preprint* astro-ph/9204001)
[35] Gehrels N *et al.* 2005 *Nature* **437** 851–854 (*Preprint* arXiv:astro-ph/0505630)
[36] Lee W H, Ramirez-Ruiz E and Granot J 2005 *Astrophys. J. Lett.* **630** L165–L168 (*Preprint* arXiv:astro-ph/0506104)
[37] East W E, Paschalidis V and Pretorius F 2015 *Astrophys. J. Lett.* **807** L3 (*Preprint* 1503.07171)
[38] Paschalidis V, East W E, Pretorius F and Shapiro S L 2015 *Phys. Rev. D* **92** 121502 (*Preprint* 1510.03432)
[39] East W E, Paschalidis V, Pretorius F and Shapiro S L 2016 *Phys. Rev. D* **93** 024011 (*Preprint* 1511.01093)
[40] Paschalidis V, Ruiz M and Shapiro S L 2015 *Astrophys. J. Lett.* **806** L14 (*Preprint* 1410.7392)
[41] Chawla S, Anderson M, Besselman M, Lehner L, Liebling S L, Motl P M and Neilsen D 2010 *Physical Review Letters* **105** 111101–+
[42] Etienne Z B, Liu Y T, Paschalidis V and Shapiro S L 2012 *Phys.Rev.* **D85** 064029
[43] Etienne Z B, Paschalidis V and Shapiro S L 2012 *Phys.Rev.* **D86** 084026
[44] Kiuchi K, Sekiguchi Y, Kyutoku K, Shibata M, Taniguchi K and Wada T 2015 *Phys. Rev. D* **92** 064034 (*Preprint* 1506.06811)
[45] Abramowicz M A and Fragile P C 2013 *Living Reviews in Relativity* **16** URL http://www.livingreviews.org/lrr-2013-1
[46] Beckwith K, Hawley J F and Krolik J H 2008 *Astrophys. J.* **678** 1180–1199
[47] Vlahakis N and Königl A 2003 *Astrophys. J.* **596** 1080–1103
[48] Faber J A and Rasio F A 2012 *ArXiv e-prints* (*Preprint* 1204.3858)
[49] Baiotti L and Rezzolla L 2016 *ArXiv e-prints* (*Preprint* 1607.03540)
[50] Ruiz M, Lang R N, Paschalidis V and Shapiro S L 2016 *Astrophys. J. Lett.* **824** L6 (*Preprint* 1604.02455)
[51] Rezzolla L, Giacomazzo B, Baiotti L, Granot J, Kouveliotou C and Aloy M A 2011 *Astrophys. J. Lett.* **732** L6
[52] Kiuchi K, Kyutoku K, Sekiguchi Y, Shibata M and Wada T 2014 *Phys.Rev.* **D90** 041502 (*Preprint* 1407.2660)
[53] Endrizzi A, Ciolfi R, Giacomazzo B, Kastaun W and Kawamura T 2016 *ArXiv e-prints:1604.03445* (*Preprint* 1604.03445)
[54] Kawamura T, Giacomazzo B, Kastaun W, Ciolfi R, Endrizzi A, Baiotti L and Perna R 2016 *Phys. Rev. D* **94** 064012 (*Preprint* 1607.01791)
[55] Zrake J and MacFadyen A I 2013 *Astrophys. J. Lett.* **769** L29 (*Preprint* 1303.1450)
[56] Kiuchi K, Cerd-Durn P, Kyutoku K, Sekiguchi Y and Shibata M 2015 *Phys. Rev.* **D92** 124034 (*Preprint* 1509.09205)
[57] Palenzuela C 2013 *Mon. Not. R. Astron. Soc.* **431** 1853–1865 (*Preprint* 1212.0130)
[58] Dionysopoulou K, Alic D, Palenzuela C, Rezzolla L and Giacomazzo B 2013 *Phys. Rev. D* **88** 044020

IOP Conf. Series: Journal of Physics: Conf. Series **837** (2017) 012011 doi:10.1088/1742-6596/837/1/012011

Simulating the Local Interstellar Medium

Jonathan D. Slavin

Smithsonian Astrophysical Observatory, 60 Garden Street, Cambridge, MA 02138 USA

E-mail: jslavin@cfa.harvard.edu

Abstract. The Local Interstellar Medium (LISM) in many ways exemplifies the diffuse ISM of the Galaxy and star forming galaxies in general. Though devoid of molecular gas, it includes warm gas, hot, supernova-heated gas and even cold neutral gas clouds. The complex of local interstellar clouds (CLIC), which includes the cloud that directly surrounds the heliosphere, is made up of warm, low density, partially ionized gas. These clouds have somehow come to be embedded within the hot Local Bubble, apparently having survived the passage of the shock that heated the gas. We describe our attempts to understand this surprising situation and to explain the thermal and ionization state of the CLIC as well as the state of the Local Bubble via hydrodynamical and photoionization modeling. We also discuss the broader implications of our results for the interaction of the different temperature-density phases in the diffuse ISM of galaxies.

1. Introduction

The Local Interstellar Medium (LISM) has been defined by convention as that region within roughly 50 - 100 pc of the Sun that comprises the so-called Local Bubble (a.k.a. the Local Cavity) and all that is contained within it. This bubble has been detected in a number of ways including Na I and Ca II absorption lines and dust extinction mapping. Shown in Figure 1 is a projection onto the Galactic plane of the 3D map of the bubble made by [1] using Na I absorption lines. The irregularly shaped bubble is essentially empty of neutral Na out to 50 - 150 pc in the plane. At high Galactic latitudes where the sampling is sparser, the size of the bubble is less well defined but clearly extends farther than in the plane, to distances well beyond 100 pc.

Important progress was made in understanding the LISM when it was realized that the diffuse soft x-ray background, first observed by sounding rockets [2], is coming from hot gas mainly contained within the Local Bubble. In Figure 2 we show the ROSAT All Sky Survey map of the emission in 1/4 keV band. While there has been some controversy regarding the nature of the emission [see 3, 4, 5], it now seems clear that most of the 1/4 keV emission does originate from hot, $T \sim 10^6$ K, low density, $n \sim 0.005$ cm^{-3} gas within the neutral gas cavity that surrounds us.

Though the Local Bubble is nearly devoid of the cold neutral gas that contains neutral Na, it is known that there is warm and partially ionized gas close to the Sun. The Complex of Local Interstellar Clouds (CLIC) is a collection of about 15 of low density, $n \sim 0.3$ cm^{-3}, warm, $T \approx 7000$ K clouds located within 15 pc of the Sun. These clouds have been identified by absorption lines with distinct velocity components along several lines of sight, each of which is consistent with a single velocity vector [6, 7]. The ratios of several ion column densities, e.g. Mg I/Mg II and C II*/C II provide evidence of their partial ionization and warm temperatures. In

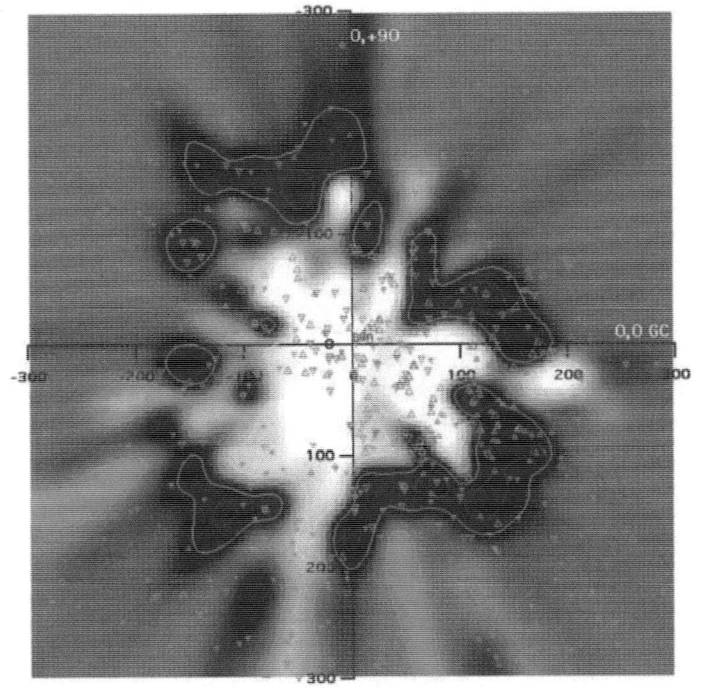

Figure 1. The Local Bubble as inferred from Na I absorption line measurements projected onto the Galactic plane [1, used by permission].

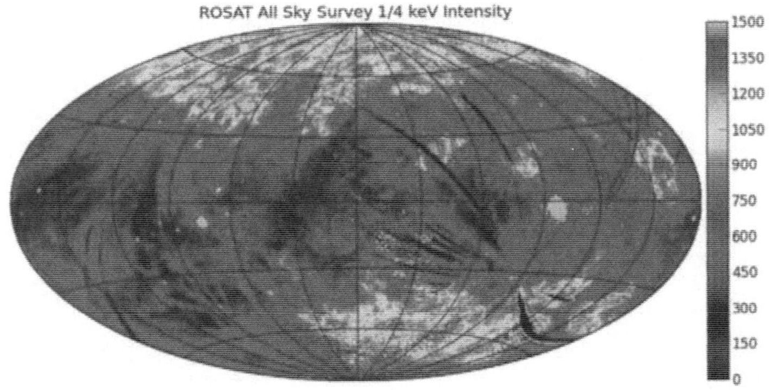

Figure 2. ROSAT 1/4 keV map of the diffuse soft x-ray background. The emission, especially at low Galactic latitude, is believed to come primarily from hot gas contained within the Local Bubble. The units of the map are 10^{-6} counts s^{-1} arcmin^{-2}.

Figure 3 we show a schematic view of the clouds as reconstructed from the data and projected onto the Galactic plane [from 4]. In Figure 4 we show some recent absorption line data from HST observations toward Sirius B, only 2.64 pc away. As can be seen there are two velocity components in this direction, one of which corresponds with the cloud that surrounds the heliosphere, known as the Local Interstellar Cloud (LIC). A recently proposed alternative possibility to the multiple cloud picture is that all of the CLIC cloud material is part of a single monolithic cloud that has a variety of flows within it [8]. There is, in addition, a cold ($T \approx 20$ K) and dense cloud known as the Local Leo Cold Cloud located roughly 20 pc from the Sun, which has a variety of interesting properties [9, 10, 11], though space does not permit discussion of that cloud here.

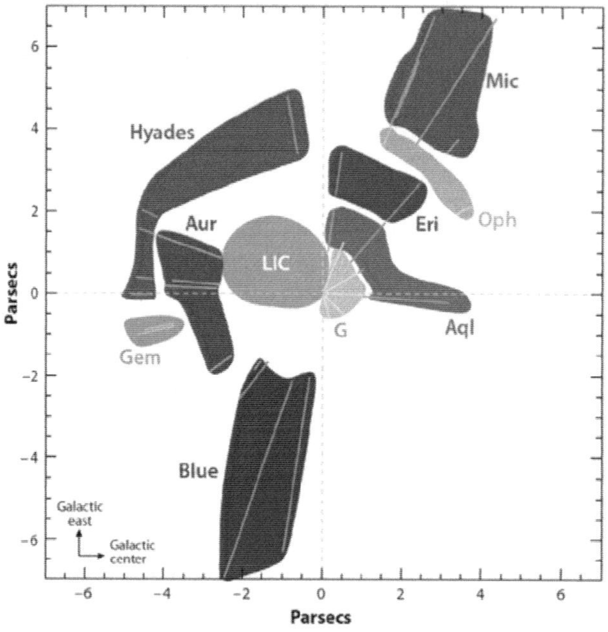

Figure 3. Schematic view of the CLIC, projected onto the Galactic plane, based on absorption line data toward nearby stars [from 4]. The location of Sirius ($d = 2.64$ pc, $\ell = 227.2$, $b = -8.9$) on the plot is (x, y) = (-1.77, -1.91), inside the Blue Cloud.

The various clouds of the CLIC (or velocity components of the LIC) share a similar overall velocity that is away from the Galactic center and the Scorpius-Centaurus OB association. Within the complex there is significant dispersion in velocity around the mean velocity, but for the single cloud that surrounds the heliosphere, the velocities for the more than 80 lines of sight are remarkably consistent with a single velocity vector which is in turn consistent with the velocity of neutral interstellar helium that has been detected flowing through the heliosphere [12]. In addition, determinations of the turbulent broadening of absorption lines, based on the assumption that line widths are determined by the combination of a constant turbulent broadening and a thermal broadening (which of course depends on the mass of the ion), find typical values of only $\sim 1 - 3$ km s^{-1}. These facts argue that at least a portion of the LIC is quite quiescent at present.

2. Mysteries of LISM

As detailed above, we now have a fairly clear picture of the nature of the LISM from observations. What is much less clear is how the LISM came to be in its current configuration. Among the open questions on the LISM are:

- How did the CLIC come to be inside the Local Bubble?
- How was the Local Bubble created? A supernova? Multiple supernovae?
- What is the ionization source for the CLIC? What is heating the clouds?

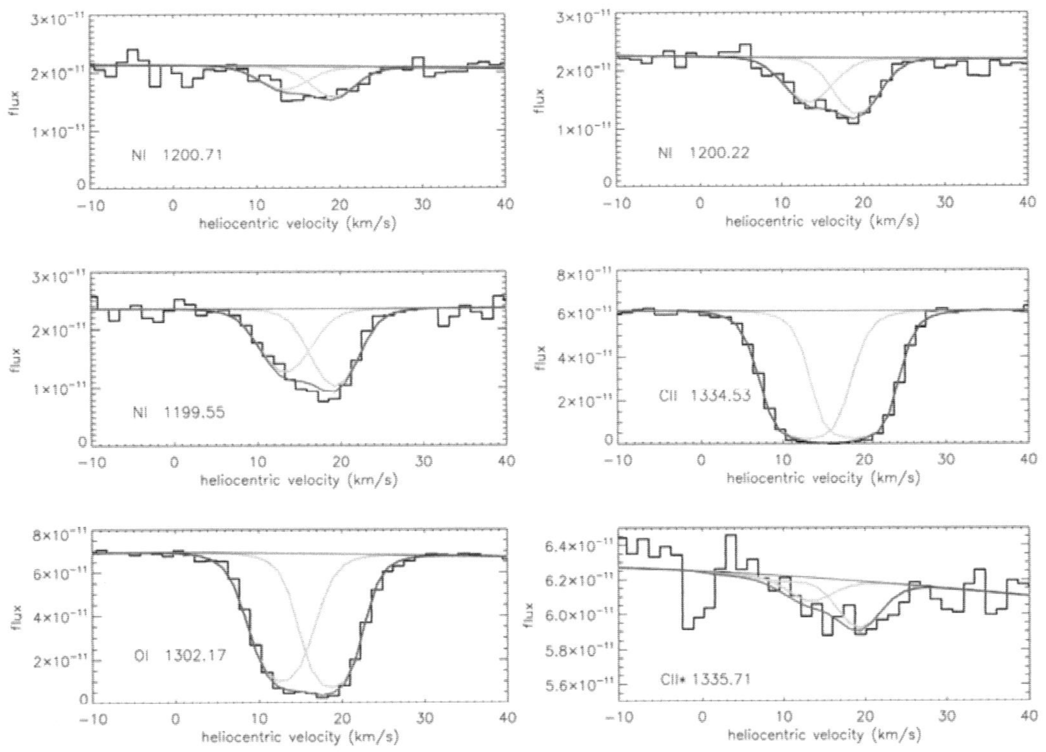

Figure 4. Absorption lines observed in the HST spectrum of Sirius B. The profiles are fitted here with a smooth continuum (blue) and two velocity components (green) corresponding to the LIC velocity and the so-called Blue Cloud velocity.

- What can the LISM tell us about the energetics and dynamics of the diffuse ISM in the Galaxy and galaxies more generally?

The first question is important because the low density of the nearby clouds is closer to that associated with the intercloud medium. If, as we suspect, the Local Bubble was created by one or more supernovae, we would have expected any low density material to be swept by the shock into the expanding shell.

As mentioned above, the nearby clouds are partially ionized. Photoionization by the radiation field from nearby hot stars has been investigated [13] and found to be insufficient for either the ionization (particularly for He) or the heating of the gas. It was suggested [14] that the ionization could be due to non-equilibrium recombination caused, for example, by the passage of a shock. The problem with this idea is that overionized gas tends to cool too quickly. That is, if you ionize the gas via a shock that is slow enough that cooling sets in behind it, then the gas cools much more quickly than it recombines dropping in temperature well below 7000 K while it is still highly ionized, $X(\mathrm{H}^+) > 0.5$. As an example, the gas behind a 100 km s^{-1} shock into a density of 0.25 cm^{-3} will cool in only $\sim$ 200 yr, while it will take close to 8×10^4 yr for it to recombine. To get something like the currently derived conditions for the LIC, $X(\mathrm{H}^+) \sim 0.25$ and $T \sim 7000$ K would require a substantial heating source. If that source is the radiation field, then there is ionization associated with the heating and we still need to identify that additional source of ionizing photons. The only other identified source of ionization is turbulent heating, but the low level turbulence in the CLIC makes it questionable if such a source is viable.

In our modeling of the ionization of the LIC [15] we included radiation from the boundary of the cloud under the assumption that it was evaporating via thermal conduction. With this source in addition to the radiation field from nearby hot stars and diffuse emission from the hot gas in the Local Bubble, we were able to match the observed ionization level. However, it is not certain that the cloud is evaporating since the expected column densities of highly ionized species in the evaporative boundaries, e.g. C IV and O VI have not been observed. The possibly more complex dynamics and energetics that is hinted at by the spread in velocities of the CLIC may be playing a role. The situation as it stands clearly demands more sophisticated modeling.

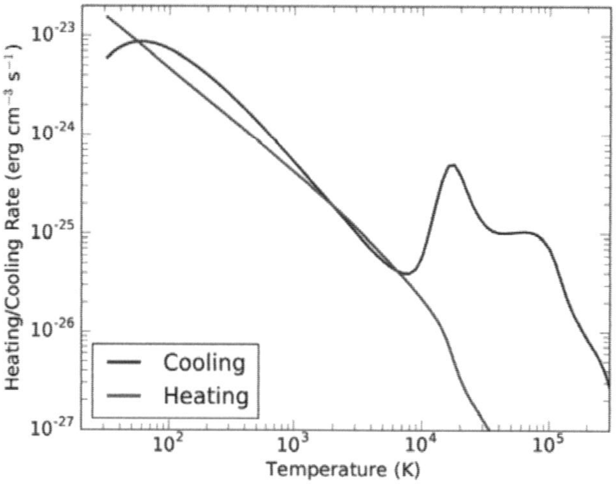

Figure 5. Heating and cooling rates vs. temperature for conditions similar to those in the LISM. Note the three intersections, which allow for thermal equilibrium. Only the low temperature and high temperature equilibria are stable.

3. Physical Processes in the ISM

To increase our understanding of the ISM in general and the LISM in particular requires inclusion of all the most important physical processes. Among those that are central to the energetics of the diffuse ISM are radiative cooling and heating via photoionization. In the diffuse ISM, particularly in the warm medium, cooling results mainly from collisional excitation of forbidden lines, though radiative recombination and free-free emission also contribute. Those processes are all two-body mechanisms and thus the cooling rate goes as density squared. Heating, in contrast, is mostly caused by photoionization of H and He, though in some circumstances photoelectric ejection from dust grains is also important. In Figure 5 we show the heating and cooling curves for LISM conditions. Note that the curves intersect in three places leading to three possible thermal equilibrium solutions. The middle intersection, however, results in an unstable equilibrium in which any displacement leads to movement away from the equilibrium temperature and toward either the warm equilibrium near 7000 K or the cold one near 70 K. The thermal pressure assumed in making this plot sets the overall scaling of the two curves. A high pressure will raise the cooling curve relative to the heating curve and result in only the single equilibrium at low temperature. A low pressure on the other hand will result in only the warm temperature equilibrium. These ideas are illustrated in Figure 6 in which we show the thermal equilibrium curve, which is the locus of points for which heating equals cooling given a range of pressure values. For combinations of pressure and density above the curve cooling exceeds heating and the gas will move down towards the curve with its path determined by the dynamics, i.e. whether the cooling is so fast as to be isochoric (constant density), isobaric, adiabatic or some combination. Similarly gas with pressure and density putting it below the curve will have a heating rate that exceeds the cooling and will heat up, moving upward toward the equilibrium curve. The speed with which the equilibrium curve is approached can vary by large factors however, with dense gas that's far from equilibrium having much shorter equilibration times than lower density gas that is relatively close to equilibrium.

We have proposed that the CLIC was at one time colder and denser that it is currently so that it could be left behind when the shock that heated lower density gas to X-ray emitting temperatures passed by. In this picture then the clouds are shocked and overpressured and subsequently expand. If the expansion overshoots, the clouds will then fall below the equilibrium curve and could then get heated to the warm temperatures we see in the CLIC currently.

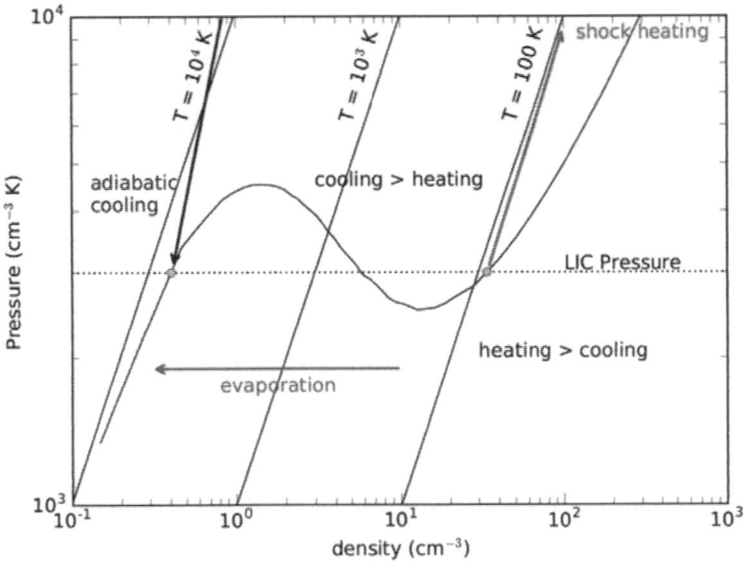

Figure 6. Phase diagram for interstellar gas in the local ISM. The blue curve is the locus of points for which heating and cooling balance. As indicated, for the region below the curve heating exceeds cooling, while above the curve cooling exceeds heating. The orange circles correspond to equilibrium points at the labeled "LIC Pressure" (here approximated as 3000 cm^{-3} K) for a cold, dense phase and for a warm, low density phase. Example paths that could be followed for gas that is cooling adiabatically, shocked by a slow shock or evaporated via thermal conduction are shown.

Thermal conduction can also be very important in multi-phase environments in the ISM, especially those that include hot gas in which the conductivity is expected to be very high. There are ongoing questions about the efficacy of thermal conduction in diffuse plasmas, but we do not have space to discuss those here. At this point we are exploring the consequences of thermal conduction that is only limited by so-called saturation of the heat flux. Under such conditions a cold or warm cloud, $T < 10^4$ K, that is embedded in a hot medium, $T > 10^5$ K, is expected to evaporate as the heat flux into the cloud drives a mass outflow. Solutions have been found to the cloud evaporation problem under various conditions [e.g. 16, 17, 18, 19] though these are typically quite idealized. In particular the cloud is generally assumed to be spherical or planar and at rest in the surrounding medium. Nevertheless they do illustrate that thermal conduction can lead to the destruction of clouds and play an important role in the evolution of the ISM.

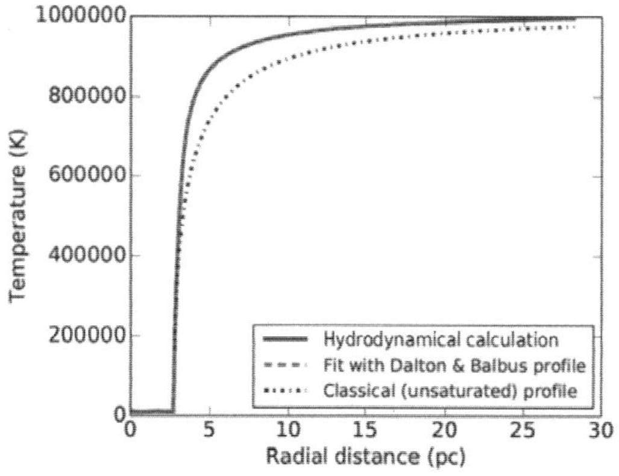

Figure 7. Temperature profile for numerical calculation of an evaporating cloud. The results of the calculation (green) are overplotted with a Dalton & Balbus [19] type analytical profile (red dashed curve). A "classical" Cowie & McKee profile which has the same asymptotic temperature is also plotted (blue dot-dashed line) for comparison.

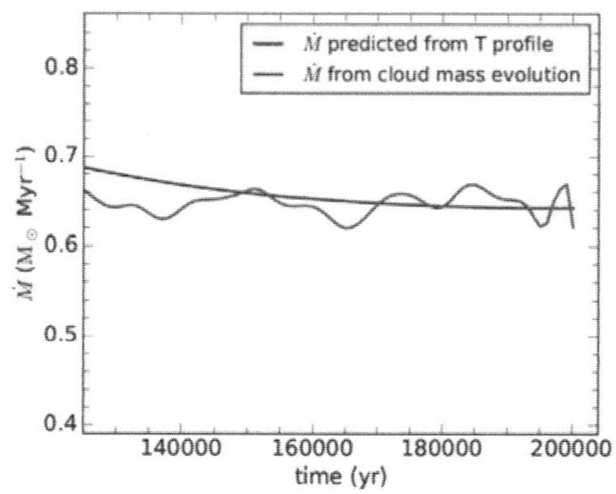

Figure 8. Mass loss rate for numerical calculation of cloud evaporation. The rate is calculated from that predicted for the temperature profile given the Dalton & Balbus [19] solution (blue line) and compared with the rate found from the decrease of cloud mass over time (green line). After an initial adjustment period, the rates match very closely.

As we discuss below, we have undertaken a series of investigations of the LISM employing the FLASH code. As part of these we have made use of the built-in capacity to include electron thermal conduction with saturation of heat flux in the code. We have tested FLASH's thermal conduction unit by doing cloud evaporation calculations. For these we put a warm spherical cloud in a hot surrounding medium in 2D cylindrical symmetry. In Figure 7 we show the temperature profile for the evaporating cloud along with a fit to the profile using the analytical solution from Dalton & Balbus [19]. They found an analytical solution for the temperature profile under the assumption that radiative cooling is not important but including the effects of heat flux saturation. As can be seen from the figure the match is extremely good. For comparison we also show the profile one would derive under the assumption of no heat flux saturation as in the "classical" solution of Cowie & McKee [16]. In Figure 8 we compare the cloud mass loss rates derived from fitting the temperature profiles with Dalton & Balbus type curves and the rate from calculating the change in cloud mass. Again, the match, after an initial non-steady evaporation period, is excellent. These results give us confidence that the implementation of thermal conduction with heat flux saturation in FLASH is accurate.

4. Numerical modeling of the LISM

We are currently involved in taking the next steps to modeling the LISM employing the magnetohydrodynamical code FLASH (version 4.3). Our simulations to date start with a single or multiple cold (T = 100 K), dense (n = 25 cm^{-3}) clouds embedded in a warm (T = 10^4 K), lower density (0.25 cm^{-3}) intercloud medium. We set off a supernova explosion close, 10 to 20 pc away from, the cloud(s). The physical processes included are thermal conduction including heat flux saturation, and cooling and heating using look-up tables for the coefficients. We have not included the magnetic field as yet, but that is the next step planned. We have used AMR with a resolution at highest refinement of 0.02 pc and done the calculations in 2D assuming cylindrical symmetry.

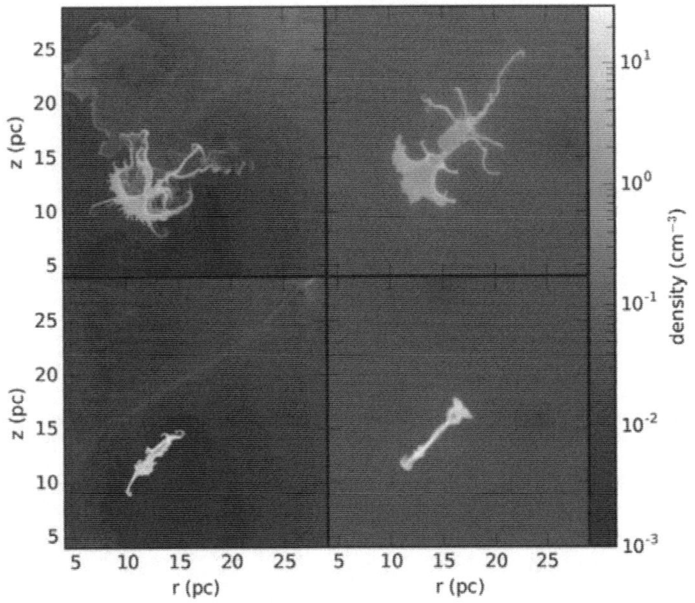

Figure 9. Comparison of runs with a single cloud overrun by a SNR shock. Moving clockwise from upper left the runs include: (*upper left*) neither cooling nor conduction, (*upper right*) conduction but no cooling, (*lower right*) both cooling and conduction and (*lower left*) cooling but no conduction.

4.1. Single Cloud Runs

In figure 9 we show a comparison of results for a series of single cloud runs designed to test the effects of cooling and thermal conduction. Heating was not included, in these runs. The figure shows the density for the cloud at a time 1.75×10^5 yr after the SN explosion. The cloud, originally centered at $(r, z) = (10.61, 10.61)$ (so it is 15 pc from the origin), first encounters the shock at $t \approx 6500$ yr after the explosion. The figure clearly shows the effects of cooling and conduction, not only on the cloud, but also on the remnant in which it is embedded. Without cooling or conduction, the cloud is shredded by Kelvin-Helmholtz and Richtmyer-Meshkov instabilities. The density in the medium surrounding the cloud is very low. When thermal conduction is operating, the instabilities are substantially suppressed, partly because of the evaporative outflow that is generated and partly because the shear layer has been smoothed out. In addition, we see that the density in the surrounding medium is higher because conduction has flattened the temperature gradient in the bulk of the SNR leading to a more uniform density distribution as well [see, e.g. 20]. In the runs that include radiative cooling the cloud is much smaller and denser because the cloud that propagates through the cloud is slow enough to be radiative. The smaller cross section for the cloud then leads to less shredding via hydrodynamical instabilities. Based on these runs, it appears that cooling is the more important of the physical processes determining the dynamical evolution of the cloud.

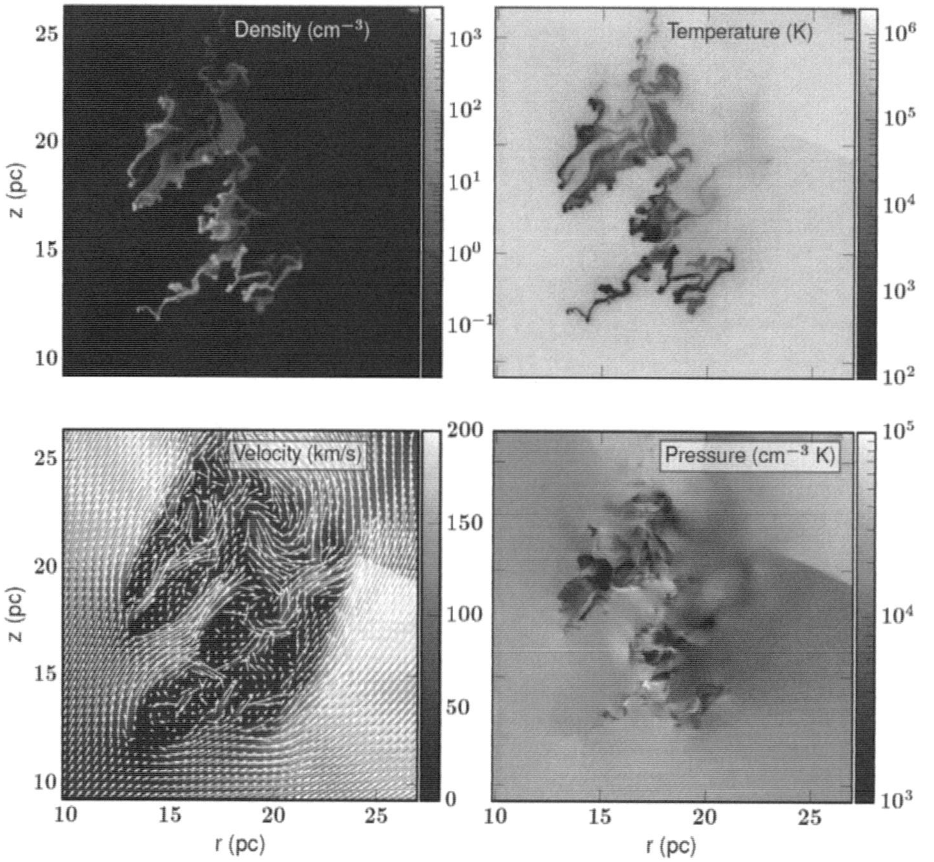

Figure 10. Simulation results for multiple cloud run. Clockwise from upper left is density, temperature, pressure and total velocity. Note the filamentary form of the cloud. The velocity field is complex and the temperature, density and pressure vary considerably through the cloud material. The current day CLIC appears more quiescent than this simulated cloud complex.

4.2. Multiple Cloud Runs

To test the effects of multiple clouds we have carried out simulations that start with 15 clouds in close proximity to each other. For this case we have included heating as well as cooling so that we can assess if there is a rebound in the temperature after the shocks pass through the clouds. In Figure 10 we show the results for a run at a time 2×10^5 yr after the explosion. The cloud complex was located initially around $(r, z) = (15, 15)$ pc so it is clear that it has moved out from the center a small distance. The high degree of variation in temperature, density and pressure as well as the complex and very non-uniform velocity field contrast with the current day CLIC wherein it appears that there are several different vectors but there is coherence in the velocity of large patches of the sky. It will be important to examine whether and under which conditions the cloud complex can dissipate the energy imparted by the shock. We expect that the magnetic field will play an important role in that evolution.

5. Conclusions

The LISM is a very interesting region of the ISM that provides opportunities that cannot be found elsewhere for understanding the physical processes that operate in the ISM. Using numerical simulations that include all the most important physical processes at work, we have begun to explore plausible scenarios for the origins and evolution of the LISM in general and the CLIC in particular.

Acknowledgments

This work was support by NASA Astrophysics Theory Grant No. NNX15AI60G. The author thanks the organizers of ASTRONUM 2016 for the invitation to present this work. The author also thanks his collaborators for helpful conversations and: Priscilla Frisch for work on the LIC, Cecile Gry who did the fits to the HST data and Nicolas Lehner who was PI on the HST observing proposal. The data was based on observations made with the NASA/ESA Hubble Space Telescope, and obtained from the Space Telescope Science Institute, which is operated by the Association of Universities for Research in Astronomy, Inc., under NASA contract NAS 5-26555. The observations are associated with program #GO-12981.01-A.

References

[1] Welsh B Y, Lallement R, Vergely J L and Raimond S 2010 *A&A* **510** A54
[2] McCammon D, Burrows D N, Sanders W T and Kraushaar W L 1983 *ApJ* **269** 107
[3] Welsh B Y and Shelton R L 2009 *Ap&SS* **323** 1
[4] Frisch P C, Redfield S and Slavin J D 2011 *ARA&A* **49** 237
[5] Galeazzi M, Chiao M, Collier M R, Cravens T, Koutroumpa D, Kuntz K D, Lallement R, Lepri S T, McCammon D, Morgan K, Porter F S, Robertson I P, Snowden S L, Thomas N E, Uprety Y, Ursino E and Walsh B M 2014 *Nature* **512** 171
[6] Redfield S and Linsky J L 2008 *ApJ* **673** 283
[7] Redfield S and Linsky J L 2015 *ApJ* **812** 125
[8] Gry C and Jenkins E B 2014 *A&A* **567** A58
[9] Meyer D M, Lauroesch J T, Heiles C, Peek J E G and Engelhorn K 2006 *ApJL* **650** L67
[10] Peek J E G, Heiles C, Peek K M G, Meyer D M and Lauroesch J T 2011 *ApJ* **735** 129
[11] Meyer D M, Lauroesch J T, Peek J E G and Heiles C 2012 *ApJ* **752** 119
[12] McComas D J, Bzowski M, Fuselier S A, Frisch P C, Galli A, Izmodenov V V, Katushkina O A, Kubiak M A, Lee M A, Leonard T W, Möbius E, Park J, Schwadron N A, Sokół J M, Swaczyna P, Wood B E and Wurz P 2015 *ApJS* **220** 22
[13] Vallerga J 1998 *ApJ* **497** 921 ISSN 0004-637X
[14] Lyu C H and Bruhweiler F C 1996 *ApJ* **459** 216
[15] Slavin J D and Frisch P C 2008 *A&A* **491** 53 ISSN 0004-6361
[16] Cowie L L and McKee C F 1977 *The Astrophysical Journal* **211** 135 ISSN 0004-637X
[17] McKee C F and Cowie L L 1977 *ApJ* **215** 213 ISSN 0004-637X
[18] Slavin J D 1989 *ApJ* **346** 718 ISSN 0004-637X
[19] Dalton W W and Balbus S A 1993 *The Astrophysical Journal* **404** 625 ISSN 0004-637X
[20] Slavin J D and Cox D P 1992 *ApJ* **392** 131 ISSN 0004-637X

Semi-convective layer formation

[1]**F. Zaussinger,** [2]**F. Kupka,** [1]**Ch. Egbers,** [1]**M. Neben,** [1]**S. Hücker,** [1]**C. Bahr,** [1]**M. Schmitt**

[1] BTU Cottbus-Senftenberg, Departement for Aerodynamics and Fluid Mechanics, Cottbus, Germany
[2] Wolfgang Pauli Institute, c/o Faculty of Mathematics of the Univ. of Vienna, Vienna, Austria

E-mail: florian.zaussinger@b-tu.de

Abstract. Semi-convective mixing, as an example of double-diffusive convection, is of general importance in multi-component fluid mixing processes. In astrophysics it occurs when the mean molecular weight gradient caused by a mixture of light material on top of heavier one counteracts the convective instability caused by a temperature gradient. Direct numerical simulations of double-diffusive fluid flows in a realistic stellar or planetary parameter space are currently non-feasible. Hence, a model describing incompressible semi-convection was developed, which allows to investigate semi-convective layer formation. A detailed parameter study with varying Rayleigh number and stability parameter has been performed for the giant planet case. We conclude that semi-convective layering may not play that important role as suggested in earlier works for the planetary case.

1. Introduction

Various situations are found in nature and in engineering problems where an unstably stratified fluid column is partially stabilized by a counteracting gradient. The double-diffusive instability (DD) belongs to these scenarios. As the name suggests, two different diffusivities are needed to describe this instability: the diffusivity κ_{T} of a fast diffusing component, e.g. temperature T, and the diffusivity κ_{S} of, say, a slowly diffusing solute such as salt in water or a component of a gas with different mean molecular weight than the other component(s) of the gas (such as helium in comparison to hydrogen in a gas mixture such as that one found in many stars). The ratio of both diffusivities is known as the Lewis number Le and for the cases of interest here, $\mathrm{Le} = \kappa_S/\kappa_T \ll 1$. Apart from the diffusivities the stratification is an important component of DD. Situations where the temperature stratification is unstable against small perturbations and the solute stabilises the convective buoyancy force are known as the *diffusive regime*. In astrophysics this case is known as *semi-convection*. On the other hand, if the solute is unstably stratified and the temperature gradient is positive in the vertical direction, so-called *salt-fingers* can occur. Consequently, both instabilities are determined mainly by the ratio of the thermal and the solute Brunt-Väisälä frequencies N_T^2 and N_S^2. The stability ratio R_ρ compares the impact of the solute stratification on the thermal buoyancy. The diffusive regime is thereby defined as $R_\rho = -N_S^2/N_T^2$, the salt-finger regime as $R_\rho = -N_T^2/N_S^2$. In spite of this classification the value of R_ρ related to unity is important: $R_\rho > 1$ implies that the stratification is much more stable than in the opposite case of $R_\rho < 1$. In particular, the parameter space of $1 < R_\rho < \mathrm{Le}^{-0.5}$

is characterized by layering processes which lead to so-called *staircases*. These staircases are well mixed layers separated by sharp interfaces with large gradients of T and S. The common challenge of modelling layered DD processes is to calculate the thermal and the solute flux over these interfaces. The fluxes in turn are needed to calculate the effective diffusivities, which can be used to estimate merging timescales and hence further evolutionary properties (mixing times, overshooting) of the star. The investigated numerical model is based on a modified experimental setup inspired by Turner [15]. The purpose of the numerical study is to investigate evolutionary aspects of the formation of the staircases and their spatial evolution.

1.1. DD in astrophysics

The challenging stellar parameter regime of $\mathrm{Pr}, \mathrm{Le} \ll 1$ makes it impossible to investigate the realistic stellar fluid flow with direct numerical simulations. Merryfield [7] performed the first simulations for $\mathrm{Pr} = 0.1$ and $\mathrm{Le} = 0.04$ for a setup otherwise akin to the convective core boundary of massive stars. He found coherent structures, but layer formation was missing, which was explained with a too low numerical resolution. Biello [2] investigated semi-convection in the same parameter space. He improved Merryfield's model setup and observed layering. Bascoul [1] found semi-convective layering for both regimes, the astrophysical one and the geophysical one. He simulated an expanding core by applying a time-dependent flux boundary condition at the bottom. Zaussinger et al. [16], [17] compared compressible and incompressible semi-convection experiments and confirmed Spruit's estimates for a parameter space accessible to direct numerical simulation ($0.01 \leqslant \mathrm{Le} \leqslant \mathrm{Pr} \leqslant 7$ with Ra up to about 10^6). Kupka et al. [10] recently published results concerning comparisons of semi-convection in oceans and in stars. Even extrapolations from the oceanographic to the stellar parameter space have been presented. Moll and Garaud [4] recently published a new model for double diffusive mixing, based on an extensive series of 3D simulations. As a result a modified power law for estimating the thermal and solute fluxes was derived. Layered semi-convection has been investigated by Chabrier and Baraffe [3] for the giant planet case, where the Prandtl- and the Lewis number are surprisingly high. The present study is oriented towards investigating their approximations.

2. Governing Equations

The double-diffusive system is described by four non-linear equations for the incompressible case. The scaled set of equations are based on the Boussinesq approximation, where the underlying equations are the mass conservation, the Navier-Stokes equation, the solute equation and the temperature equation. The density is assumed to be constant, apart from the buoyancy term. This approximation is justified since we consider small layers (in terms of the vertical extent), moderate flow speeds (compared to the sound speed) and small fluctuations of the dynamical variables around their (constant) background state.

$$\begin{aligned}
&\nabla \cdot \vec{u} = 0 \\
&\frac{\mathrm{D}\,\tilde{\mathrm{u}}}{\mathrm{D}\,\mathrm{t}} = -\frac{1}{\rho_0}\nabla p + \nu\Delta\vec{u} + \rho\vec{g} \\
&\frac{\mathrm{D}\,\mathrm{T}}{\mathrm{D}\,\mathrm{t}} = \kappa_T \Delta T \\
&\frac{\mathrm{D}\,\mathrm{S}}{\mathrm{D}\,\mathrm{t}} = \kappa_S \Delta S
\end{aligned} \tag{1}$$

IOP Conf. Series: Journal of Physics: Conf. Series **837** (2017) 012012 doi:10.1088/1742-6596/837/1/012012

Here we use the subsequent scaling denoted by *, which follows the idea that semi-convective processes happens on the thermal time scale,

$$\begin{aligned}
\mathrm{t} &= t^* \cdot \tfrac{H^2}{\kappa_T} \\
\mathrm{u} &= u^* \cdot \tfrac{\kappa_T}{H} \\
\mathrm{p} &= p^* \cdot \tfrac{\rho_0 \kappa_T \nu}{H^2} \\
\mathrm{T} &= T_0 + \Delta T T^* \\
\mathrm{S} &= S_0 + \Delta S S^*.
\end{aligned} \tag{2}$$

The buoyancy is modelled in terms of the extended Boussinesq approximation. The density, normalized to a background density $\rho_0 = 1$, hence is given by

$$\rho = 1 - \alpha(\mathrm{T} - \mathrm{T}_0) + \beta(\mathrm{S} - \mathrm{S}_0) \tag{3}$$

We follow standard notation, where the velocity is denoted by $\vec{u}(\vec{x}, t) = (u(\vec{x}, t), v(\vec{x}, t), w(\vec{x}, t))$, and $T(\vec{x}, t)$ is the (potential) temperature with units chosen to scale between 0 and 1, $S(\vec{x}, t)$ is the solute concentration scaled in similar manner such that S is between 0 and 1, and $p(\vec{x}, t)$ is the pressure. The height H, the thermal diffusion coefficient κ_T, the solute diffusion coefficient κ_S, the background density ρ_0, the kinematic viscosity ν, the vertical temperature contrast ΔT, the vertical saline contrast ΔS, the thermal expansion coefficient α, the saline expansion coefficient β and the gravitational acceleration g_z are needed to describe the double diffusive system. The dimensionless set of equations is obtained by applying the mentioned scaling (2),

$$\begin{aligned}
\nabla \cdot \vec{u} &= 0 \\
\mathrm{Pr}^{-1} \frac{\mathrm{D}\,\tilde{\mathrm{u}}}{\mathrm{D}\,\mathrm{t}} &= -\nabla p + \Delta \vec{u} - \mathrm{Ra_T} \cdot T + \mathrm{Ra_S} \cdot S \\
\frac{\mathrm{D}\,\mathrm{T}}{\mathrm{D}\,\mathrm{t}} &= \Delta T \\
\frac{\mathrm{D}\,\mathrm{S}}{\mathrm{D}\,\mathrm{t}} &= \mathrm{Le} \Delta \mathrm{S}
\end{aligned} \tag{4}$$

after having introduced the thermal Rayleigh number

$$\mathrm{Ra_T} = \tfrac{\alpha g \Delta T H^4}{\kappa_T \nu},$$

the solute Rayleigh number

$$\mathrm{Ra_S} = \tfrac{\beta g \Delta S H^4}{\kappa_T \nu},$$

the Prandtl number $\mathrm{Pr} = \frac{\nu}{\kappa_T}$, the Lewis number $\mathrm{Le} = \frac{\kappa_S}{\kappa_T}$ and the stability parameter $R_\rho = \frac{\mathrm{Ra_S}}{\mathrm{Ra_T}}$. The material derivative is defined as $D/Dt = \partial/\partial t + \vec{u} \cdot \nabla$. We have dropped the scaling indicator * here for better readability.

The height of the domain is set to $1 < H < 5$ and the thermal diffusion time scale $\tau = \frac{H^2}{\kappa_T}$ defines one unit of time. Especially the correct height of the box is difficult to find, since the expected height of the single layers vary in time. However, a rough estimate gives $H_{\mathrm{layer}} = R_\rho^{-1}$ for a macroscopic and well developed layer. Consequently, the dynamical behaviour of double diffusive convection is described by the four dimensional space spanned by $(\mathrm{Pr}, \mathrm{Le}, \mathrm{Ra_T}, \mathrm{R}_\rho)$. The boundaries are assumed to be impermeable for the temperature (top: $T = 0$, bottom: $T = H$) and the solute (top: $S = 0$, bottom: $S = H$). This ensures a thermal and a solute diffusive flux of $\frac{\Delta T}{H} = \frac{\Delta S}{H} = 1$. Impenetrable, stress free boundary conditions are set for the velocity, hence $\frac{\partial v}{\partial z} = 0$ and $u = 0$ with z denoting the vertical direction. The horizontal boundary conditions are periodic. We define the standard box by setting $H = 1$, $\Delta T = 1$ and $\Delta S = 1$. In order to use higher boxes with, e.g., the scaling factors aH and $a\Delta T$ as well as $a\Delta S$, the Rayleigh number as function of the height needs to be rescaled by

$$\mathrm{Ra[aH]} = \mathrm{Ra_{ref}} \cdot \mathrm{a}^4. \tag{5}$$

The reference Rayleigh number $\mathrm{Ra_{ref}}$ refers always to a box with height $H = 1$ and gradients in temperature and the solute as defined. However, it is very convenient to compare different box geometries with $\mathrm{Ra_{ref}}$. The simulation time changes accordingly $(aH)^2$.

2.1. Numerical setup

The governing equations are solved numerically with the ANTARES software suite, which treats advective terms with a 5th-order WENO type scheme on a rectangular grid. A comprehensive overview on the numerical details of this code is presented in Muthsam et al. [9] and Zaussinger and Spruit [16]. The interior domain consists of several layers 'peeled off' from the boundary of the underlaying convective zone. Hence a non-uniform grid is not needed. The main focus in choosing the grid size lies on the resolution of the solute boundary layers, which are the steepest. Not resolving them is a major limitation and hence would alter extrapolations in the numerically non-feasible stellar case significantly. Consequently, the thermal and the solute boundary layers are resolved with at least three points to guarantee correctly calculated fluxes. The minimum resolution can easily be estimated with the thermal Rayleigh number, by estimating the thermal boundary layer thickness δ_T,

$$\frac{\delta_T}{H} \approx \sqrt[4]{\frac{1}{\mathrm{Ra_T} \cdot \mathrm{Pr}}} \tag{6}$$

and by using

$$\delta_S = \sqrt{\mathrm{Le}}\, \delta_T, \tag{7}$$

the solute boundary thickness is calculated in turn. The layer formation itself is triggered by a linear temperature and solute gradient, and a small perturbation in the solute.

2.2. Model parameters and box geometry

The four dimensional parameter space spanned by the quantities $(\mathrm{Pr}, \mathrm{Le}, \mathrm{Ra_T}, R_\rho)$ is too large to be covered by numerical simulations without excessive computational expenses. The astrophysically relevant restriction of $\mathrm{Le} < \mathrm{Pr}$ reduces the computational amount significantly. This is reflected by the Schmidt number, $\mathrm{Sc} = \frac{\nu}{\kappa_S}$, which is typically in the oder of $\mathrm{Sc} \approx 100$. In the following the 'loosely' astrophysical (giant planet) based case of $\mathrm{Pr} = 1$ and $\mathrm{Le} = 0.01$ is investigated in more detail.

Since layer formation needs a certain amount of convective cells, the aspect ratio is an important factor. Based on a couple of initial tests, the aspect ratio is set to 2, which results in a typical box geometry of height $H = 5$ and width $L = 10$. This leads to thermal Rayleigh numbers of $\mathrm{Ra_T} > 10^7$, where the smallest scales are resolved by a standard Smagorinsky subgrid model. In order to reduce confusion all values are given in terms of the reference height $H = 1$ and have to be scaled according to equation (5).

3. The giant planet case with $\mathrm{Pr} = 1$ and $\mathrm{Le} = 0.01$

This case represents values, which are comparable to a giant gas planet with $10^{-2} < \mathrm{Pr} < 1$ and $\mathrm{Le} = 10^{-2}$. The study is covered by 72 numerical simulations, where the parameter space is spanned by the Rayleigh number and the stability parameter (see table 1). The dynamical changes have typically occurred within $\tau = 0.1$, which reduces the computational costs significantly. The numerical time step is of the order of $\Delta\tau = 10^{-6}$ and gives a computational

	R_ρ					
Ra_{ref}	1	$1.1-1.3$	1.4	1.5	1.6	$1.7-3.5$
$5.0 \cdot 10^3$	C	LU	D	D	D	D
$1.0 \cdot 10^4$	C	LU	L4	D	D	D
$1.5 \cdot 10^4$	C	LU	L5	D	D	D
$2.0 \cdot 10^4$	C	LU	L6	L6	D	D
$5.0 \cdot 10^4$	C	LU	L6	L7	D	D
$1.0 \cdot 10^5$	C	LU	L6	L8	D	D
$2.0 \cdot 10^5$	C	LU	L7	L9	D	D

Table 1. Semi-convective regimes for varying $\mathrm{Ra}_{\mathrm{ref}}$ and R_ρ after $\tau = 0.05$. No layering, full overturns (C), layers unstable with tendency to full overturning (LU), diffusive or conductive (D) and layered regime with number X of staircases (LX).

effort of two days on 36 processors for each simulation. However, this is very cheap compared to other setups. High resolution 3D simulations with more than 400 points in each direction are very expensive, especially, when non-linearities begin to dominate the processes and thus lead to restrictive, advection dominated time steps. Since the stability parameter is less then the predicted maximum value of $R_{\mathrm{max}} = \mathrm{Le}^{-0.5} = 10$, layer formation is expected in $1 < R_\rho < R_{\mathrm{max}}$.

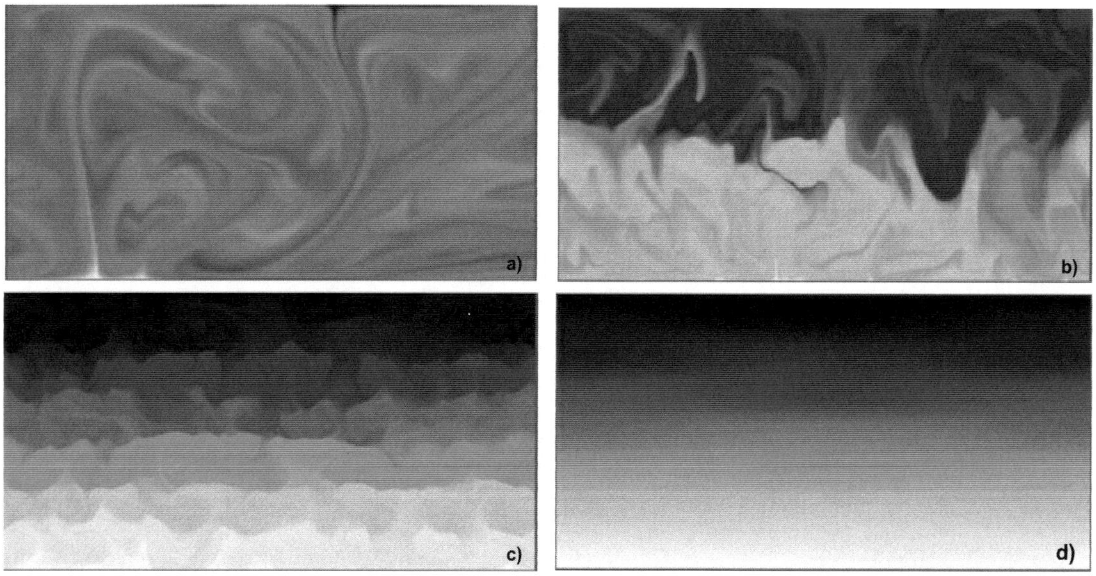

Figure 1. Solute (yellow high content, black low content) for varying stability parameters. $Ra_T = 3 \cdot 10^7$, $Pr = 1$, $Le = 0.01$: a) $R_\rho = 1$ fully convective, b) $R_\rho = 1.3$ unstable layering, c) $R_\rho = 1.4$ stable layering, d) $R_\rho = 3$ conductive.

Chabrier and Baraffe [3] and Zaussinger et al. [16], [17] estimated $10^5 - 10^6$ layers per pressure scale height H_p after a short initial layer formation period of some weeks. However, this number of layers has not to be simulated, since most relevant fluid dynamical processes can be investigated with 4–7 stairs. The layered convective regime is embedded between the fully convective regime, $R_\rho \lessapprox 1$, and the very stable conductive regime, $R_{\rho,\mathrm{max}} \gtrapprox 10$. The existence of this upper stability limit $R_{\rho,\mathrm{max}}$ is known from, e.g., Stevenson [14] and in the astrophysical context from Spruit [13].

However, the actual range of layered convection for the presented values of Rayleigh numbers is fairly small, $1.4 < R_\rho < 1.5$, which coincides with theoretical estimates from Mirouh [8] and Leconte [12] that a minimum stability parameter $R_{\rho,\mathrm{min}}$ separates the layered regime from the turbulent diffusive state. We found this value for $R_{\rho,\mathrm{min}} \approx 1.5$ (see figure 1c). The stability change from layered convection to turbulent diffusive convection is very sharp and needs to be investigated by numerical simulations for each new set of parameters. Since the super adiabatic gradient in the interior of a giant gas planet like Jupiter is extremely small, $\nabla - \nabla_{\mathrm{ad}} \approx 10^{-8}$ (see [3]), it has to be a happy coincidence that ∇_μ ranges exactly between $10^{-8} - 10^{-7}$ to trigger layering over the whole pressure scale height H_p in the small layering windows. This leads to the assumption that layering is a rare phenomenon and rather local than global. We found that the amount of layers increase with the Rayleigh number and is only limited by the vertical extent of the box and by $R_{\rho,\mathrm{max}}$. Since merging processes between convective cells destabilise the stack, the layered regime is only a short-term phenomenon. Merging effects and local entrainment decompose the stack, which leads to the question, how important the (temporal) layered semi-convective state is for the planetary evolution. A detailed study on time scales is currently in preparation.

4. Conclusion

Realistic numerical simulations of double diffusive stacks are a formidable task, since the small diffusivities and especially the extreme gap between the convective and the evolutionary time scale of the giant planet have to be covered. It seems to be impossible to simulate a whole pressure scale height with all scales resolved. But this is not necessary, since extrapolations from numerically feasible parameters and theoretical assumptions give enough tools to investigate the astrophysically relevant regime. The results show that the geophysical relevant parameter space can be applied to the astrophysical case, even when some properties like different time scales or the heights are numerically challenging.

Acknowledgements

We thank Robert Rode from the Graduate Research School (GRS) at the BTU Cottbus-Senftenberg for funding the costs for the Astronum 2016. Furthermore, we thank Andreas Krebs from the 'Center for Flow and Transport, Modelling and Measurement ($CFTM^2$)' for managing the computational resources used for the numerical simulations. Work of F. Kupka has been supported through Austrian Science Fund (FWF) grants P25229-N27 and P29172-N27.

References

[1] G. Bascoul: Double-diffusive convection in stars, PhDT, University of Cambridge, priv. comm., (2007)
[2] J. Biello, Layer formation in semiconvection, PhDT, University of Chicago, priv. comm., (2001)
[3] G. Chabrier, I. Baraffe, Heat Transport in Giant (Exo)planets: A New Perspective, ApJ Letters, 661-1 (2007)
[4] R. Moll, P. Garaud, S. Stellmach, A New Model for Mixing By Double-Diffusive Convection (Semi-Convection). III. Thermal and Compositional Transport Through Non-Layered ODDC, ApJ, 823, (2016)
[5] H.E. Huppert, J.S. Turner: Double-diffusive convection, JFM, 106, 199-329 (1981)
[6] H. E. Huppert, D. R. Moore. Nonlinear double-diffusive convection. JFM, 78, 821-854 (1976)
[7] W. J. Merryfield, Hydrodynamics of semiconvection, ApJ, 444, 318-337 (1995)
[8] Mirouh, G. M., Garaud, P., Stellmach, S., Traxler, A. L., Wood, T. S. A new model for mixing by double-diffusive convection (semi-convection). I. The conditions for layer formation. ApJ, 750, 61 (2012).
[9] H. Muthsam et al., ANTARES - A Numerical Tool for Astrophysical RESearch with applications to solar granulation, NewA, 15, 460-475, (2010)
[10] F. Kupka, Semi-convection in the ocean and in stars: A multi-scale analysis, MetZ, 24-33, 343358 (2015)
[11] J. Leconte, G. Chabrier, A new vision of giant planet interiors: Impact of double diffusive convection, A&A, 540, A20 (2012)

[12] J. Leconte, G. Chabrier, Layered convection as the origin of Saturns luminosity anomaly, Nature GeoSc, 6, 347350 (2013)
[13] H. Spruit, Semiconvection: theory, A&A, 552, A76 (2013)
[14] D. J. Stevenson, Formation of the giant planets, Planet. Space Sci., 30, 755 (1982)
[15] J. S. Turner, The behaviour of a stable salinity gradient heated from below, JFM, 33, 183-200 (1968)
[16] F. Zaussinger, H. Spruit, Semiconvection: numerical simulations, A&A, 554, A119 (2013)
[17] F. Zaussinger, F. Kupka, H. J. Muthsam, Semi-convection, LNP, Vol. 865, 219-237 (2013)

Charge-exchange source terms in magnetohydrodynamic plasmas

Anthony M. DeStefano, Jacob Heerikhuisen
Department of Space Science, University of Alabama in Huntsville, USA
E-mail: amd0020@uah.edu

Abstract. In the modeling of space plasma environments, source terms are often used to couple separate species of particles and/or fluids. There have been many techniques developed over the years to make such coupling more tractable while maintaining maximum physical fidelity. In our current application we use the formalism of the Boltzmann collision integral to compute source terms due to charge-exchange events in the heliosphere. The charge-exchange cross sections often encountered in heliospheric interactions can be fit to laboratory data, but in most cases cannot be directly integrated over analytically. Therefore, researchers often employ various levels of approximation, either semi-analytic or numerical. We explore several assumptions to the charge-exchange source term integrals, namely using Maxwellian velocity spaces for like-mass species and either hard-sphere, power-law, or exact forms of the cross section.

1. Introduction

Modeling of the interaction between the solar wind (SW) and local interstellar medium (LISM) dates back to the dawn of the space age, [1–4]. As knowledge of properties of the SW and LISM improved, the models correspondingly began to include more sophisticated physics. The LISM is a partially ionized gas [5] where LISM hydrogen ions and neutrals are weakly coupled by photoionization and charge-exchange [6, 7]. The crossing of the termination shock in 2004 and 2007 by NASA's Voyager 1 & 2 spacecraft [8, 9], respectively, and the transition into interstellar space by Voyager 1 in 2012 [10] have provided invaluable data to the heliospheric modeling community. In addition, the Interstellar Boundary Explorer (IBEX) [11], launched in October 2008, has been making observations of energetic neutral atoms (ENAs), which discovered the IBEX ribbon and has shed light on the structure of the heliosphere, interstellar matter, and several other scientific areas (see [12] for a complete list of discoveries). The accepted theory of the generation of the IBEX ribbon is the secondary ENA mechanism [13, 14], which includes charge-exchange in the micro physics between the LISM plasma and ENAs.

It was shown qualitatively by Wallis [7] that resonant charge-exchange between hydrogen atoms and protons greatly influences the structure and location of the termination shock and heliopause. Since that time it has commonly been accepted that this charge-exchange mechanism is a critical ingredient for models of the heliosphere. Other collision mechanisms, both inelastic and elastic, may also contribute in modifying the global structure of the heliosphere. As pointed out by Williams et al. [15], momentum exchange for H-H and H-p collisions are equally as frequent as charge-exchange, with mean free paths of $\sim$200 AU outside the heliosphere. This may degrade the distinction between the separate populations of hydrogen atoms (interstellar, ENA, etc.) and act to thermalize them to a common temperature. In this paper we will solely focus on resonant charge-exchange with same-mass, Maxwellian distributed velocity space species.

Two popular simulation methods used to model the heliosphere are the fluid/multi-fluid and fluid-kinetic simulations, where the fluids are either hydrodynamic (HD), magnetohydrodynamic (MHD), or both. The ions are always HD/MHD fluids, while the neutrals may be a single fluid, ensemble of fluids, or kinetic particles. The self consistent problem of the SW-LISM interaction, including resonant charge-exchange, was considered by [16, 17] using a fluid approach, see also [18–21]. Examples of a multi-fluid approach include [15, 22, 23]. The work of Malama [24] introduced the kinetic treatment of neutral hydrogen via the Monte Carlo method, leading to the paper of Baranov & Malama [25] solving the problem using charge-exchange source terms to couple particle neutrals to a fluid plasma. However, the effects of the interplanetary and interstellar magnetic fields were ignored. A decade or so later, the modeling groups of [26–28] adopted the Monte Carlo kinetic approach along with a MHD description of the proton/electron mixture. For the sake of brevity, in this paper we will limit discussion to source terms that couple the fluid ions and fluid neutrals via resonant charge-exchange.

In the past, authors have only compared the analytic solutions of the hard-sphere or power-law source term approximations to their asymptotic forms (e.g., [19, 29, 30]). However, comparisons to numerical evaluations of the exact source term integrals have not been done thus far, and are necessary in determining the effectiveness of these different approximations for various parameter domains. The purpose of this paper will be to analyze the relative errors of these approximations to the exact source term integrals for a range of effective temperatures as well as specific cases of parameters particular to the heliosphere.

In §2 we will outline the results of using two Maxwellian fluids (one ion and one neutral) in the collision source term integrals while keeping the form of the charge-exchange cross section generalized. Then, in §3 we will proceed to derive expressions for power-law and in §4 for hard-sphere (constant over integral) cross sections, as a special case of the power-law formalism. Proceeding to §5.1 we will compare the numerical results of using the exact, power-law, and hard-sphere forms of the charge-exchange cross section for the momentum and thermal energy source terms for a range of effective temperatures. §5.2 will be dedicated to specific cases of temperatures and bulk velocities for protons and hydrogen atoms in the heliosphere. Finally, we will summarize in §6.

2. Resonant charge-exchange source terms

To begin, we will assume elastic, resonant, charge-exchange collisions between protons and hydrogen atoms. The following calculations will be done for the mass, momentum, and energy sources imparted on the proton population, but analogous equations can be found for the hydrogen population by swapping the p and H subscripts. Under the theory of the Boltzmann collision integral for isotropic cross sections we can write (e.g., [31]),

$$\begin{Bmatrix} Q_n \\ \mathbf{Q}_M \\ Q_{E_{tot}} \end{Bmatrix} = \iint d^3v_p d^3v_H |\mathbf{v}_H - \mathbf{v}_p| \sigma_{ex}(|\mathbf{v}_H - \mathbf{v}_p|) f_p f_H \triangle\xi_{Hp}, \tag{1}$$

where the source term moments are,

$$\triangle\xi_{Hp} = \begin{cases} m_H - m_p \approx 0 & \text{mass} \\ m_H\mathbf{v}_H - m_p\mathbf{v}_p = m(\mathbf{v}_H - \mathbf{v}_p) & \text{momentum} \\ \frac{1}{2}m_H v_H^2 - \frac{1}{2}m_p v_p^2 = \frac{1}{2}m(v_H^2 - v_p^2) & \text{total energy} \end{cases} \tag{2}$$

and σ_{ex} is the charge-exchange cross section (e.g., [32] or [33]). The mass density source equates to zero because the p-H charge-exchange reaction is symmetric. Throughout this paper we will assume that the proton and hydrogen masses are the same, viz. $m \equiv m_H = m_p$. The p-H charge-exchange cross section is parametrized in [32] by the following expression,

$$\sigma_{ex}(E) = (a_1 - a_2 \ln E)^2 \left(1 - e^{-a_3/E}\right)^{4.5} \quad \text{for } E \in [0.005, 250] \text{ keV}, \tag{3}$$

where $a_1 = 4.15$, $a_2 = 0.531$, $a_3 = 67.3$ keV, and E in units of keV with σ_{ex} in units of 10^{-16} cm^2. The distribution functions of the protons and hydrogen atoms, f_p and f_H, are assumed to be Maxwellian (i.e., independently in local thermal equilibrium) such that,

$$f_p f_H = \frac{n_p n_H}{\pi^3 v_{tp}^3 v_{tH}^3} \exp\left[-\frac{(\mathbf{v}_p - \mathbf{u}_p)^2}{v_{tp}^2} - \frac{(\mathbf{v}_H - \mathbf{u}_H)^2}{v_{tH}^2}\right], \tag{4}$$

where n_p & n_H are the number densities, $\mathbf{u}_p$ & $\mathbf{u}_H$ the bulk flow velocities, and v_{tp} & v_{tH} the thermal speeds defined by the temperature as $v_{ti} = \sqrt{2k_B T_i/m_i}$ for each species, k_B is the Boltzmann constant and T_i and m_i are the temperatures and masses of the species. The macroscopic variables n_i, $\mathbf{u}_i$, and T_i are implied to be functions of position and time, independent of the phase space variables $\mathbf{v}_i$, for $i = p, H$. Next, we define the center of mass velocity $\mathbf{V}$ and relative velocity $\mathbf{g}$ (see e.g., [29, 31]) such that

$$\mathbf{V} = \frac{m_p \mathbf{v}_p + m_H \mathbf{v}_H}{m_p + m_H} \approx \frac{1}{2}(\mathbf{v}_p + \mathbf{v}_H) \tag{5}$$

$$\mathbf{g} = \mathbf{v}_H - \mathbf{v}_p \tag{6}$$

or writing the original velocities in terms of the new definitions,

$$\mathbf{v}_H = \mathbf{V} + \frac{1}{2}\mathbf{g} \tag{7}$$

$$\mathbf{v}_p = \mathbf{V} - \frac{1}{2}\mathbf{g}. \tag{8}$$

Note that the Jacobian of this transformation is equal to 1. To proceed with evaluating Eq. (1) we first integrate over the center of mass velocity. The following definitions aid with simplifying the notation:

$$\triangle \mathbf{u} = \mathbf{u}_H - \mathbf{u}_p \tag{9}$$

$$\mathbf{U} = \frac{\mathbf{u}_p v_{tH}^2 + \mathbf{u}_H v_{tp}^2}{v_t^2} \tag{10}$$

$$\triangle v_t^2 = v_{tH}^2 - v_{tp}^2 \tag{11}$$

$$v_t^2 = v_{tH}^2 + v_{tp}^2 \,. \tag{12}$$

After the change of variables of Eqs. (5) and (6), the exponent of Eq. (4) reads

$$-\frac{v_t^2}{(v_{tp} v_{tH})^2}\left[V^2 + \mathbf{V}\cdot\mathbf{X} + b\right] \tag{13}$$

where

$$\mathbf{X} = -\frac{\triangle v_t^2}{v_t^2}\mathbf{g} - 2\mathbf{U} \tag{14}$$

$$b = \frac{1}{4}g^2 + \mathbf{g}\cdot\mathbf{U} + \frac{u_p^2 v_{tH}^2 + u_H^2 v_{tp}^2}{v_t^2}. \tag{15}$$

We choose $\mathbf{X}$ to be aligned along the z-axis and the perpendicular component of $\mathbf{g}$, with respect to $\mathbf{X}$, along the x-axis in a spherical coordinate system. This choice makes the integration over the azimuthal angle ϕ trivial, as well as for the polar angle $\cos\theta = \mu$, where $\hat{\mathbf{V}} \cdot \hat{\mathbf{X}} = \mu$. The integral over $\mathbf{g}$ is done in a similar manner; we integrate via a spherical coordinate system with the z-axis aligned along $\triangle\mathbf{u}$, simplifying the azimuthal integration.

Since the charge-exchange cross section is only a function of the magnitude of $\mathbf{g}$ we can compute all but the last integral over g. We then arrive at

$$\mathbf{Q}_M = \frac{2mn_pn_H}{\Gamma(3/2)v_t^3}\int_0^\infty dg\sigma_{ex}(g)g^4 e^{-\frac{g^2+\triangle u^2}{v_t^2}} i_1\left(\frac{2g\triangle u}{v_t^2}\right)\triangle\hat{\mathbf{u}} \tag{16}$$

$$\begin{aligned} Q_{E_{tot}} &= \mathbf{Q}_M \cdot \mathbf{U} + \frac{mn_pn_H\triangle v_t^2}{\Gamma(3/2)v_t^5}\int_0^\infty dg\sigma_{ex}(g)g^5 e^{-\frac{g^2+\triangle u^2}{v_t^2}} i_0\left(\frac{2g\triangle u}{v_t^2}\right) \\ &= Q_{E_K} + Q_{E_T} \end{aligned} \tag{17}$$

where Γ is the gamma function and i_n is the modified spherical Bessel function of the first kind, for the momentum and total energy charge-exchange source terms, in agreement with Eqs. (25) and (27) of McNutt et al.'s work [29]. Notice that the first and second term of $Q_{E_{tot}}$ represents the kinetic and thermal energy source denoted by Q_{E_K} and Q_{E_T}, respectively.

From Eq. (17) we can understand the physical meaning of $\mathbf{U}$, namely, the kinetic energy source is given by the component of the momentum source in the direction of $\mathbf{U}$. There is a bias towards the bulk flow of the cooler species for the direction of $\mathbf{U}$. In this sense, the hotter species has less of an effect on the overall kinetic energy source, albeit a greater influence on the thermal energy source.

3. Power-law cross section approximation of source terms

Given cross sections that vary slowly with the collision energy (i.e., relative speed g), we can model the cross section as having a power-law form given by,

$$\sigma_{ex}(g) \sim \sigma_0\left(\frac{g^2}{g_0^2}\right)^\nu \tag{18}$$

similar to Eq. (1) of [30], where g_0 is the effective relative speed, $\sigma_0 = \sigma_{ex}(g_0)$ the cross section, and ν the power-law index. Inserting Eq. (18) into Eqs. (16)-(17) and using the substitution $z = g^2\triangle u^2/v_t^4$, we arrive at the intermediate step,

$$\mathbf{Q}_M = \frac{mn_pn_H\sigma_0}{\Gamma(3/2)v_t^3g_0^{2\nu}}\left(\frac{v_t^2}{\triangle u}\right)^{2\nu+5} e^{-\frac{\triangle u^2}{v_t^2}}\int_0^\infty dz z^{\nu+3/2}e^{-z\frac{v_t^2}{\triangle u^2}} i_1(2\sqrt{z})\triangle\hat{\mathbf{u}} \tag{19}$$

$$Q_{E_{tot}} = \mathbf{Q}_M \cdot \mathbf{U} + \frac{mn_pn_H\sigma_0\triangle v_t^2}{\Gamma(3/2)v_t^5g_0^{2\nu}}\left(\frac{v_t^2}{\triangle u}\right)^{2\nu+5} e^{-\frac{\triangle u^2}{v_t^2}}\int_0^\infty dz z^{\nu+2}e^{-z\frac{v_t^2}{\triangle u^2}} i_0(2\sqrt{z}). \tag{20}$$

To proceed we use the following confluent hypergeometric function substitutions, which can be derived by matching the power series of the confluent hypergeometric and hyperbolic functions

$${}_0F_1(-;3/2;t) = \frac{\sinh(2\sqrt{t})}{2\sqrt{t}} = i_0(2\sqrt{t}) \tag{21}$$

$${}_0F_1(-;5/2;t) = \frac{3}{4t}\left[\cosh(2\sqrt{t}) - \frac{\sinh(2\sqrt{t})}{2\sqrt{t}}\right] = \frac{\Gamma(5/2)}{\Gamma(3/2)\sqrt{t}}i_1(2\sqrt{t}), \tag{22}$$

the integral

$$\int_0^\infty t^{\alpha-1}\exp(-at)_0F_1(-;b;t)dt = a^{-\alpha}\Gamma(\alpha)_1F_1(\alpha;b;1/a), \tag{23}$$

derived from the general form in Eq. 7.522.5 of Gradshteyn & Ryzhik [34], and Kummer's transformation, Eq. 13.1.27 of Abramowitz & Stegun [35],

$$e^{-t}{}_1F_1(a;b;t) = {}_1F_1(b-a;b;-t). \tag{24}$$

Applying these substitutions and transformation we obtain,

$$\mathbf{Q}_M = A_\nu \triangle \mathbf{u} G_{1,\nu}\left(\frac{\triangle u}{v_t}\right) \tag{25}$$

$$Q_{E_{tot}} = Q_{E_K} + Q_{E_T} = \mathbf{Q}_M \cdot \mathbf{U} + \frac{1}{2}A_\nu \triangle v_t^2 G_{2,\nu}\left(\frac{\triangle u}{v_t}\right), \tag{26}$$

where the coefficient A_ν and functions $G_{1,\nu}(x)$ and $G_{2,\nu}(x)$ are,

$$A_\nu = \frac{m n_p n_H \sigma_0 v_t}{(g_0/v_t)^{2\nu}} \tag{27}$$

$$G_{1,\nu}(x) = \frac{\Gamma(\nu+3)}{\Gamma(5/2)}{}_1F_1\left[-\nu-\frac{1}{2};\frac{5}{2};-x^2\right] \tag{28}$$

$$G_{2,\nu}(x) = \frac{\Gamma(\nu+3)}{\Gamma(3/2)}{}_1F_1\left[-\nu-\frac{3}{2};\frac{3}{2};-x^2\right], \tag{29}$$

and ${}_1F_1$ the Kummer confluent hypergeometric function. Note we recover Eq. (23) of [30].

It can be observed that if both the proton and hydrogen species have the same bulk velocity there will be no net exchange in momentum. However, for different temperatures there will be exchange in energy transfered from the hotter species to the colder species. On the other hand, if the species have the same temperature but different bulk flow velocities, the sign of the momentum and energy exchange is determined by the sign of $\triangle\mathbf{u}$. Finally, if both the bulk flows and temperatures are the same, the species are in equilibrium with each other and there is no net exchange of momentum or energy, even though charge-exchange events will occur.

In numerical applications, it is beneficial to use simplified forms of Eqs. (25)-(26) by matching the first order term of the Taylor and asymptotic series for small and large values of $x = \triangle u/v_t$ to a model function,

$$\left[ax^2 + b^{1/(\nu+\frac{n}{2})}\right]^{\nu+\frac{n}{2}}. \tag{30}$$

By doing so, and factoring the coefficients such that $a = 1$, it can be shown that,

$$G_{1,\nu}(x) \approx \left[x^2 + \left(\frac{\Gamma(\nu+3)}{\Gamma(5/2)}\right)^{1/(\nu+\frac{1}{2})}\right]^{\nu+\frac{1}{2}} \tag{31}$$

$$G_{2,\nu}(x) \approx \left[x^2 + \left(\frac{\Gamma(\nu+3)}{\Gamma(3/2)}\right)^{1/(\nu+\frac{3}{2})}\right]^{\nu+\frac{3}{2}}, \tag{32}$$

where Eq. (31) is analogous to Eq. (31) of [30]. Therefore, Eqs. (31)-(32) are said to be asymptotic to Eqs. (28)-(29) for $x \ll 1$ or $x \gg 1$. From a practical standpoint, using a single constant for the power-law index ν fails for most regimes of the p-H cross section, even when using a clever effective relative speed $\bar{v}$, e.g., [30]. To combat this issue, it is useful to adopt a variable power-law given by,

$$\nu = \frac{1\text{ keV}}{2\triangle E}\ln\left[\frac{\sigma_{ex}(E_0 e^{\triangle E/1\text{keV}})}{\sigma_{ex}(E_0 e^{-\triangle E/1\text{keV}})}\right]. \quad (33)$$

Eq. (33) is derived (see Appendix) from the central difference method by taking the Taylor expansion of the logarithm of the cross section with respect to the logarithm of the collision energy $E_0 = \frac{1}{2}m(\triangle u)^2$. The energies $\triangle E$ and E_0 are in units of keV, and $\triangle E$ is some small change in energy, typically < 0.001 keV for our purposes. The cross section in A_ν should then be evaluated at $\triangle u$ and not $\bar{v}$. As an illustration, Figure 1 is an application of Eq. (33) using σ_{ex} of Eq. (3).

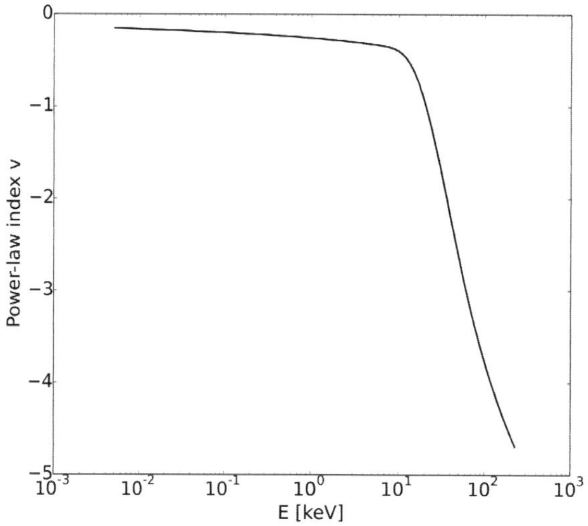

Figure 1: A power-law index fit to the p-H charge-exchange cross section [32] using Eq. (33). The exponential cutoff becomes noticeable for energies greater than ~ 10 KeV.

3.1. Physical interpretation of power-law cross section

An interesting consequence of using a power-law cross section, as defined in Eq. (18), is that the charge-exchange cross section can be thought of as the result of scattering by repulsion of a central force, given by $F = K_{ab}r^{-p}$, where K_{ab} is some constant that depends on the physical application (see Eq. 9.15 of [31] and pg. 172 of [36] for specific examples) and r is the distance between particles a and b. A typical central force found in nature is the Coulomb interaction, were $p = 2$ and $K_{ab} = q_a q_b/(4\pi\epsilon_0)$, where q_i are the charges of each species and ϵ_0 is the permittivity of free space.

From the Appendix of Section 9 of [31] we see that $\nu = -2/(p-1)$ relates the power-law index ν of the cross section to the central force power index p. Common central force laws are Coulomb interactions ($p = 2$, $\nu = -2$), Maxwell molecules ($p = 5$, $\nu = -1/2$), and hard-sphere interactions ($p \to \infty$, $\nu = 0$). The p-H charge-exchange cross section in Eq. (3) has a relatively hard power-law fit in the ranges of 5 eV ($p = 14$, $\nu = -0.154$) to 10 keV ($p = 6$, $\nu = -0.4$) reaching steeper than Coulomb-like interactions for 250 keV ($p = 1.42$, $\nu = -4.76$). See Figure 1 for the full range of energies within the domain of the parameterized fit of Eq. (3).

4. Hard-sphere cross section approximation of source terms

For the hard-sphere approximation, we take $\nu = 0$ in Eqs. (25)-(26), and simplify to obtain

$$G_{1,0}(x) = \frac{8}{3\sqrt{\pi}} {}_1F_1\left[-\frac{1}{2}; \frac{5}{2}; -x^2\right] = \frac{\exp(-x^2)}{\sqrt{\pi}}\left(1 + \frac{1}{2x^2}\right) + \mathrm{erf}(x)\left(x + \frac{1}{x} - \frac{1}{4x^3}\right) \quad (34)$$

$$G_{2,0}(x) = \frac{4}{\sqrt{\pi}} {}_1F_1\left[-\frac{3}{2}; \frac{3}{2}; -x^2\right] = \frac{\exp(-x^2)}{\sqrt{\pi}}\left(x^2 + \frac{5}{2}\right) + \mathrm{erf}(x)\left(x^3 + 3x + \frac{3}{4x}\right), \quad (35)$$

where erf is the error function, comparable to Eqs. (49)-(50) of [29]. See also [6] and [19]. The momentum and total energy charge-exchange source terms for a constant, hard-sphere cross section are then

$$\mathbf{Q}_M = A_0 \triangle \mathbf{u} G_{1,0}\left(\frac{\triangle u}{v_t}\right) \quad (36)$$

$$Q_{E_{tot}} = Q_{E_K} + Q_{E_T} = \mathbf{Q}_M \cdot \mathbf{U} + \frac{1}{2} A_0 \triangle v_t^2 G_{2,0}\left(\frac{\triangle u}{v_t}\right), \quad (37)$$

where $A_0 = m n_p n_H \sigma_0 v_t$ and $\sigma_0 = \sigma_0(\bar{v})$, to be evaluated at the effective relative speed $\bar{v}$. The definition of $\bar{v}$ is not unique, so we will take some time to discuss some variations of this definition in §4.1. We may also simplify $G_{1,0}$ and $G_{2,0}$ by plugging $\nu = 0$ into Eqs. (31)-(32), obtaining

$$G_{1,0}(x) \approx \sqrt{x^2 + \frac{64}{9\pi}} \quad (38)$$

$$G_{2,0}(x) \approx \left[x^2 + \left(\frac{16}{\pi}\right)^{1/3}\right]^{3/2}, \quad (39)$$

which are asymptotic to Eqs. (34) and (35) for $x \ll 1$ or $x \gg 1$.

4.1. Effective relative speed

In the case of hard-sphere collisions, the cross section is constant, and can thus be taken outside of the integral during evaluation. In this case it must then be computed at an effective relative speed, $\bar{v}$. One simple way to define $\bar{v}$ is via the average interaction rate (e.g., [19, 36]),

$$\bar{v} = \frac{1}{n_p n_H} \iint d^3v_p d^3v_H |\mathbf{v}_H - \mathbf{v}_p| f_p f_H \quad (40)$$

and for Maxwellian velocity distributions for both species it can be shown that,

$$\bar{v} = v_t G_{0,0}(x) \quad (41)$$

$$G_{0,0}(x) = \left[\frac{\exp(-x^2)}{\sqrt{\pi}} + \left(\frac{1}{2x} + x\right)\mathrm{erf}(x)\right] \approx \sqrt{\frac{4}{\pi} + x^2} \quad (42)$$

where $x = \triangle u / v_t$, $\triangle u = |\mathbf{u}_H - \mathbf{u}_p|$, and $v_t^2 = v_{tH}^2 + v_{tp}^2$. This definition of $\bar{v}$ is used for the effective relative speed for the number, momentum, and total energy source in [19], in contrast to the next definition (see Eqs. (44)-(46)).

For the hard-sphere approximation the cross section is Taylor expanded, as explained by [29], around an effective relative speed g_0,

$$\sigma_{ex}(g) = \sigma_0 + \left(\frac{d\sigma_{ex}}{dg}\right)_{g=g_0} (g - g_0) + \cdots \quad (43)$$

where g_0 is found such that the integral over the second term vanishes, for each moment 1, $\mathbf{g}$, and g^2, thereby eliminating first order errors. We assume that higher than first order terms are insignificant compared to lower order terms of the Taylor series expansion. Therefore, omitting the details (see Appendix B of [37] for a similar discussion) we find that the effective relative speeds for the number, momentum, and total energy sources are, using $x = \triangle u/v_t$

$$\bar{v} = \frac{3\sqrt{\pi}}{4} v_t \frac{{}_1F_1\left[-1; \frac{3}{2}; -x^2\right]}{{}_1F_1\left[-\frac{1}{2}; \frac{3}{2}; -x^2\right]} = v_t \frac{x^2 + \frac{3}{2}}{G_{0,0}(x)} \approx v_t \sqrt{x^2 + \frac{9\pi}{16}} \tag{44}$$

$$\bar{v}_m = \frac{15\sqrt{\pi}}{16} v_t \frac{{}_1F_1\left[-1; \frac{5}{2}; -x^2\right]}{{}_1F_1\left[-\frac{1}{2}; \frac{5}{2}; -x^2\right]} = v_t \frac{x^2 + \frac{5}{2}}{G_{1,0}(x)} \approx v_t \sqrt{x^2 + \frac{225\pi}{256}} \tag{45}$$

$$\bar{v}_E = \frac{15\sqrt{\pi}}{16} v_t \frac{{}_1F_1\left[-2; \frac{3}{2}; -x^2\right]}{{}_1F_1\left[-\frac{3}{2}; \frac{3}{2}; -x^2\right]} = v_t \frac{x^4 + 5x^2 + \frac{15}{4}}{G_{2,0}(x)} \approx v_t \sqrt{x^2 + \frac{225\pi}{256}}. \tag{46}$$

The right hand sides of both Eqs. (45) and (46) are the same due to the fact that in the analytic equations, the leading order term of the Taylor series at $x = 0$ are identical. The asymptotic forms of Eqs. (42) and (44)-(46) are valid for $x \ll 1$ or $x \gg 1$, i.e., when $\triangle u \ll v_t$ or $\triangle u \gg v_t$.

5. Comparison of the exact, power-law, and hard-sphere cross section in source term integrals

5.1. Range of temperature sums

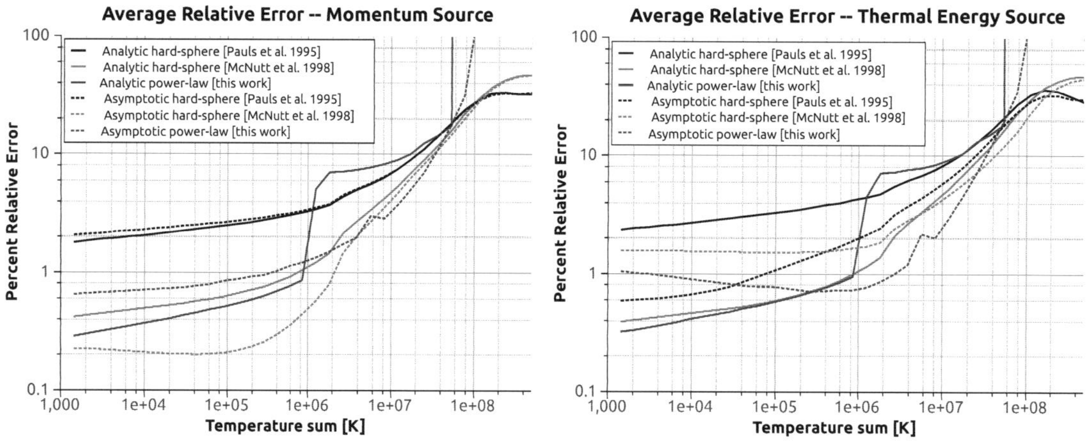

Figure 2: The average relative error of the momentum source (left) and the thermal energy source (right) of various cross-section approximations found in [19], [29] and this work, as a function of the temperature sum, T_{sum} (see Eq. (47)). The solid curves are analytic expressions while the dotted curves are asymptotic expressions of the hard-sphere and power-law cross section approximations.

In this section we will compare the hard-sphere approximation using the effective relative speeds from either Eqs. (41) or (45)-(46), found in Pauls et al. [19] and McNutt et al. [29], respectively, with the numerical evaluation of the exact momentum and thermal energy source term integrals, given in Eq. (16) and the second term Q_{E_T} of Eq. (17). A similar comparison is made for the power-law cross section using a variable power-law index (defined in Eq. (33)) that is fit to the exact cross section in Eq. (3). These comparisons are summarized in Table 1.

Table 1: A list of equations used for each approximation of the charge-exchange cross section.

Reference	Pauls et al. 1995 [19]	McNutt et al. 1998 [29]	This work
Approximation	Hard-sphere	Hard-sphere	Variable power-law
Analytic	Eqs. (34)-(35), (41)a	Eqs. (34)-(35), (45)b-(46)b	Eqs. (28)-(29), (33)
Asymptotic	Eqs. (38)-(39), (41)b	Eqs. (38)-(39), (45)c-(46)c	Eqs. (31)-(32), (33)

The average relative errors for bulk collision energies $E_0 = \frac{1}{2}m(\triangle u)^2$ in the range of $E_0 \in [5 \text{ eV}, 250 \text{ keV}]$ as a function of the temperature sum,

$$T_{sum} = \frac{mv_t^2}{2k_B} = T_H + T_p, \tag{47}$$

are illustrated in Figure 2 for the cases in Table 1. For low collision energies E_0 the exponential cut-off of the charge-exchange cross section (see Eq. (3)) does not play a role. However, for larger collision energies the situation changes; the temperature sum acts as a scaling factor for the cross section argument and for sufficiently large temperature sums, $T_{sum} \sim 10^6$ K, all approximations begin to suffer due to the non-linear and non-power-law nature of the exact cross section. In other words, the Maxwellian distribution becomes very broad for large thermal speeds and begins to *feel* the higher order terms of the Taylor series of the cross section such that the constant, linear, or power-law approximations become inadequate. This is especially true near the exponential cut-off in the cross section above ~ 10 keV (see Eq. (3)).

Examining Figure 2, the power-law formalism completely fails for large temperature sums $T_{sum} \geq 5 \times 10^7$ K, because of the convergence issues resulting from large, negative power-law indices from high collisional energies. Physically, this translates to inverse power-law forces steeper than Coulomb interactions, which would require a finite cutoff of the integral bounds for convergence. For $T_{sum} < 1 \times 10^6$ K, the analytic expressions for the hard-sphere, using the effective relative velocity in [29], and variable power-law approximations produce the best accuracy compared with the exact expressions. Because of the computationally expensive hypergeometric functions used in the variable power-law expressions, it is suggested that the manageable forms of the hard-sphere approximation of [29] or the asymptotic variable power-law forms be used alternatively.

If generalized Lorentzian velocity distributions (i.e., kappa distributions) are used instead of Maxwellian distributions, then the power-law and hard-sphere assumptions of the charge-exchange cross section do not necessarily hold. For example, Heerikhuisen et al. [38] showed differences between assuming either Maxwellian or kappa velocity space distributed protons charge-exchanging with hydrogen atoms in the inner heliosheath (IHS). They showed that a constant, hard-sphere assumption of the cross section produces inaccurate charge-exchange rates, even for moderate temperatures, because the wings (i.e., power-law suprathermal protons) of the kappa distribution extend dozens of thermal speeds when $\kappa \leq 2$.

5.2. Specific cases in the heliosphere

In the heliospheric environment there are a wide variety of thermal and flow conditions for the ions and neutral atoms. We outline the parameters used for three test cases found in the supersonic SW, IHS, and outer heliosheath (OHS) in Table 2.

From Table 3 we can see that the most consistent performing approximation across all cases is the analytic form of McNutt et al. [29]. All approximations perform well in Case 1 in the

Table 2: List of temperatures and bulk velocities, evaluated at various locations along the LISM flow vector in the nose direction of the heliosphere, for protons and pickup ions (PUIs) in the SW, IHS, or OHS that may charge-exchange with neutral hydrogen atoms or hydrogen energetic neutral atoms (H ENAs).

Case	Temperature [K]		Bulk Velocity [km s^{-1}]	
	T_p	T_H	$\mathbf{u}_p$	$\mathbf{u}_H$
1. SW protons with LISM H	1×10^5 [39]	7.5×10^3 [40]	+400 [41]	−25 [40]
2. IHS PUIs with LISM H	1.6×10^7 [39]	7.5×10^3 [40]	+78 [42]	−25 [40]
3. OHS PUIs with H ENAs	4.22×10^6 [42]	2×10^5 [43]	−10 [42]	+400 [41]

Table 3: For each case described in Table 2 (both analytic and asymptotic forms) the percent relative errors are calculated with respect to the numerical evaluation of the exact momentum and total energy sources, which are the upper and lower element respectively. The temperature sum and bulk flow differences specific to each case are tabulated for quick reference.

	Percent relative error of charge-exchange source terms							
Case	Hard-sphere, [19]		Hard-sphere, [29]		Variable power-law		Parameters	
$\mathbf{Q}_m$ $Q_{E_{tot}}$	Ana.	Asymp.	Ana.	Asymp.	Ana.	Asymp.	T_{sum} [K]	$\Delta\mathbf{u}$ [km s^{-1}]
1.	0.338 0.574	0.396 0.121	0.152 0.150	0.030 0.244	0.032 0.032	0.124 0.124	1.08×10^5	−425
2.	20.5 20.8	20.4 20.1	3.66 3.66	3.53 3.79	22.7 22.8	22.8 22.5	1.60×10^7	−103
3.	9.28 9.15	9.40 29.3	2.63 3.85	0.501 17.6	1.36 0.047	2.52 19.3	4.42×10^6	410

supersonic SW, because the temperatures and bulk velocities result in lower collisional energies (see Figure 2). In the IHS for Case 2, the hard-sphere forms found in Pauls et al. [19], more specifically the effective relative speed expression, and variable power-law expressions, derived in this work, suffer because of the large temperature sums. The effective relative speeds defined by McNutt et al. [29] cancel out the first order errors from the hard-sphere approximation (see §4.1), and therefore are sufficient for IHS PUI temperatures. For Case 3, most of the asymptotic forms fail to match the exact source terms while the analytic power-law equations operate the best, although they are computationally expensive to use due to the hypergeometric functions involved.

6. Summary

In §2 we derived simplified, exact source term expressions for a generalized, isotropic cross section assuming charge-exchange collisions between Maxwellian fluids. We then assumed two approximations of the cross section: power-law in §3, and hard-sphere in §4. By using the p-H charge-exchange cross section from Eq. (3) we showed that all approximations evaluated in this paper introduce noticeable errors for $T_{sum} > 10^6$ K. Power-law source terms fail to converge for $T_{sum} \geq 5 \times 10^7$ K, because of the steeper than Coulomb-like power-law index ν. For $T_{sum} < 10^6$ K, the most accurate and simple source term expressions are the analytic hard-sphere forms of McNutt et al. [29]. This is solely due to the fact that the effective relative speeds $\bar{v}$ are chosen to cancel first order errors introduced by a constant cross section, in contrast to $\bar{v}$ defined in Pauls et al. [19]. For heliospheric applications of Maxwellian fluid ions and neutrals, the analytic source terms derived in McNutt et al. [29] perform consistently, in accordance with Table 3. Finally, we remark that for non-resonant cross sections (e.g., proton-helium charge-exchange p+He $\rightarrow$ H+He$^+$ [33]) the hard-sphere and power-law assumptions would be invalid, and exact source term integrals must be derived for non-symmetric reactions with dissimilar masses.

Acknowledgments
A.D. acknowledges support by NASA Headquarters under the NASA Earth and Space Science Fellowship Program - Grant NNX15AT10H. A.D. would also like to thank the reviewers for useful suggestions in addition to friends and family for invaluable feedback and support through prayer. J.H. acknowledges support from NASA Grant NNX16AG83G.

Appendix
Derivation of Eq. (33): The power-law index ν can be extracted from the cross section σ_{ex}, Eq. (3), as a function of collision energy E by computing the slope of σ_{ex} on a log-log scale. We first make the substitutions $y = \ln(\sigma_{ex} \times 10^{16}\ \mathrm{cm}^{-2})$ and $x = \ln(E \times 1\ \mathrm{keV}^{-1})$ in order to convert to a log-log scale. Then, ν may be approximated using the central difference method for the derivative,

$$\nu = \frac{dy}{dx} \sim \frac{y(x+h) - y(x-h)}{2h} \text{ as } h \rightarrow 0. \tag{A.1}$$

We take $h = \triangle E \times 1\ \mathrm{keV}^{-1}$ and exponentiate the arguments of y (to convert back to a linear scale in energy) so that we arrive at

$$\nu = \frac{1\ \mathrm{keV}}{2\triangle E} \ln\left[\frac{\sigma_{ex}(Ee^{\triangle E/1\mathrm{keV}})}{\sigma_{ex}(Ee^{-\triangle E/1\mathrm{keV}})}\right] \tag{A.2}$$

where E is in units of keV.

References

[1] Davis L 1955 *PhRv* **100** 1440–1444
[2] Parker E N 1961 *ApJ* **134** 20
[3] Axford W I, Dessler A J and Gottlieb B 1963 *ApJ* **137** 1268
[4] Baranov V B and Krasnobaev K V 1971 *Kosmicheskie Issledovaniia* **9** 620–622
[5] Reynolds R J 1989 *ApJ* **345** 811–814
[6] Holzer T E 1972 *JGR* **77** 5407
[7] Wallis M K 1975 *Nature* **254** 202
[8] Stone E C, Cummings A C, McDonald F B, Heikkila B C, Lal N and Webber W R 2005 *Sci* **309** 2017–2020
[9] Stone E 2008 *37th COSPAR Scientific Assembly* (*COSPAR Meeting* vol 37) p 3046
[10] Stone E C, Cummings A C, McDonald F B, Heikkila B C, Lal N and Webber W R 2013 *Sci* **341** 150–153

[11] McComas D J, Allegrini F, Bochsler P, Bzowski M, Christian E R, Crew G B, DeMajistre R, Fahr H, Fichtner H, Frisch P C *et al.* 2009 *Sci* **326** 959

[12] Schwadron N A, McComas D J, Christian E R, Desai M I, Funsten H O, Fuselier S A, Möbius E, Reno M, Scherrer J and Zirnstein E 2016 *AIP Conf. Series* (*AIP Conf. Series* vol 1720) p 080002

[13] Heerikhuisen J, Pogorelov N V, Zank G P, Crew G B, Frisch P C, Funsten H O, Janzen P H, McComas D J, Reisenfeld D B and Schwadron N A 2010 *ApJL* **708** L126–L130

[14] McComas D J, Dayeh M A, Allegrini F, Bzowski M, DeMajistre R, Fujiki K, Funsten H O, Fuselier S A, Gruntman M, Janzen P H *et al.* 2012 *ApJS* **203** 1

[15] Williams L L, Hall D T, Pauls H L and Zank G P 1997 *ApJ* **476** 366–384

[16] Baranov V B, Ermakov M K and Lebedev M G 1981 *SvAL* **7** 206–209

[17] Baranov V B, Ermakov M K and Lebedev M G 1982 *Akademiia Nauk SSSR Izvestiia Mekhanika Zhidkosti i Gaza* 122–128

[18] Steinolfson R S, Pizzo V J and Holzer T 1994 *GeoRL* **21** 245–248

[19] Pauls H L, Zank G P and Williams L L 1995 *JGR* **100** 21595–21604

[20] Fahr H J, Kausch T and Scherer H 2000 *A&A* **357** 268–282

[21] Pogorelov N V and Zank G P 2005 *AdSpR* **35** 2055–2060

[22] Zank G P, Pauls H L, Williams L L and Hall D T 1996 *JGR* **101** 21639–21656

[23] Pogorelov N V, Zank G P and Ogino T 2006 *ApJ* **644** 1299–1316

[24] Malama Y G 1991 *Ap&SS* **176** 21–46

[25] Baranov V B and Malama Y G 1993 *JGR* **98** 15

[26] Heerikhuisen J, Florinski V, Zank G P and Pogorelov N V 2006 *Numerical Modeling of Space Plasma Flows* (*ASPC* vol 359) ed Zank G P and Pogorelov N V p 251

[27] Izmodenov V V, Malama Y G, Ruderman M S, Chalov S V, Alexashov D B, Katushkina O A and Provornikova E A 2009 *SSRv* **146** 329–351

[28] Opher M, Bibi F A, Toth G, Richardson J D, Izmodenov V V and Gombosi T I 2009 *Nature* **462** 1036–1038

[29] McNutt Jr R L, Lyon J and Goodrich C C 1998 *JGR* **103** 1905

[30] McNutt R L, Lyon J and Goodrich C C 1999 *JGR* **104** 14803–14810

[31] Burgers J M 1969 *Flow Equations for Composite Gases* (New York: Academic Press)

[32] Lindsay B G and Stebbings R F 2005 *JGR (Space Physics)* **110** A12213

[33] Barnett C F, Hunter H T and Kirkpatrick M I 1990 *Atomic Data For Fusion: Collision of H, H_2, He and Li Atoms and Ions with Atoms and Molecules* (TN: Oak Ridge National Laboratories)

[34] Gradshteyn I S and Ryzhik I M 1965 *Table of integrals, series and products* (New York: Academic Press)

[35] Abramowitz M and Stegun 1964 *Handbook of Mathematical Functions With Formulas, Graphs and Mathematical Tables* (United States Department of Commerce, National Bureau of Standards)

[36] Chapman S and Cowling T G 1960 *The Mathematical Theory of Non-Uniform Gases* (London and New York, Cambridge Univ. Press)

[37] Scherer K, Fichtner H, Fahr H J, Bzowski M and Ferreira S E S 2014 *A&A* **563** A69

[38] Heerikhuisen J, Zirnstein E and Pogorelov N 2015 *JGR* **120** 1516–1525

[39] Zank G P, Heerikhuisen J, Pogorelov N V, Burrows R and McComas D 2010 *ApJ* **708** 1092–1106

[40] McComas D J, Bzowski M, Frisch P, Fuselier S A, Kubiak M A, Kucharek H, Leonard T, Möbius E, Schwadron N A, Sokół J M *et al.* 2015 *ApJ* **801** 28

[41] McComas D J, Elliott H A, Schwadron N A, Gosling J T, Skoug R M and Goldstein B E 2003 *GeoRL* **30** 1517

[42] Zirnstein E J, Heerikhuisen J, Zank G P, Pogorelov N V, McComas D J and Desai M I 2014 *ApJ* **783** 129

[43] Heerikhuisen J, Gamayunov K V, Zirnstein E J and Pogorelov N V 2016 *ApJ* **831** 137

IOP Conf. Series: Journal of Physics: Conf. Series **837** (2017) 012014 doi:10.1088/1742-6596/837/1/012014

Modeling Coronal Mass Ejections with the Multi-Scale Fluid-Kinetic Simulation Suite

N. V. Pogorelov,[1,2] S. N. Borovikov,[2] I. A. Kryukov,[3], S. T. Wu,[2] M. S. Yalim,[2] P. C. Colella,[4] and B. Van Straalen[4]

[1]Department of Space Science, University of Alabama in Huntsville, Huntsville, AL 35805, U.S.A.

[2]Center for Space Plasma and Aeronomic Research, University of Alabama in Huntsville, Huntsville, AL 35805, U.S.A.

[3]Institute for Problems in Mechanics, Russian Academy of Sciences, Moscow, Russia

[4]Applied Numerical Algorithms Group, Lawrence Berkeley National Laboratory, MS 50A-1148, Berkeley, CA 94720, U.S.A.

E-mail: `nikolai.pogorelov@uah.edu`

Abstract. The solar eruptions and interacting solar wind streams are key drivers of geomagnetic storms and various related space weather disturbances that may have hazardous effects on the space-borne and ground-based technological systems as well as on human health. Coronal mass ejections (CMEs) and their interplanetary counterparts, interplanetary CMEs (ICMEs), belong to the strongest disturbances and therefore are of great importance for the space weather predictions. In this paper we show a few examples of how adaptive mesh refinement makes it possible to resolve the complex CME structure and its evolution in time while a CME propagates from the inner boundary to Earth. Simulations are performed with the Multi-Scale Fluid-Kinetic Simulation Suite (MS-FLUKSS).

1. Introduction

The solar wind (SW) emerging from the Sun is the main driving mechanism of solar events which may lead to geomagnetic storms that are the primary causes of space weather disturbances. Such disturbances affect the magnetic environment of Earth and may have hazardous effects on the space-borne and ground-based technological systems as well as human health. For this reason, accurate modeling of the background SW is a necessary part of space weather forecasting. Geomagnetic storms are caused both by SW stream interactions and by the largest solar coronal disturbances called coronal mass ejections (CMEs) and their interplanetary counterparts, interplanetary coronal mass ejections (ICMEs). Therefore, modeling stream interaction background and ICME propagating through it on the basis of observational data is a key area of research in solar and heliospheric physics. Such modeling should necessarily be built on mathematically- and physically-consistent connections between eruptive events, magnetic phenomena on the Sun, and SW structures in the solar atmosphere and inner heliosphere (IHS). Substantial success has been achieved in global numerical modeling of the IHS, see, e.g., [1, 2, 3, 4, 5, 6, 7, 8, 9, 10, 11, 12, 13, 14, 15, 16, 17, 18, 19, 20, 21, 22, 23, 24, 25, 26, 27, 28, 29, 30, 31, 32, 33, 34, 35, 36, 37, 38, 39, 40, 41]. The quest to create a nearly real-time model of the solar atmosphere and IHS not only requires appropriate codes, but also a

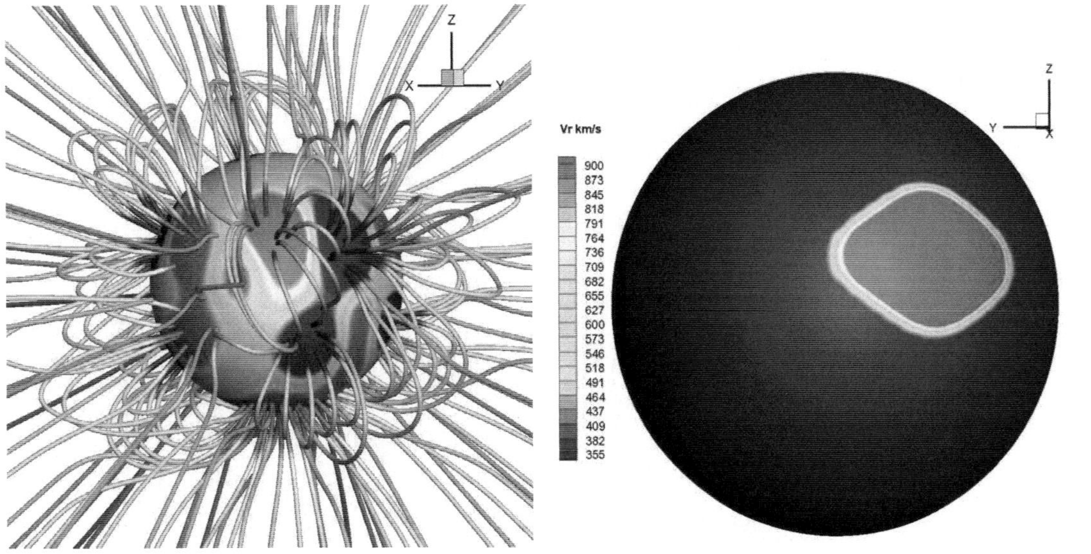

Figure 1. (*Left panel.*) The initial distribution of the heliospheric magnetic field. (*Right panel.*) The CME "blob" structure in radial velocity at 10 $R_\odot$ as seen from Earth.

formulation of time-dependent boundary conditions (b.c.'s) that would incorporate remote and *in situ* observations of the Sun in a self-consistent way (see, e.g., [42, 38, 43, 44, 45, 46]).

Such boundary conditions are implemented in a Multi-Scale Fluid-Kinetic Simulation Suite (MS-FLUKSS) collaboratively developed by the UAH and LBNL teams [47, 48, 49, 50]. MS-FLUKSS is a collection of problem-oriented routines incorporated into the Chombo adaptive mesh refinement (AMR) framework [51]. We added the MHD, multi-fluid components, and kinetic Boltzmann components, and the possibility of using non-orthogonal meshes, while the parallelization, load balancing, and other data choreography of the AMR code remains the responsibility of Chombo. We also have implemented a hybrid parallelization for efficient use of the plasma data set in the kinetic module [49]. A new version of our AMR code allows us to perform simulations with the fourth order of accuracy in time and space and uses cubed spheres [52, 53, 54, 55, 56] to generate meshes around the Sun. MS-FLUKSS solves MHD equations with the volumetric heating source terms. It can also resolve the transport of neutral atoms kinetically, by solving the Boltzmann equation. Beyond the Alfvénic surface out into the interplanetary space, we also take into account the influence of interstellar neutral atoms, nonthermal pickup ions, and SW turbulence. In addition, we have implemented special algorithms to track exactly the surfaces that passively propagate with the SW (heliospheric current sheet, CME driving surface, etc) once their initial position is known. Combining these features with AMR, we are able to analyze transient heliospheric structures with high precision. MS-FLUKSS is a parallel (hybrid, MPI/OpenMP implementation) code which has been applied successfully to a number of heliospheric problems and SW interaction with the local interstellar medium [57, 58, 59, 60, 61, 62, 63, 64, 65, 66, 67].

In this paper, we present a few examples of our application of MS-FLUKSS to modeling CMEs. In all these simulations, the inner boundary conditions are specified on a sphere where the radial component of the SW velocity vector is greater than the fast magnetosonic speed and provided externally. This means that we do not use our own coronal model (see the paper of Yalim et al. in this volume). As such, we do not use these simulations to reproduce observations at Earth, but rather investigate the capacity of MS-FLUKSS to model discontinuities that

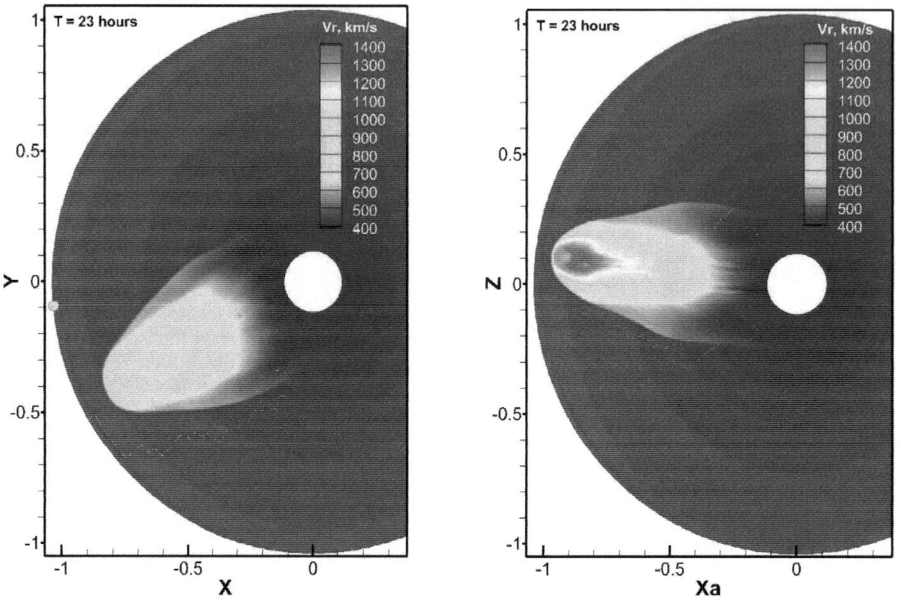

Figure 2. The radial velocity component distribution (*left panel*) in the ecliptic plane and (*right panel*) in the plane formed by the z-axis and the CME axis for a Jan 23, 2012 CME 23 hours after eruption. Distances are given in astronomical units.

accompany CME propagation through the interplanetary space.

2. Modeling results

Firstly, we show our solution for the CME event that occurred on Jan 23, 2012. The boundary conditions are designed at 10 $R_\odot$. We use a simple Parker solution, which extends from 1 $R_\odot$ to 1 AU, as the background SW in this simulation. This background solution is further modified by the insertion of a CME into it at 10 $R_\odot$. It is assumed that the temperature and density at the coronal base is 1.85×10^6 K and 3.35×10^{-4} kg/km^3, respectively. In this case, $V_\odot = 8.16$ km/s. This results in the critical sphere at 4.08 $R_\odot$ and the corresponding velocity on it equal to 154 km/s. The initial distribution of magnetic field between the inner boundary and 1 AU is obtained with the PFSS method (see Fig. 1, *left panel*). Magnetic field in the solar corona is assumed to be potential and radial at some distance from the Sun (at the source surface). The distribution is approximated by associated Legendre functions. The coefficients in the expansion are derived from the line-of-sight magnetic field measurement through the Fe I λ5250 line observations at the Wilcox Solar Observatory. The CME is inserted at 1.5 $R_\odot$ in the direction N6W27 with respect to the Earth location using the "blob" approach [68, 69, 70, 71] (the blob radius is 0.33 $R_\odot$). The maximum velocity, density, and temperature in the blob are the following: 2241 km/s, 2×10^{-3} kg/km^3, and 8×10^6 K. Such a blob of high density and pressure is created and launched with high velocity in the outward radial direction, being superimposed with the background SW flow. In Fig. 1 (*right panel*), we show this CME as seen on the surface of 10 $R_\odot$ from Earth. To obtain the solution between the inner sphere and 1 AU, we use a spherical grid with the base level resolution 256^3 and one AMR level (the refinement ratio is 4) around the CME. The computational mesh is adapted to the CME profile to capture fine features of its structure including the forward shock wave, compressed ambient SW, and tangential discontinuity that separates the ejected plasma from the compressed SW.

IOP Conf. Series: Journal of Physics: Conf. Series **837** (2017) 012014 doi:10.1088/1742-6596/837/1/012014

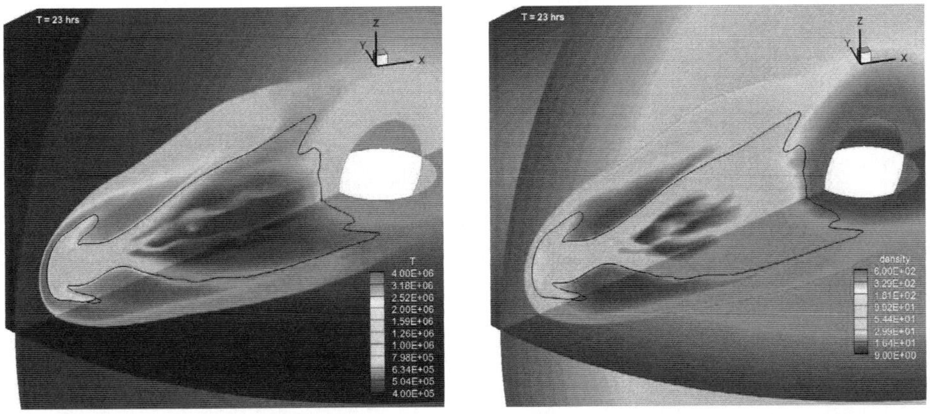

Figure 3. Plasma temperature (left panel) and density (right panel) distributions in the meridional and ecliptic planes for a Jan 23, 2012 CME 23 hours after eruption. The black line in both figures shows the surface driving the CME, which is identified by following the evolution of the "blob" surface with a level-set method.

Figure 2 shows the radial velocity distribution in the ecliptic plane (*left panel*) and in the plane formed by the Sun's rotation axis (the z-axis) and the line defined by the CME axis (*right panel*) 23 hrs after its origin. In Fig. 3, we show the plasma temperature and density distributions for the same CME. To demonstrate the evolution of this CME in time, in Figs. 4 and 5, respectively, we show the distributions of plasma temperature in the ecliptic plane and in the plane formed by the z-axis and the Earth position 23 and 34 hrs after eruption. Additionally, in Fig. 6, we show the linear distributions of the radial velocity component (*left panel*) and density (*right panel*) as functions of time at Earth.

Since the initial shape of the blob is well defined as a tangential discontinuity and knowing that such discontinuities propagate at SW velocity, we can follow the evolution of this surface (the CME driver) as a function of time. It is seen from Fig. 3, that the "mushroom" shape of this CME resembles that of an astrophysical jet injected into the surrounding interstellar medium [72]. Since tangential discontinuities are typically unstable, one would expect that further increase of space resolution will reveal more instabilities.

Another simulation is related to a CME that occurred on 4 November, 1997. The boundary conditions at 10 $R_\odot$ are taken from [73] and are obtained with the Solar-Interplanetary Conservation Element/Solution Element (SIP-CESE) MHD model [46]. The boundary distributions on the inner sphere are shown in Fig. 7 (the density and the radial velocity component) and Fig. 8 (the radial and toroidal magnetic field components) at different moments of time: $t = 0$, $t = 90$ min, and $t = 150$ min. The physical time interval was placed around the CME injection time ($t = 0$) and covered 12 hours.

The boundary conditions were propagated to 0.45 AU with MS-FLUKSS and saved there. Afterwards, the latter were used to establish the solution to 12 AU. The evolution of the SW radial velocity component and density are shown in Figs. 9 and 10, respectively, for the same moments of time as above. It is seen that AMR allows us to resolve the discontinuities accompanying the CME with high accuracy. The solution presented here is in agreement with that obtained in [73]. Worth noticing is the latitudinal asymmetry in the CME propagation.

Finally, in Fig. 11, we show the radial velocity distribution to distance of 12 AU 4 and 23 days after the eruption. Our solution at 12 AU may be used to drive time-dependent interaction of the SW and LISM. Such simulations will be shown elsewhere.

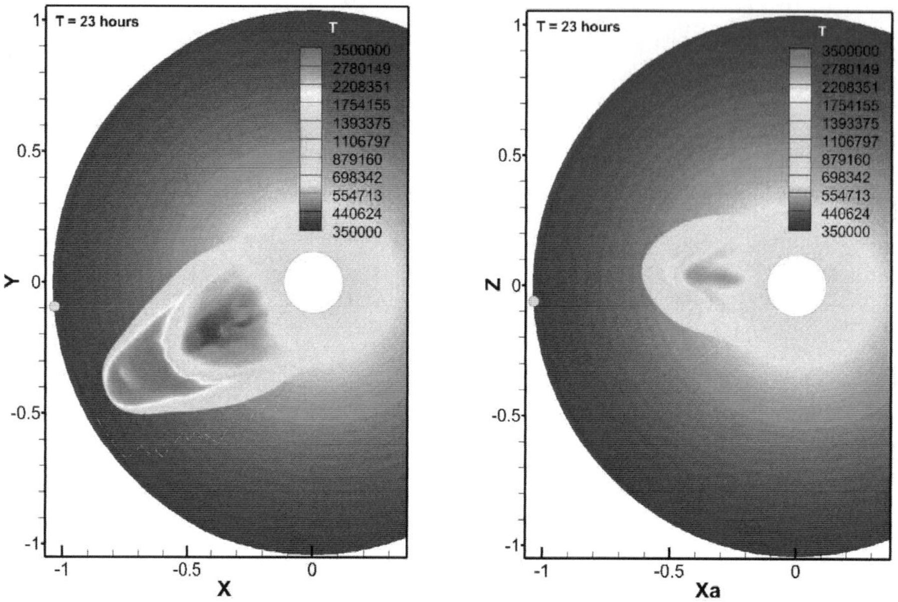

Figure 4. Plasma temperature in the ecliptic plane (*left panel*) and the plane formed by the z-axis and the Earth location (*right panel*) for a Jan 23, 2012 CME 23 hours after eruption. Distances are given in astronomical units.

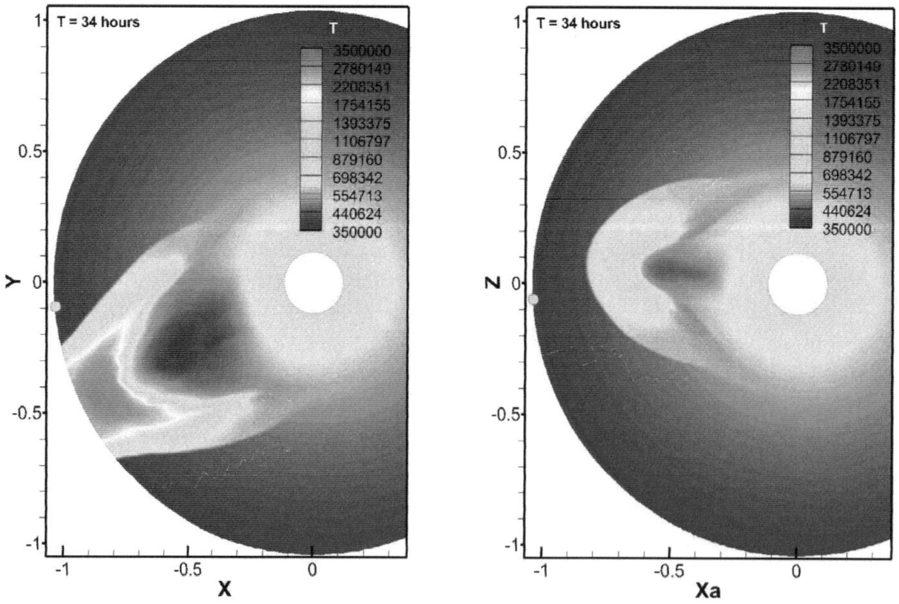

Figure 5. The same as in Fig. 4, but 34 hrs after eruption. Distances are given in astronomical units.

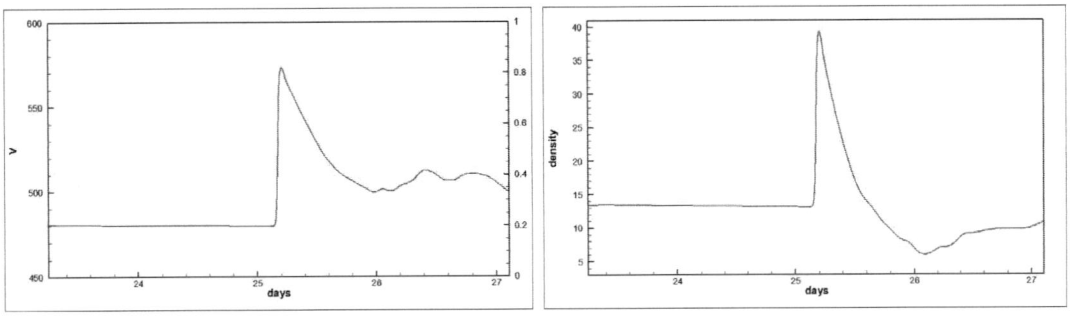

Figure 6. Plasma density (*right panel*) and velocity (*left panel*) as functions of time at Earth for the CME event on Jan 23, 2012.

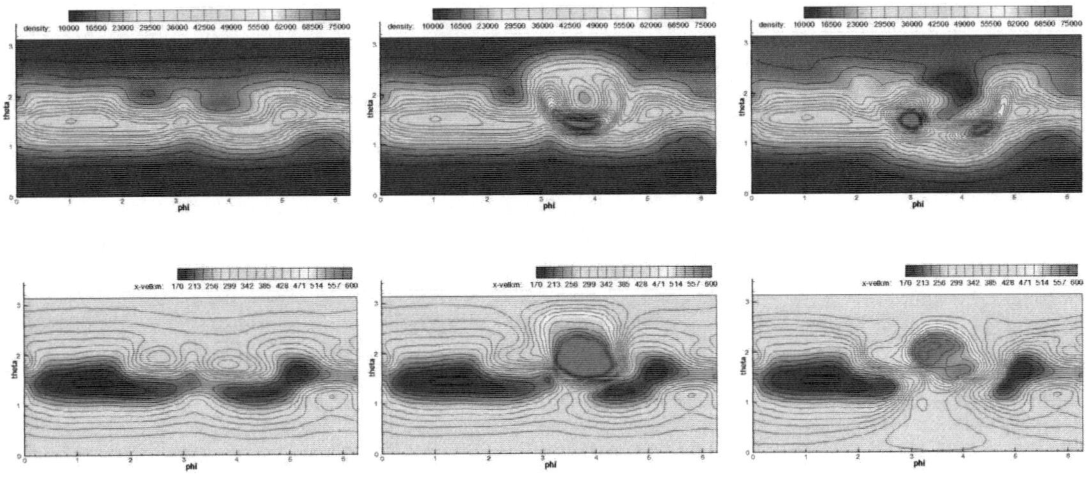

Figure 7. Time-dependent boundary conditions at 10 $R_{\odot}$: plasma density (*top row*) and the radial velocity component (*bottom row*) at $t = 0$, $t = 90$ min, and $t = 150$ min after the CME launch (from the left to the right) on 4 Nov 1997.

3. Conclusions

This paper is a brief summary of our efforts to model CME propagation with MS-FLUKSS. It shows our ability to resolve discontinuities in the SW flow sharply enough to have meaningful predictions at 1 AU and beyond. We did not take into account turbulence and separation between thermal and non-thermal ions in this paper. This clearly should be done at distances beyond 5–10 AU, when these processes become important. Such simulations are of importance for the New Horizons mission and for the future IMAP mission. Regardless of the system of equations to be used near the solar surface, one needs to formulate enough physical boundary conditions to ensure the consistency of the mathematical model of the solar corona. Such boundary conditions can be provided only by the SDO/HMI vector magnetograms. Paper [41] shows the application of characteristic boundary conditions when applying vector magnetograms to modeling active regions on the solar surface.

As shown in [74], the quality of SW predictions at Earth provided by existing first-principles models is improving, but still lags behind data assimilation approaches. On the other hand, it is hard to imagine a data-assimilation model that would be able to reproduce the plasma

IOP Conf. Series: Journal of Physics: Conf. Series **837** (2017) 012014 doi:10.1088/1742-6596/837/1/012014

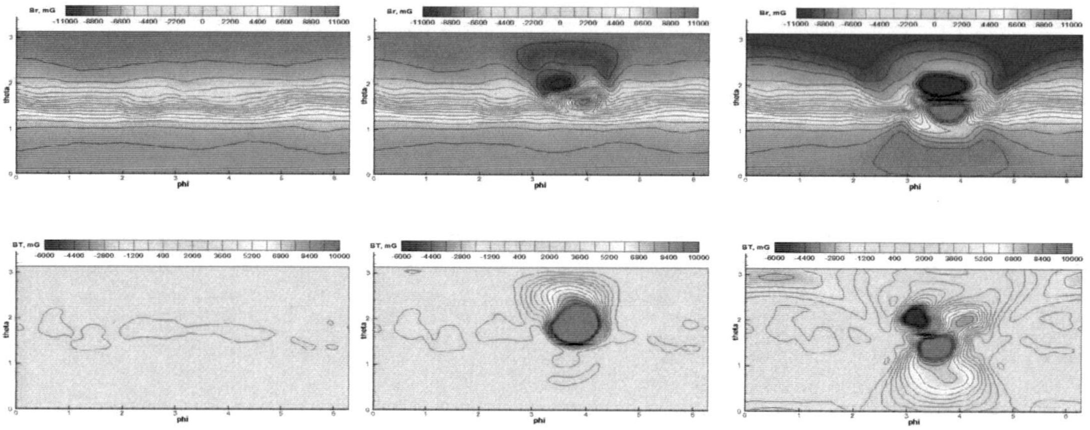

Figure 8. Time-dependent boundary conditions at 10 $R_\odot$: the radial (*top row*) and toroidal (*bottom panel*) magnetic field components at $t = 0$, $t = 90$ min, and $t = 150$ min after the CME launch (from the left to the right) on 4 Nov 1997.

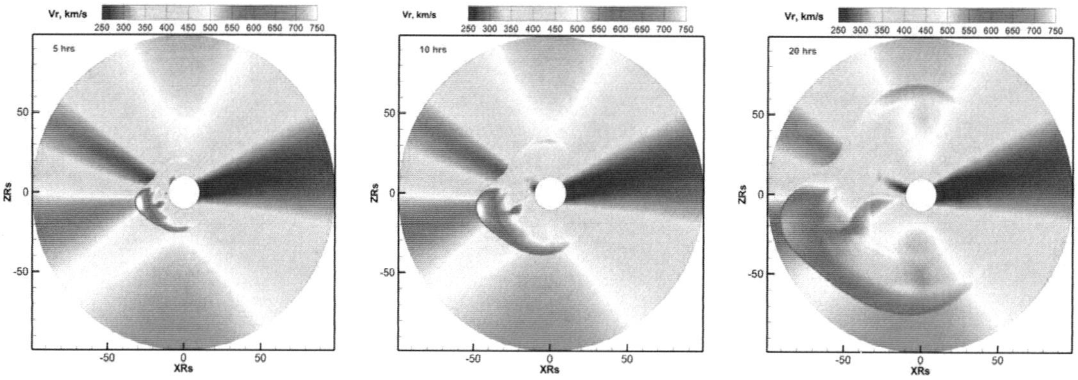

Figure 9. From the left to the right, the distribution of the radial SW velocity component 5, 10, and 20 hrs after the CME launch on 4 Nov 1997. Distances are given in units of $R_\odot$.

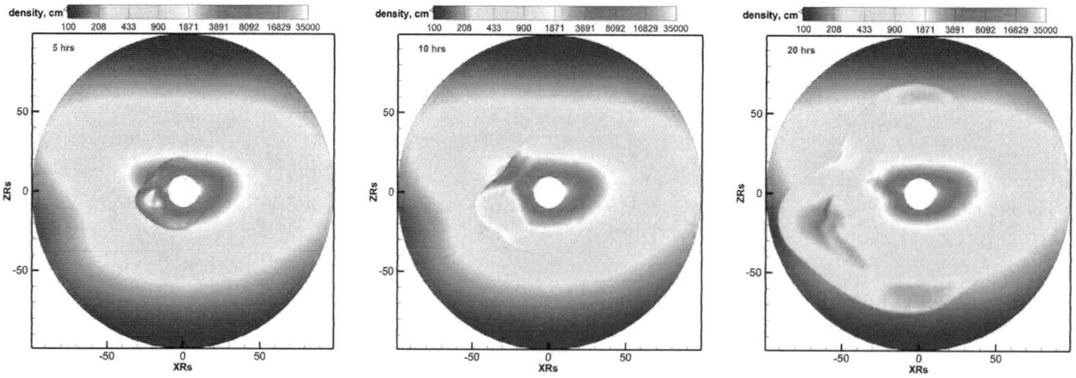

Figure 10. From the left to the right, the distribution of plasma density 5, 10, and 20 hrs after the CME launch on 4 Nov 1997. Distances are given in units of $R_\odot$.

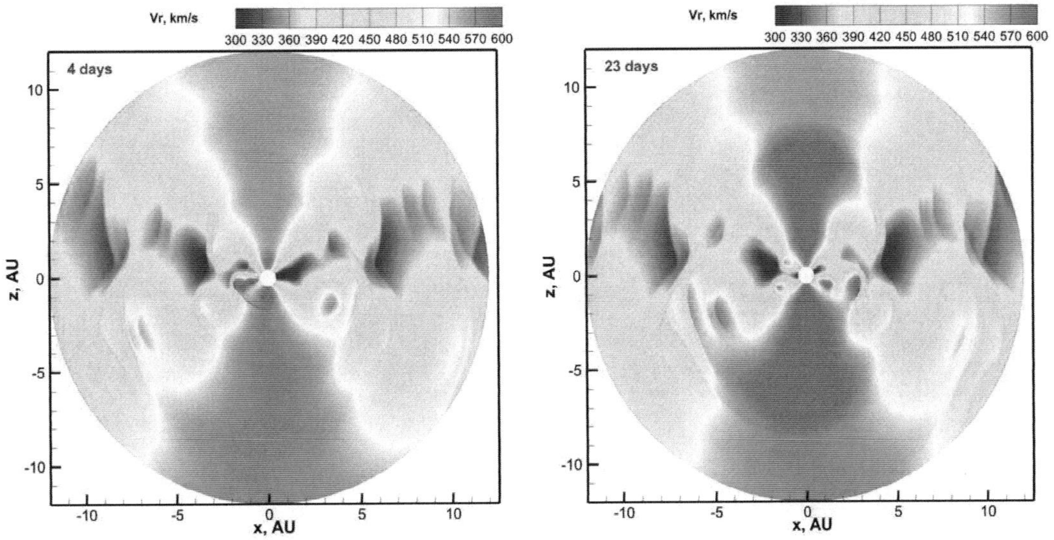

Figure 11. Distributions of the radial velocity component 4 days (*left panel*) and 23 days (*right panel*) after the CME launch on 4 Nov 1997.

and magnetic field properties at Solar Orbiter and Solar Probe Plus, Earth, and other planets at once [59, 75]. The heliospheric community is seeking for an MHD simulation model that is largely data-driven and utilizes characteristic boundary conditions. Given the current and future observational capabilities offered through spacecraft missions around the world, it is possible to apply such models to study the problems of CME origination, propagation and interaction central to heliophysics science, utilizing all available and critical observations. These quantitative measurements range from the photospheric magnetograms to in-situ measurements of plasma and magnetic field properties, which provide the inner boundary conditions, constraints and validations to models. A completely data-driven approach might enable physics-based simulation to improve numerical solutions by making them more realistic and improving our understanding of the underlying physical processes that lead to CME formation, eruption, and dynamic and morphological changes.

Acknowledgments
The authors are grateful to X. Feng, F. Shen, and Y. F. Zhou for providing the time-dependent boundary conditions for these simulations.

This work was supported, in part, by NASA grants NNX14AJ53G, NNX14AF41G, NNX14AF43G, NNX15AN72G, and NNX16AG83G, and NSF SHINE project AGS-1358386. We acknowledge NSF PRAC award ACI-1144120 and related computer resources from the Blue Waters sustained-petascale computing project. Supercomputer time allocations were also provided on SGI Pleiades by NASA High-End Computing Program award SMD-15-5860 and on Stampede by NSF XSEDE project MCA07S033.

Bibliography
[1] Detman T R, Intriligator D S, Dryer M, Sun W, Deehr C S and Intriligator J 2011 *Journal of Geophysical Research (Space Physics)* **116** A03105
[2] van der Holst B, Sokolov I V, Meng X, Jin M, Manchester IV W B, Tóth G and Gombosi T I 2014 *Astrophys. J.* **782** 81

[3] Feng X, Zhang S, Xiang C, Yang L, Jiang C and Wu S T 2011 *Astrophys. J.* **734** 50
[4] Feng X, Ma X and Xiang C 2015 *Journal of Geophysical Research (Space Physics)* **120** 10
[5] Hayashi K, Zhao X P and Liu Y 2008 *Journal of Geophysical Research (Space Physics)* **113** A07104
[6] Hayashi K 2013 *Journal of Geophysical Research (Space Physics)* **118** 6889–6906
[7] Intriligator D S, Detman T, Gloecker G, Gloeckler C, Dryer M, Sun W, Intriligator J and Deehr C 2012 *Journal of Geophysical Research (Space Physics)* **117** A06104
[8] Lee E, Lukin V S and Linton M G 2014 *A&A* **569** A94
[9] Leake J E, Linton M G and Török T 2013 *Astrophys. J.* **778** 99
[10] Leake J E, Linton M G and Antiochos S K 2014 *Astrophys. J.* **787** 46
[11] Linker J A, Mikić Z, Biesecker D A, Forsyth R J, Gibson S E, Lazarus A J, Lecinski A, Riley P, Szabo A and Thompson B J 1999 *J. Geophys. Res. Space Phys.* **104** 9809–9830
[12] Linker J A, Lionello R, Mikić Z, Titov V S and Antiochos S K 2011 *Astrophys. J.* **731** 110
[13] Linker J A, Caplan R M, Downs C, Lionello R, Riley P, Mikic Z, Henney C J, Arge C N, Kim T and Pogorelov N 2016 *Journal of Physics Conference Series* **719** 012012
[14] Lionello R, Linker J A and Mikić Z 2009 *Astrophys. J.* **690** 902–912
[15] Lionello R, Török T, Titov V S, Leake J E, Mikić Z, Linker J A and Linton M G 2016 *Astrophys. J. Lett.* **831** L2
[16] Lugaz N and Roussev I I 2011 *Journal of Atmospheric and Solar-Terrestrial Physics* **73** 1187–1200
[17] Manchester W B, Ridley A J, Gombosi T I and Dezeeuw D L 2006 *Advances in Space Research* **38** 253–262
[18] Manchester IV W B, van der Holst B, Tóth G and Gombosi T I 2012 *Astrophys. J.* **756** 81
[19] Merkin V G, Lionello R, Lyon J G, Linker J, Török T and Downs C 2016 *Astrophys. J.* **831** 23
[20] Odstrcil D, Riley P and Zhao X P 2004 *Journal of Geophysical Research (Space Physics)* **109** A02116
[21] Odstrcil D, Pizzo V J, Arge C N, Bissi M M, Hick P P, Jackson B V, Ledvina S A, Luhmann J G, Linker J A, Mikic Z and Riley P 2008 Numerical Simulations of Solar Wind Disturbances by Coupled Models *Numerical Modeling of Space Plasma Flows* (*Astronomical Society of the Pacific Conference Series* vol 385) ed Pogorelov N V, Audit E and Zank G P p 167
[22] Odstrcil D and Pizzo V J 2009 *Solar Phys.* **259** 297–309
[23] Oran R, van der Holst B, Landi E, Jin M, Sokolov I V and Gombosi T I 2013 *Astrophys. J.* **778** 176
[24] Oran R, Landi E, van der Holst B, Lepri S T, Vásquez A M, Nuevo F A, Frazin R, Manchester W, Sokolov I and Gombosi T I 2015 *Astrophys. J.* **806** 55
[25] Riley P, Linker J A and Mikić Z 2001 *J. Geophys. Res. Space Phys.* **106** 15889–15902
[26] Riley P, Linker J A and Mikić Z 2002 *Journal of Geophysical Research (Space Physics)* **107** 1136
[27] Riley P, Linker J A, Mikić Z, Odstrcil D, Zurbuchen T H, Lario D and Lepping R P 2003 *Journal of Geophysical Research (Space Physics)* **108** 1272
[28] Riley P, Linker J A, Mikić Z, Lionello R, Ledvina S A and Luhmann J G 2006 *Astrophys. J.* **653** 1510–1516
[29] Riley P, Lionello R, Linker J A, Mikic Z, Luhmann J and Wijaya J 2011 *Solar Phys.* **274** 361–377
[30] Riley P, Linker J A and Arge C N 2015 *Space Weather* **13** 154–169
[31] Riley P, Lionello R, Linker J A, Cliver E, Balogh A, Beer J, Charbonneau P, Crooker N, DeRosa M, Lockwood M, Owens M, McCracken K, Usoskin I and Koutchmy S 2015 *Astrophys. J.* **802** 105
[32] Riley P and Richardson I G 2013 *Solar Phys.* **284** 217–233
[33] Roussev I I, Gombosi T I, Sokolov I V, Velli M, Manchester IV W, DeZeeuw D L, Liewer P, Tóth G and Luhmann J 2003 *Astrophys. J. Lett.* **595** L57–L61
[34] Sokolov I V, van der Holst B, Oran R, Downs C, Roussev I I, Jin M, Manchester IV W B, Evans R M and Gombosi T I 2013 *Astrophys. J.* **764** 23
[35] Usmanov A V and Goldstein M L 2006 *Journal of Geophysical Research (Space Physics)* **111** A07101
[36] Usmanov A V, Matthaeus W H, Breech B A and Goldstein M L 2011 *Astrophys. J.* **727** 84
[37] Wang A H, Wu S T, Tandberg-Hanssen E and Hill F 2011 *Astrophys. J.* **732** 19
[38] Wu S T, Wang A H, Liu Y and Hoeksema J T 2006 *Astrophys. J.* **652** 800–811
[39] Wu C C, Fry C D, Wu S T, Dryer M and Liou K 2007 *Journal of Geophysical Research (Space Physics)* **112** A09104
[40] Wu S T, Wang A H, Gary G A, Kucera A, Rybak J, Liu Y, Vrśnak B and Yurchyshyn V 2009 *Advances in Space Research* **44** 46–53
[41] Jiang C, Wu S T, Feng X and Hu Q 2016 *Nature Communications* **7** 11522
[42] Mikić Z, Linker J A, Schnack D D, Lionello R and Tarditi A 1999 *Physics of Plasmas* **6** 2217–2224
[43] Hayashi K 2005 *Astrophys. J. Suppl.* **161** 480–494
[44] van der Holst B, Manchester IV W B, Frazin R A, Vásquez A M, Tóth G and Gombosi T I 2010 *Astrophys. J.* **725** 1373-1383
[45] Yang L P, Feng X S, Xiang C Q, Liu Y, Zhao X and Wu S T 2012 *Journal of Geophysical Research (Space Physics)* **117** A08110

[46] Feng X, Yang L, Xiang C, Jiang C, Ma X, Wu S T, Zhong D and Zhou Y 2012 *Solar Phys.* **279** 207–229
[47] Borovikov S N, Kryukov I A and Pogorelov N V 2009 Adaptive Mesh Refinement on Curvilinear Grids *Numerical Modeling of Space Plasma Flows: ASTRONUM-2008* (*Astronomical Society of the Pacific Conference Series* vol 406) ed Pogorelov N V, Audit E, Colella P and Zank G P p 127
[48] Pogorelov N V, Borovikov S N, Florinski V, Heerikhuisen J, Kryukov I A and Zank G P 2009 Multi-scale Fluid-Kinetic Simulation Suite: A Tool for Efficient Modeling of Space Plasma Flows *Numerical Modeling of Space Plasma Flows: ASTRONUM-2008* (*Astronomical Society of the Pacific Conference Series* vol 406) ed Pogorelov N V, Audit E, Colella P and Zank G P p 149
[49] Borovikov S N, Heerikhuisen J and Pogorelov N V 2013 Hybrid Parallelization of Adaptive MHD-Kinetic Module in Multi-Scale Fluid-Kinetic Simulation Suite *Numerical Modeling of Space Plasma Flows (ASTRONUM2012)* (*Astronomical Society of the Pacific Conference Series* vol 474) ed Pogorelov N V, Audit E and Zank G P p 219
[50] Pogorelov N V, Borovikov S N, Bedford M C, Heerikhuisen J, Kim T K, Kryukov I A and Zank G P 2013 Modeling Solar Wind Flow with the Multi-Scale Fluid-Kinetic Simulation Suite *Numerical Modeling of Space Plasma Flows (ASTRONUM2012)* (*Astronomical Society of the Pacific Conference Series* vol 474) ed Pogorelov N V, Audit E and Zank G P p 165
[51] Colella P, Bell J, N K, Ligocki T, Lijewski M and van Straalen B 2007 *Journal of Physics Conference Series* **78** 012013
[52] Colella P, Dorr M, Hittinger J, Martin D F and McCorquodale P 2009 *Journal of Physics Conference Series* **180** 012010
[53] Colella P, Dorr M, Hittinger J, McCorquodale P W and Martin D F 2009 High-Order Finite-Volume Methods on Locally-Structured Grids *Numerical Modeling of Space Plasma Flows: ASTRONUM-2008* (*Astronomical Society of the Pacific Conference Series* vol 406) ed Pogorelov N V, Audit E, Colella P and Zank G P p 207
[54] McCorquodale P and Colella P 2011 *Communications in Applied Mathematics and Computational Science* **6** 1-25
[55] Guzik S M, McCorquodale P and Colella P 2012 *AIAA Paper 2012-0574*
[56] Ivan L, De Sterck H, Susanto A and Groth C P T 2015 *Journal of Computational Physics* **282** 157–182
[57] Borovikov S N, Pogorelov N V and Ebert R W 2012 *Astrophys. J.* **750** 42
[58] Gamayunov K V, Zhang M, Pogorelov N V, Heerikhuisen J and Rassoul H K 2012 *Astrophys. J.* **757** 74
[59] Kim T K, Pogorelov N V, Borovikov S N, Jackson B V, Yu H S and Tokumaru M 2014 *Journal of Geophysical Research (Space Physics)* **119** 7981–7997
[60] Kim T K, Pogorelov N V, Zank G P, Elliott H A and McComas D J 2016 *Astrophys. J.* **832** 72
[61] Kryukov I A, Pogorelov N V, Zank G P and Borovikov S N 2012 Numerical modeling of the solar wind turbulence *American Institute of Physics Conference Series* (*American Institute of Physics Conference Series* vol 1436) ed Heerikhuisen J, Li G, Pogorelov N and Zank G pp 48–54
[62] McComas D J, Allegrini F, Bochsler P and et al 2009 *Science* **326** 959
[63] McComas D J, Alexashov D, Bzowski M, Fahr H, Heerikhuisen J, Izmodenov V, Lee M A, Möbius E, Pogorelov N, Schwadron N A and Zank G P 2012 *Science* **336** 1291
[64] Pogorelov N V, Suess S T, Borovikov S N, Ebert R W, McComas D J and Zank G P 2013 *Astrophys. J.* **772** 2
[65] Pogorelov N V, Borovikov S N, Heerikhuisen J and Zhang M 2015 *Astrophys. J. Lett.* **812** L6
[66] Pogorelov N V 2016 *Journal of Physics Conference Series* **719** 012013
[67] Zank G P 2015 *Ann. Rev. Astron. Astrophys.* **53** 449–500
[68] Chané E, Jacobs C, van der Holst B, Poedts S and Kimpe D 2005 *A&A* **432** 331–339
[69] Chané E, van der Holst B, Jacobs C, Poedts S and Kimpe D 2006 *A&A* **447** 727–733
[70] Chané E, Poedts S and van der Holst B 2008 *A&A* **492** L29–L32
[71] Shen F, Feng X S, Wang Y, Wu S T, Song W B, Guo J P and Zhou Y F 2011 *Journal of Geophysical Research (Space Physics)* **116** A09103
[72] Stone J M and Norman M L 1994 *Astrophys. J.* **420** 237–246
[73] Zhou Y F, Feng X S, Wu S T, Du D, Shen F and Xiang C Q 2012 *Journal of Geophysical Research (Space Physics)* **117** A01102
[74] Jian L K, MacNeice P J, Mays M L, Taktakishvili A, Odstrcil D, Jackson B, Yu H S, Riley P and Sokolov I V 2016 *Space Weather* **14** 592–611
[75] Manoharan P, Kim T, Pogorelov N V, Arge C N and Manoharan P K 2015 *Journal of Physics Conference Series* **642** 012016

A data-driven MHD model of the global solar corona within Multi-Scale Fluid-Kinetic Simulation Suite (MS-FLUKSS)

Mehmet Sarp Yalim[2], Nikolai Pogorelov[1,2], and Yang Liu[3]

[1]Department of Space Science, The University of Alabama in Huntsville, 320 Sparkman Drive, Huntsville, AL 35805, USA

[2]Center for Space Plasma and Aeronomic Research, The University of Alabama in Huntsville, 320 Sparkman Drive, Huntsville, AL 35805, USA

[3]W. W. Hansen Experimental Physics Laboratory, Stanford University, 452 Lomita Mall, Stanford, CA 94305, USA

E-mail: msy0002@uah.edu,np0002@uah.edu,yliu@sun.stanford.edu

Abstract. We have developed a data-driven magnetohydrodynamic (MHD) model of the global solar corona which uses characteristically-consistent boundary conditions (BCs) at the inner boundary. Our global solar corona model can be driven by different observational data including Solar Dynamics Observatory/Helioseismic and Magnetic Imager (*SDO/HMI*) synoptic vector magnetograms together with the horizontal velocity data in the photosphere obtained by the time-distance helioseismology method, and the line-of-sight (LOS) magnetogram data obtained by *HMI*, Solar and Heliospheric Observatory/Michelson Doppler Imager (*SOHO/MDI*), National Solar Observatory/Global Oscillation Network Group (*NSO/GONG*) and Wilcox Solar Observatory (*WSO*). We implemented our model in the Multi-Scale Fluid-Kinetic Simulation Suite (MS-FLUKSS) – a suite of adaptive mesh refinement (AMR) codes built upon the Chombo AMR framework developed at the Lawrence Berkeley National Laboratory. We present an overview of our model, characteristic BCs, and two results we obtained using our model: A benchmark test of relaxation of a dipole field using characteristic BCs, and relaxation of an initial PFSS field driven by HMI LOS magnetogram data, and horizontal velocity data obtained by the time-distance helioseismology method using a set of non-characteristic BCs.

1. Introduction

The solar wind (SW) emerging from the Sun is the main driving mechanism of solar events which may lead to geomagnetic storms which are the primary causes of space weather disturbances that affect the magnetic environment of Earth and may have hazardous effects on the space-borne and ground-based technological systems as well as human health. Therefore, accurate modeling of the SW is very important to understand the underlying mechanisms of such storms.

As the number of space-borne and ground-based observatories increased significantly over the past decades, a number of data-driven models of SW that involve formulations of time-dependent boundary conditions (BCs) that incorporate remote and *in situ* observations of the Sun in a self-consistent way have been developed (see, e.g. [1, 2, 3, 4, 5, 6]).

The SW emerges from the Sun. In order to model the background SW in a physically-consistent way, it is essential to apply a proper boundary treatment based on observational data on the Sun. For this purpose, we have developed a data-driven magnetohydrodynamic (MHD) model of the global solar corona which applies characteristically-consistent BCs at the inner boundary of our computational domain located in the lower solar corona. Our characteristic BC formulation follows [7], is *entirely based on observations, and does not rely on any non-reflection principle, as in* [8]. It allows us to specify sufficient number of *mathematically admissible* BCs. The implementation of characteristic BCs at the lower corona is based not only on the knowledge of how many quantities should be specified, but also on the requirement that time-increments of quantities that are specified as physical BCs should be uniquely expressible in terms of the increments of proper characteristic variables.

With the availability of vector magnetograms and surface velocity components, we can impose our characteristic BC formulation to obtain a realistic time-dependent background SW model. Our model is driven by Solar Dynamics Observatory/Helioseismic and Magnetic Imager (*SDO/HMI*) synoptic vector magnetograms together with the horizontal velocity data in the photosphere obtained, e.g., with the time-distance helioseismology method [9]. In addition, our model can also be driven by other observational data including the line-of-sight (LOS) magnetogram data obtained by *HMI*, Solar and Heliospheric Observatory/Michelson Doppler Imager (*SOHO/MDI*), National Solar Observatory/Global Oscillation Network Group (*NSO/GONG*), and Wilcox Solar Observatory (*WSO*).

We obtain the initial distribution of plasma properties in the computational domain from the Potential Field Source Surface (PFSS) model [10, 11] for the magnetic field and Parker's 1D isothermal SW model [12] for the hydrodynamic variables. We model the SW acceleration by introducing volumetric heating source terms into the momentum and energy equations that incorporate expansion factors to take the coronal magnetic field distribution into account [13, 14].

We have implemented our model in the Multi-Scale Fluid-Kinetic Simulation Suite (MS-FLUKSS) [15] – a suite of adaptive mesh refinement (AMR) codes designed to solve the coupled systems of MHD, gas dynamics Euler, and kinetic Boltzmann equations [16, 17, 18, 19]. MS-FLUKSS is built upon the Chombo AMR framework developed at the Lawrence Berkeley National Laboratory. MS-FLUKSS also has modules that treat pickup ions as a separate fluid and turbulence beyond the Alfvénic surface.

In section 2, we will present an overview of our global solar corona model with particular emphasis given to our characteristic BC formulation and *SDO/HMI* vector magnetogram and horizontal velocity data. In section 3, we will present results we obtained using our model. Finally, in section 4, we will present our conclusions.

2. Data-driven MHD model of global solar corona

2.1. Governing equations

We solve the set of ideal MHD equations with volumetric heating source terms to model acceleration of SW [13, 14]. They are written in terms of conservative variables, in conservation-law form as follows:

$$\frac{\partial}{\partial t}\begin{pmatrix}\rho\\ \rho\mathbf{v}\\ \mathbf{B}\\ E\end{pmatrix}+\nabla\cdot\begin{pmatrix}\rho\mathbf{v}\\ \rho\mathbf{v}\mathbf{v}+\mathbf{I}(p+\frac{B^2}{8\pi})-\frac{\mathbf{B}\mathbf{B}}{4\pi}\\ \mathbf{v}\mathbf{B}-\mathbf{B}\mathbf{v}\\ (E+p+\frac{B^2}{8\pi})\mathbf{v}-\frac{\mathbf{B}}{4\pi}(\mathbf{v}\cdot\mathbf{B})\end{pmatrix}=\begin{pmatrix}0\\ \rho[\mathbf{g}+(\mathbf{\Omega}\times\mathbf{r})\times\mathbf{\Omega}+2(\mathbf{v}\times\mathbf{\Omega})]+\mathbf{S_M}\\ 0\\ \rho\mathbf{v}\cdot[\mathbf{g}+(\mathbf{\Omega}\times\mathbf{r})\times\mathbf{\Omega}]+S_E\end{pmatrix},\tag{1}$$

where ρ, $\mathbf{v}$, $\mathbf{B}$, p, E, and $\mathbf{g}$ are the density, velocity, magnetic field, thermal pressure, specific total energy of the plasma, and gravitational acceleration, respectively. The source terms in the momentum and energy conservation equations include the Coriolis and centrifugal forces which are present when the system is solved in a frame corotating with the Sun. Accordingly, Ω and $\mathbf{r}$ correspond to the angular velocity of the Sun and position vector, respectively.

In order to model the acceleration of the SW, we introduce a volumetric heating source term, S_E, into the energy conservation equation, and the corresponding source term, $\mathbf{S_M}$, into the conservation of momentum equations [13, 14]. They are given as follows:

$$S_E = \frac{Q_0}{f_s}\exp\Big(-\frac{r}{L_Q}\Big), \tag{2}$$

$$\mathbf{S_M} = \frac{M_0}{f_s}\Big(\frac{r}{R_\odot} - 1\Big)\exp\Big(-\frac{r}{L_M}\Big), \tag{3}$$

where L_{M}, L_{Q}, M_0, and Q_0 are the model constants given as $L_{\mathrm{M}} = L_Q = 0.9R_\odot$, $M_0 = 2.65 \times 10^{-14}$ N m^{-3}, and $Q_0 = 1.65 \times 10^{-6}$ J m^{-3}s^{-1}. Additionally, f_s is the expansion factor by which a magnetic flux tube expands in solid angle between its footpoint location on the photosphere and the source surface which is typically at $R_{\mathrm{SS}} = 2.5R_\odot$ [20]:

$$f_s = \frac{B(R_\odot)}{B(R_{SS})}\left(\frac{R_\odot}{R_{SS}}\right)^2. \tag{4}$$

These source terms take the coronal magnetic field topology into account by incorporating the expansion factor.

The expansion factor is initially computed in every cell located between the inner boundary and the source surface and kept constant throughout a simulation.

2.2. Observational data

Our model is primarily driven by full disk *HMI* synoptic vector magnetograms [21] and horizontal velocity data in the photosphere.

HMI [22] provides full disk vector magnetogram data with high cadence ($\sim$ 720 s) and no data gaps [23]. From these data, synoptic vector magnetograms are computed for each Carrington rotation for the entire *SDO* [24] mission. They have a resolution of 3600×1440 pixels and are provided by the *Joint Science Operations Center* (JSOC) under the data series name hmi.B_synoptic [21].

The horizontal velocity data are inferred by the time-distance helioseismology method [9]. The acoustic travel times are measured using *HMI* Dopplergram observations under the data series name hmi.V and this method infers solar interior properties by inverting these measurements. Thus, near-real time full disk maps of subsurface wave-speed perturbations and horizontal flow velocities are produced for depths ranging from 0 to 20 Mm, every 8 hours. Carrington synoptic maps for the subsurface properties are made from these full disk maps. These maps have a resolution of 3000×1000×6 pixels where the last dimension corresponds to the depth beneath the solar surface. We utilize the corresponding data on the photosphere.

In Figures 1 and 2, three components of the magnetic field vector and horizontal velocity components are shown on the solar surface corresponding to Carrington Rotation (CR) 2145, respectively. The horizontal velocity profiles are given within the latitude range of $\pm 60°$. Above this latitude, the inferred velocity is not very reliable.

Our model can also be driven by other observational data including *NSO/GONG*, *HMI*, *SOHO/MDI*, and *WSO* LOS magnetogram data.

2.3. Initial solution

We calculate the initial distribution of magnetic field using the PFSS model [10, 11] based either on the spherical harmonics approach [25, 26, 27] or a finite difference method by incorporating

IOP Conf. Series: Journal of Physics: Conf. Series **837** (2017) 012015 doi:10.1088/1742-6596/837/1/012015

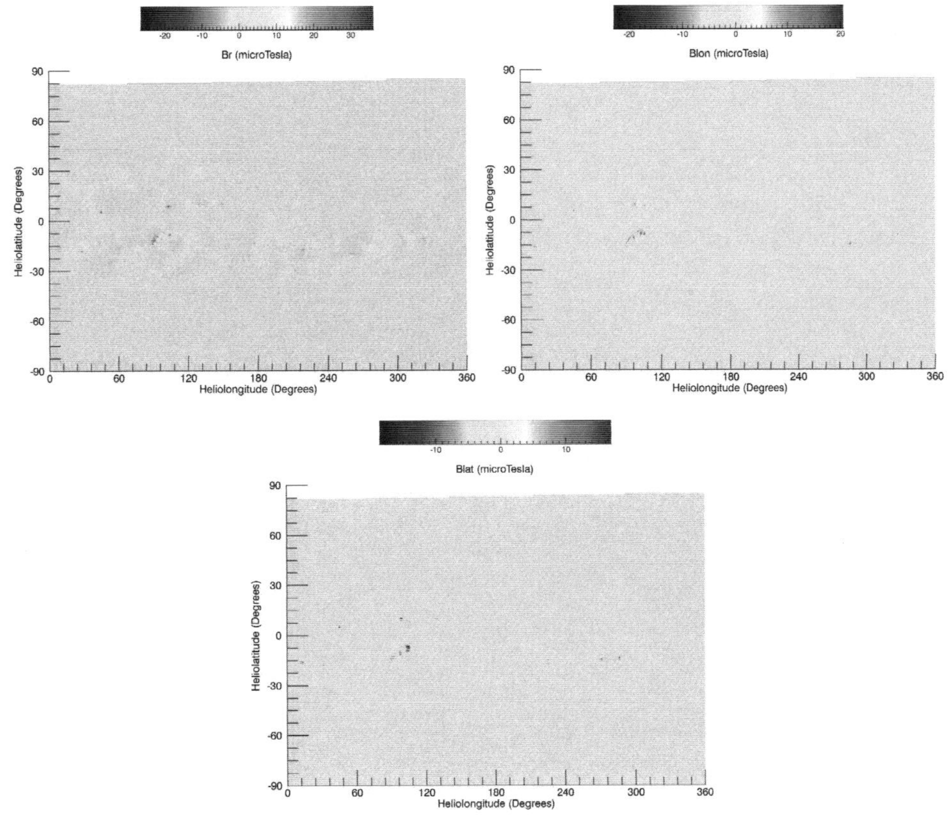

Figure 1. *HMI* vector field synoptic chart for CR 2145: (*Left panel*) radial, (*Right panel*) longitudinal, and (*Bottom panel*) latitudinal components.

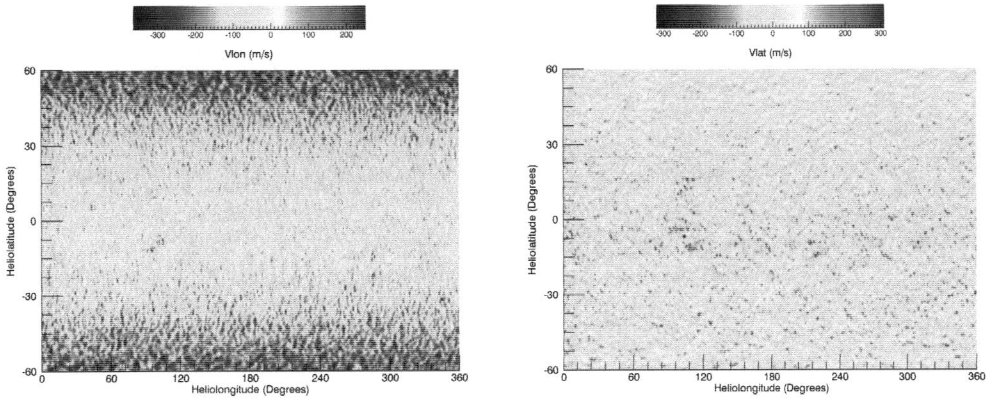

Figure 2. Synoptic chart of the horizontal velocity data for CR 2145: (*Left panel*) longitudinal, and (*Right panel*) poleward flows.

the solution provided by the Finite Difference Iterative Potential-field Solver (FDIPS) code [28] into MS-FLUKSS. FDIPS is a good alternative to the spherical harmonics approach. As the number of spherical harmonics increases, using raw magnetogram data specified on a grid that is uniform in the sine of latitude can result in inaccurate and unreliable results, especially in the

IOP Conf. Series: Journal of Physics: Conf. Series **837** (2017) 012015 doi:10.1088/1742-6596/837/1/012015

polar regions close to the Sun. As the FDIPS code accepts any magnetogram data represented in the Flexible Image Transport System (FITS) file format as input, we can also incorporate potential magnetic field solutions for all such data into MS-FLUKSS. For the rest of the plasma parameters, we compute the *initial solution* from Parker's isothermal SW model [12].

2.4. Characteristic boundary conditions

One of the important components of our data-driven MHD model is a genuinely characteristic BC formulation, which became possible only recently because of the availability of vector magnetograms. If the inner simulation boundary is in the vicinity of the photosphere ($R \approx R_\odot$), the radial component of the SW velocity is likely to be less than the Alfvén or even slow magnetosonic velocity. Not only this determines the number of physical BCs at $R \approx R_\odot$, but the characteristic variables to be specified are also uniquely defined. Since there are no analytic expressions for characteristic variables in MHD, they cannot be found from observational data. On the other hand, any admissible set of physical BCs should be resolvable for the above-mentioned physical characteristic variables. The remaining BCs should be mathematical (numerical). They should be specified, e.g., by extrapolation of the corresponding variables from inside the computational region. This approach is widely applied in aerodynamic simulations [7].

Consider the system of governing equations written in terms of the so-called primitive variables (the abbreviations are conventional).

$$\begin{aligned} \frac{\partial \rho}{\partial t} + \nabla \cdot (\rho \mathbf{v}) &= 0 \\ \frac{\partial \mathbf{v}}{\partial t} + (\mathbf{v}\cdot\nabla)\mathbf{v} &= -\frac{\nabla p}{\rho} - \frac{\mathbf{B}\times(\nabla\times\mathbf{B})}{4\pi\rho} + \mathbf{g} + \mathbf{S}_\mathrm{M} \\ \frac{\mathrm{d}p}{\mathrm{d}t} &= a^2 \frac{\mathrm{d}\rho}{\mathrm{d}t} + S_\mathrm{E} \\ \frac{\partial \mathbf{B}}{\partial t} &= \nabla \times (\mathbf{v}\times\mathbf{B}) \\ \nabla \cdot \mathbf{B} &= 0. \end{aligned} \tag{5}$$

Assuming spherical coordinates for definiteness, we rewrite this system in the quasilinear form as follows:

$$\frac{\partial \mathbf{u}}{\partial t} + \bar{A}_r \frac{\partial \mathbf{u}}{\partial r} + \bar{A}_\theta \frac{\partial \mathbf{u}}{\partial \theta} + \bar{A}_\phi \frac{\partial \mathbf{u}}{\partial \phi} = \mathbf{S}, \tag{6}$$

where $\mathbf{u}$ is the vector of primitive variables, $\bar{A}_r$, $\bar{A}_\theta$, and $\bar{A}_\phi$ are the coefficient matrices in front of the r-, θ-, and ϕ-derivatives, respectively, and $\mathbf{S}$ is the source term vector involving the gravitational force, spherical geometrical factors, and volumetric heating terms.

Let $\mathbf{H} = \bar{A}_\theta \frac{\partial \mathbf{u}}{\partial \theta} + \bar{A}_\phi \frac{\partial \mathbf{u}}{\partial \phi} - \mathbf{S}$. Then, Eq. (6) becomes

$$\frac{\partial \mathbf{u}}{\partial t} + \bar{A}_r \frac{\partial \mathbf{u}}{\partial r} + \mathbf{H} = 0. \tag{7}$$

To write out a set of compatibility relations, we multiply the above equation by the left eigenvector matrix of $\bar{A}_r$ [8, 29]:

$$\bar{L}\frac{\partial \mathbf{u}}{\partial t} + \bar{L}\bar{A}_r \frac{\partial \mathbf{u}}{\partial r} + \bar{L}\mathbf{H} = 0, \tag{8}$$

or in the component form (note that $\mathbf{l}_i \bar{A}_r = \lambda_i \mathbf{l}_i$)

$$\mathbf{l}_i \cdot \frac{\partial \mathbf{u}}{\partial t} + \lambda_i \mathbf{l}_i \cdot \frac{\partial \mathbf{u}}{\partial r} + \mathbf{l}_i \cdot \mathbf{H} = 0, \tag{9}$$

where $\mathbf{l}_i$ and λ_i are the rows of the left eigenvector matrix $\bar{L}$ and eigenvalues of $\bar{A}_r$, respectively.

On introducing a set of characteristic variables $\mathbf{W}$, such that $\delta\mathbf{W} = \bar{L}\delta\mathbf{u}$, we obtain a diagonal system ($\Lambda = \{\lambda_1, \lambda_2, \ldots, \lambda_8\}$)

$$\frac{\partial \mathbf{W}}{\partial t} + \Lambda \frac{\partial \mathbf{W}}{\partial r} + \bar{L}\mathbf{H} = 0. \tag{10}$$

No analytic expressions for $\mathbf{W}$ exist in MHD, but their increments are well defined.

At the inner (SW entrance) boundary, W_i ($i = 1, \ldots k$) corresponding to $\lambda_i > 0$ (incoming characteristics) should be specified as physical BCs, whereas the BCs for remaining characteristic variables should be numerical (i.e., they should be derived from the characteristic compatibility relations for W_j with $\lambda_j < 0$ for $j = k+1, \ldots n$). For MHD equations, $n = 8$.

Following this mathematical principle, we split the compatibility relations (i.e., Eq. (10)) into two sets corresponding to the physical (P) and numerical (N) BCs:

$$\frac{\partial}{\partial t}\begin{vmatrix} \mathbf{W}^P \\ \mathbf{W}^N \end{vmatrix} + \begin{vmatrix} \bar{L}^P \\ \bar{L}^N \end{vmatrix} \bar{A}_r \frac{\partial}{\partial r}\begin{vmatrix} \mathbf{u}^P \\ \mathbf{u}^N \end{vmatrix} + \bar{L}\mathbf{H} = 0, \tag{11}$$

where $\mathbf{W}^P$, $\mathbf{W}^N$, $\bar{L}^P$, and $\bar{L}^N$ are comprised of the characteristic variables and left eigenvectors corresponding to incoming and outgoing characteristics, respectively, and $\mathbf{u}^P$ and $\mathbf{u}^N$ are the primitive variables to be physically imposed or numerically determined.

The physical BCs are implemented by setting

$$\frac{\partial \mathbf{W}^P}{\partial t} = (\bar{L}^P)^{-1}\mathbf{C}^P(t) \text{ or } \frac{\partial \mathbf{u}^P}{\partial t} = \mathbf{C}^P(t), \tag{12}$$

where $\mathbf{C}^P(t)$ is the vector of time dependent variations of the primitive variables that are physically imposed. At each time step, $\mathbf{u}^P$ values are interpolated in time using successive observational data (e.g., vector magnetogram) with the chosen order of accuracy.

Keeping in mind Eq. (12), we can rewrite Eq. (11) in terms of the primitive variables:

$$\begin{vmatrix} \bar{I}^P \\ \bar{L}^N \end{vmatrix} \frac{\partial}{\partial t}\begin{vmatrix} \mathbf{u}^P \\ \mathbf{u}^N \end{vmatrix} + \begin{vmatrix} 0 \\ \bar{L}^N \end{vmatrix} \bar{A}_r \frac{\partial}{\partial r}\begin{vmatrix} \mathbf{u}^P \\ \mathbf{u}^N \end{vmatrix} + \begin{vmatrix} -\mathbf{C} \\ \bar{L}^N\mathbf{H} \end{vmatrix} = 0, \tag{13}$$

where $\bar{I}^P$ is a matrix of size $(P \times (N+P))$ with $I^P_{ii} = 1$ and $I^P_{ij} = 0$ for $i \neq j$, and $\mathbf{C} = \begin{vmatrix} \mathbf{C}^P(t) \\ 0 \end{vmatrix}$.

Let us define $\bar{L}_1 = \begin{vmatrix} \bar{I}^P \\ \bar{L}^N \end{vmatrix}$, and $\bar{L}_2 = \begin{vmatrix} 0 \\ \bar{L}^N \end{vmatrix}$.

After making necessary substitutions, the set of primitive variables at the inner boundary can be determined from

$$\frac{\partial}{\partial t}\begin{vmatrix} \mathbf{u}^P \\ \mathbf{u}^N \end{vmatrix} = \mathbf{C} - (\bar{L}_1)^{-1}\bar{L}_2\bar{A}_r \frac{\partial}{\partial r}\begin{vmatrix} \mathbf{u}^P \\ \mathbf{u}^N \end{vmatrix} - (\bar{L}_1)^{-1}\bar{L}_2\mathbf{H}. \tag{14}$$

It should be noted that our BCs are evolutionary (i.e., small changes in them lead to small changes in solution, see [30]), iff the left eigenvector matrix of the split system, $\bar{L}_1$, is non-singular. This determines the set of admissible physical BCs in terms of the primitive variables. It follows from our analysis that for realistic inflow condition only the plasma density and the radial component of magnetic field should always be specified as physical BCs. The choice of other physical BCs in terms of the primitive variables depends on the availability of appropriate measurements. Table 1 describes reasonable choices, as well as the number of physical and numerical BCs.

Thus, it can be seen that *SDO/HMI* data allow us to specify the necessary BCs, except for density. It was demonstrated in [31] that underestimating or overestimating density at the inner

Case	# of physical BCs	# of numerical BCs	Variables to be imposed as physical BCs	Variables to be imposed as numerical BCs
Subslow inflow ($v_r < a_s$)	5	3	ρ, B_r, v_θ, v_ϕ, and B_θ	v_r, p, and B_ϕ
Sub-Alfvénic, but superslow inflow ($a_s < v_r < a_A$)	6	2	ρ, v_θ, v_ϕ, p, B_r, and B_θ	v_r, and B_ϕ
Subfast, but super-Alfvénic inflow ($a_A < v_r < a_f$)	7	1	ρ, v_θ, v_ϕ, p, B_r, B_θ, and B_ϕ	v_r

Table 1. Different ways to specify the characteristic BCs. The slow magnetosonic, Alfvénic, and fast magnetosonic velocities are denoted as a_s, a_A, and a_f, respectively.

boundary does not affect the coronal solution. Hence, we can assume uniform density at the inner boundary since we do not have measurements.

We discretize our characteristic compatibility equations (see Eq. (14)) using TVD, Roe-type schemes. For the components of the vector magnetogram and horizontal velocity data which can be imposed as physical BCs (see Table 1), the data values are imposed at the boundary surface and they are used in combination with values in the first and second layers of inner cells to extrapolate the primitive variables into the first and second ghost cell layers, respectively. Since our characteristic BC formulation is in inertial reference frame, the synoptic maps for vector magnetogram and horizontal velocity data are also rotated taking the solar rotation into account. For the rest of the variables, their time variation is determined by solving characteristic compatibility equations (see Eq. (14)) in the ghost cells. These equations involve spatial derivatives of plasma variables which are solved by finite difference approximations taking into account the plasma parameter values in the first and second inner and ghost cell layers.

3. Results

3.1. Relaxation of a dipole magnetic field to a quasi-steady state

We present the results of a benchmark test that we performed to validate our characteristic BC formulations. In this test, we impose a dipole magnetic field with magnetic dipole moment equal to the solar dipole moment, 3.5×10^{29} N m T^{-1}, and Parker's isothermal SW model as initial conditions for the magnetic field and hydrodynamic plasma parameters, respectively. Then, we relax this configuration to a quasi-steady state using our characteristic BCs. We also performed another simulation with exactly the same grid, numerical schemes and initial conditions, but this time, using a set of non-characteristic BCs (i.e., $n = 1.5 \times 10^8$ cm^{-3}, $B_r = (B_r)_{\mathrm{Dipole}}$, $B_\theta = (B_\theta)_{\mathrm{Dipole}}$, $v_\theta = v_\phi = 0$, while B_ϕ, v_r, and p have zero derivative in the radial direction) instead. Notice that both sets of BCs have the same physical BCs corresponding to the subslow inflow case in Table 1, whereas the rest of the variables are either obtained from the compatibility equations or their radial derivatives are assumed to be zero. The comparison of the obtained solutions is shown in Figure 3.

The computational domain has a span of $1.03R_\odot \leq r \leq 20R_\odot$, $0 \leq \theta \leq \pi$, and $0 \leq \phi \leq 2\pi$. The computational grid is $128 \times 56 \times 40$ in r, θ and ϕ directions, respectively, and is radially stretched. We use a TVD, Roe-type scheme and forward Euler method for spatial and temporal discretizations, respectively. The $\nabla \cdot \mathbf{B} = 0$ constraint is enforced following [32].

In Figure 3, dipole magnetic field lines that relaxed to a quasi-steady state are shown together with the plasma flow speed contours using characteristic and non-characteristic BCs. Moreover, radial velocity variations are also shown along the y-axis for both sets of BCs.

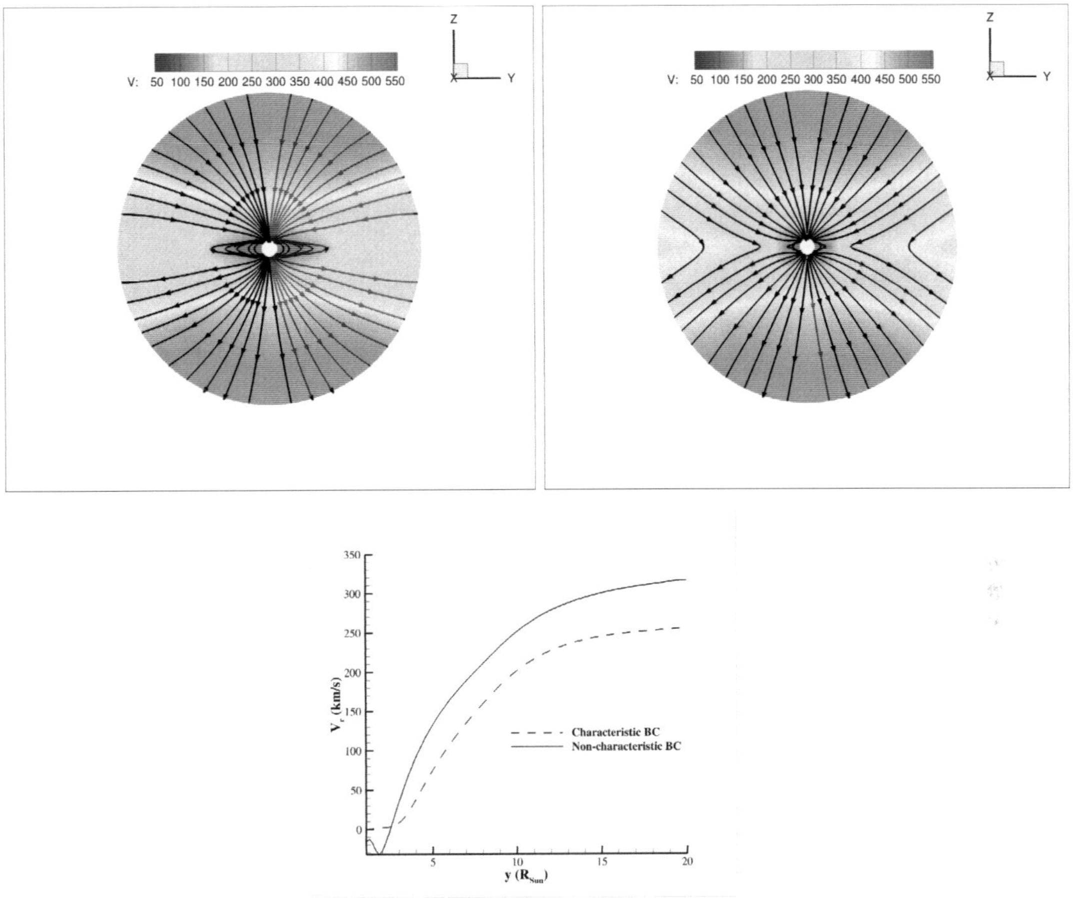

Figure 3. Dipole magnetic field lines relaxed to a quasi-steady state are shown together with the plasma flow speed contours using: (*Left panel*) characteristic BCs and (*Right panel*) one particular choice of non-characteristic BCs, and (*Bottom panel*) radial velocity variations along y-axis for characteristic and non-characteristic BCs.

As seen from Figure 3, using the characteristic BCs leads to a correct magnetic field configuration whereas the non-characteristic approach shows artificial X-points in the vicinity of the Sun. Furthermore, non-characteristic BCs produce a substantial inflow on the solar equatorial plane which does not seem right.

3.2. Relaxation of the PFSS magnetic field to a quasi-steady state

In this section, we present results related to the relaxation of an initial PFSS magnetic field topology generated by using the *HMI* LOS magnetogram data observed on 2 January 2014 to a quasi-steady state. We obtained the PFSS solution by applying the spherical harmonics approach for which we took the spherical harmonics coefficients from the *pfss_viewer* software in IDL SolarSoft, and using 10 modes.

The computational domain has a span of $1.03R_{\odot} \leq r \leq 20R_{\odot}$, $0 \leq \theta \leq \pi$, and $0 \leq \phi \leq 2\pi$. The computational grid is $128{\times}60{\times}120$ in r, θ and ϕ directions, respectively, and is radially stretched. The set of non-characteristic BCs applied at the inner boundary is given as follows: $n = 1.5 \times 10^8$ cm^{-3}, $T = 1.3 \times 10^6$ K, B_r is determined from the *HMI* LOS magnetogram data,

IOP Conf. Series: Journal of Physics: Conf. Series **837** (2017) 012015 doi:10.1088/1742-6596/837/1/012015

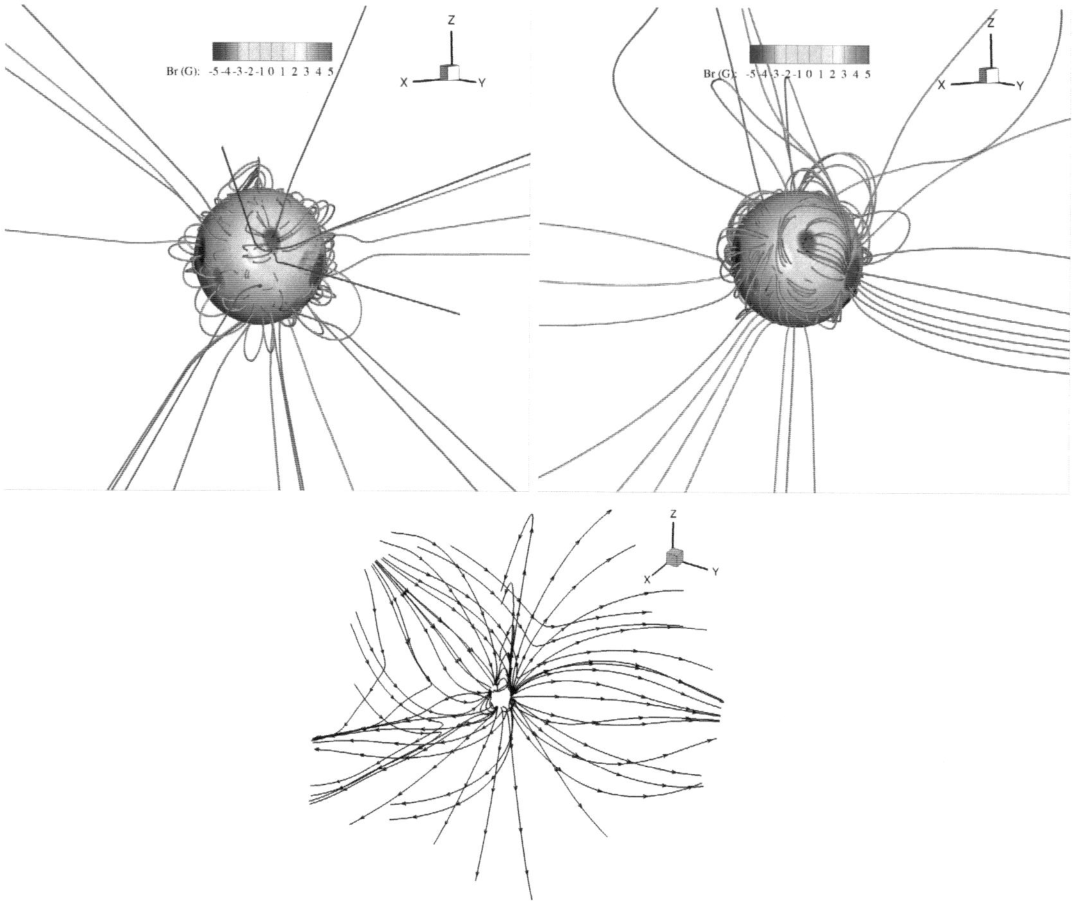

Figure 4. Magnetic field lines in the vicinity of the inner boundary at $1.03R_\odot$ with radial magnetic field line contours for (*Left panel*) initial PFSS solution generated by using the *HMI* LOS magnetogram data observed on 2 January 2014 and (*Right panel*) quasi-steady solution, and (*Bottom panel*) full 3D view of the magnetic field line topology at the quasi-steady state.

v_θ and v_ϕ are the horizontal velocity data shown in Figure 2, while B_θ, B_ϕ, and v_r have zero derivative in the radial direction. The horizontal velocity data inferred by the time-distance helioseismology method have 3000×1000 pixels resolution. We applied bilinear interpolation to interpolate the data to our computational grid at the inner boundary in order to be able to utilize in our BCs. It should be noted that differential rotation and meridional flow profiles are removed from the horizontal velocity data. Therefore, we impose empirical formulae for differential rotation [33] and meridional flow [34] at the entire inner boundary in addition to the data. The simulation is performed in the frame of reference corotating with the Sun.

In Figure 4, we show the magnetic field line distributions in the vicinity of the inner boundary with radial magnetic field line contours for the initial PFSS and quasi-steady solutions, and a full 3D view of the magnetic field line topology. In Figure 5, heliospheric current sheets corresponding to the initial PFSS and quasi-steady solutions are presented.

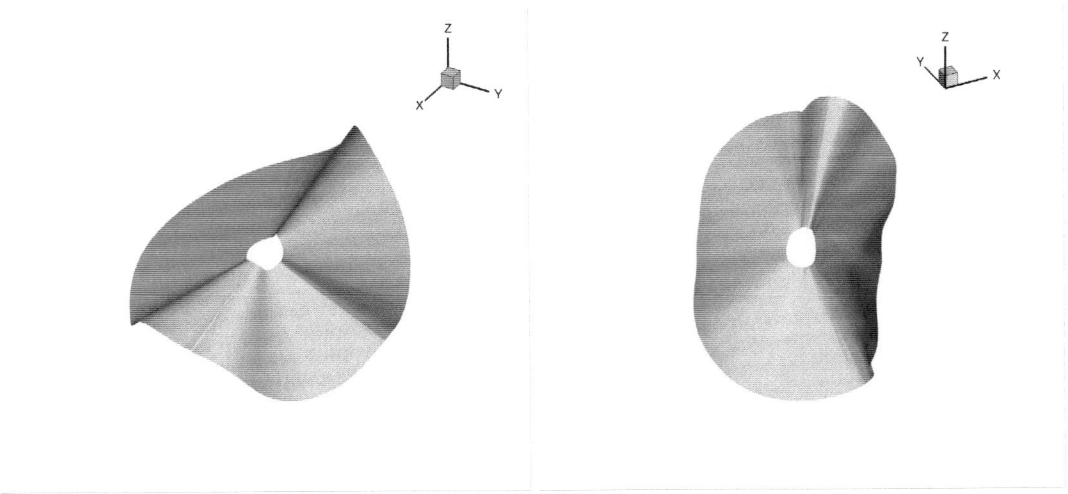

Figure 5. Heliospheric current sheet for (*Left panel*) initial PFSS solution generated by using the *HMI* LOS magnetogram data observed on 2 January 2014 and (*Right panel*) quasi-steady solution.

4. Conclusions

In this paper, we presented an overview of our data-driven MHD model of SW *based entirely on observations and not relying on any non-reflection principle* in the solar corona.

For this purpose, we developed a set of genuinely characteristic BCs which gives remarkable results already in a benchmark test.

Our model can be driven by a variety of observational data including *HMI* vector magnetogram and horizontal velocity data, and LOS magnetogram data from *HMI*, *MDI*, *GONG*, and *WSO*. In order to demonstrate this, we presented results related to the simulation of relaxation of a PFSS magnetic field topology to a quasi-steady state which is driven by *HMI* LOS magnetogram and horizontal velocity data using a set of non-characteristic BCs.

We will present results related to time-dependent simulations of our global corona model in our next publication.

Acknowledgments

The authors acknowledge the support from the NASA grant NNX14AF41G and NSF SHINE grant AGS-1358386. We also acknowledge NSF PRAC award OCI-1144120 and related computer resources from the Blue Waters sustained-petascale computing project. Supercomputer allocations were also provided on SGI Pleiades by NASA High-End Computing Program award SMD-15-5860 and on Stampede by NSF XSEDE project MCA07S033.

References

[1] Mikić Z, Linker J A, Schnack D D, Lionello R and Tarditi A 1999 *Phys. Plasmas* **6** 2217–24
[2] Wu S T, Wang A H, Liu Y and Hoeksema J T 2006 *Astrophys. J.* **652** 800–11
[3] Hayashi K 2005 *Astrophys. J. Supp.* **161** 480–94
[4] van der Holst B, Manchester IV W B, Frazin R A, Vasquez A M, Toth G and Gombosi T I 2010 *Astrophys. J.* **725** 1373–83
[5] Yang L P, Feng X S, Xiang C Q, Liu Y, Zhao X and Wu S T 2012 *J. Geophys. Res - Space Phys.* **117** A08110
[6] Feng X S, Yang L P, Xiang C Q, Liu Y, Zhao X P and Wu S T 2012 *Numerical Modeling of Space Plasma Flows* vol 459 eds N V Pogorelov et al. (ASP Conference Series) p 202
[7] Hirsch C 1994 *Numerical Computation of Internal and External Flows, Volume 2: Computational Methods for Inviscid and Viscous Flows* (John Wiley & Sons)

[8] Thompson K W 1987 *J. Comp. Phys.* **68** 1–24
[9] Zhao J, Couvidat S, Bogart R S, Parchevsky K V, Birch A C, Duvall T L, Beck J G, Kosovichev A G and Scherrer P H 2012 *Sol. Phys.* **275** 375–90
[10] Altschuler M D and Newkirk G 1969 *Sol. Phys.* **9** 131–49
[11] Schatten K H, Wilcox J M and Ness N F 1969 *Sol. Phys.* **6** 442–55
[12] Parker E N 1958 *Astrophys. J.* **128** 664–76
[13] Nakamizo A, Tanaka T, Kubo Y, Kamei S, Shimazu H and Shinagawa H 2009 *J. Geophys. Res - Space Phys.* **114** A07109
[14] Feng X, Yang L, Xiang C, Wu S T, Zhou Y and Zhong D 2009 *Astrophys. J.* **723** 300–19
[15] Pogorelov N V, Borovikov S N, Heerikhuisen J, Kim T K, Kryukov I A and Zank G P 2014 *XSEDE'14 Proc. of the 2014 Annual Conference on Extreme Science and Engineering Discovery Environment* (ACM: New York)
[16] Borovikov S N, Kryukov I A and Pogorelov N V 2009 *Numerical Modeling of Space Plasma Flows* vol 406 eds N V Pogorelov et al. (ASP Conference Series) p 127
[17] Pogorelov N V, Borovikov S N, Florinski V, Heerikhuisen J, Kryukov I A and Zank G P 2009 *Numerical Modeling of Space Plasma Flows* vol 406 eds N V Pogorelov et al. (ASP Conference Series) p 149
[18] Borovikov S N, Heerikhuisen J and Pogorelov N V 2013 *Numerical Modeling of Space Plasma Flows* vol 474 eds N V Pogorelov et al. (ASP Conference Series) p 219
[19] Pogorelov N V, Borovikov S N, Bedford M C, Heerikhuisen J, Kim T K, Kryukov I A and Zank G P 2013 *Numerical Modeling of Space Plasma Flows* vol 474 eds N V Pogorelov et al. (ASP Conference Series) p 165
[20] Wang Y-M and Sheeley Jr N R 1997 *Geophys. Res. Lett.* **24** 3141–44
[21] Liu Y, Hoeksema J T, Sun X and Hayashi K 2017 *Sol. Phys.* **292** 29
[22] Schou J et al 2012 *Sol. Phys.* **275** 229–59
[23] Hoeksema J T et al 2014 *Sol. Phys.* **289** 3483–3530
[24] Pesnell W D, Thompson B J and Chamberlin P C 2012 *Sol. Phys.* **275** 3–15
[25] Hoeksema J T 1984 *Structure and Evolution of the Large Scale Solar and Heliospheric Magnetic Fields* PhD Thesis (Stanford University)
[26] Wang Y-M and Sheeley Jr N R 1992 *Astrophys. J.* **392** 310–19
[27] Schrijver C J and DeRosa M L (2003) *Sol. Phys.* **212** 165–200
[28] Toth G, van der Holst B and Huang Z 2011 *Astrophys. J.* **732** 102
[29] Thompson K W 1990 *J. Comp. Phys.* **89** 439–61
[30] Kulikovskii A G, Pogorelov N V and Semenov A Y 2001 *Mathematical Aspects of Numerical Solution of Hyperbolic Systems* (Boca Raton: Chapman & Hall/CRC Press)
[31] Lionello R, Linker J A and Mikić Z 2009 *Astrophys. J.* **690** 902–12
[32] Powell K G, Roe P L, Linde T J, Gombosi T I and De Zeeuw D L 1999 *J. Comp. Phys.* **154** 284–309
[33] Komm R W, Howard R F and Harvey J W 1993 *Sol. Phys.* **143** 19–39
[34] Komm R W, Howard R F and Harvey J W 1993 *Sol. Phys.* **147** 207–23

Advancing parabolic operators in thermodynamic MHD models: Explicit super time-stepping versus implicit schemes with Krylov solvers

R. M. Caplan, Z. Mikić, J. A. Linker, and R. Lionello
Predictive Science Inc., 9990 Mesa Rim Road Suite 170, San Diego, CA 92121
E-mail: caplanr@predsci.com, mikicz@predsci.com, linkerj@predsci.com, lionel@predsci.com

Abstract. We explore the performance and advantages/disadvantages of using unconditionally stable explicit super time-stepping (STS) algorithms versus implicit schemes with Krylov solvers for integrating parabolic operators in thermodynamic MHD models of the solar corona. Specifically, we compare the second-order Runge-Kutta Legendre (RKL2) STS method with the implicit backward Euler scheme computed using the preconditioned conjugate gradient (PCG) solver with both a point-Jacobi and a non-overlapping domain decomposition ILU0 preconditioner. The algorithms are used to integrate anisotropic Spitzer thermal conduction and artificial kinematic viscosity at time-steps much larger than classic explicit stability criteria allow. A key component of the comparison is the use of an established MHD model (MAS) to compute a real-world simulation on a large HPC cluster. Special attention is placed on the parallel scaling of the algorithms. It is shown that, for a specific problem and model, the RKL2 method is comparable or surpasses the implicit method with PCG solvers in performance and scaling, but suffers from some accuracy limitations. These limitations, and the applicability of RKL methods are briefly discussed.

1. Introduction

Complex physical systems often have processes that act on widely different time-scales. For magnetohydrodynamic (MHD) models of the solar corona, these processes include plasma flow, slow and fast magneto-sonic waves, resistivity, kinematic viscosity, and thermal conduction [1]. When implementing the models with numerical methods, accurately capturing the dynamics of these wide-ranging time-scales can make the simulations computationally intractable due to the fine time-resolutions required. Also, when using classic explicit finite difference algorithms (such as the Euler method), there exists numerical time-step stability restrictions (typically a function of grid cell size) that can limit the time-step further.

In order to make the simulations computationally feasible, they need to be integrated at time-steps which exceed the fastest time-scales. This is possible because often, a system can be integrated without accurately capturing the detailed evolution of the faster time-scales and still obtain useful solutions. However, integrating at large time-steps can violate the explicit numerical stability requirements, and if too large, can cause the model to loose convergence and become very inaccurate. Therefore, accuracy considerations must be taken into account when integrating past the fast time-scales, and numerical methods that can overcome the explicit limits are required.

In our model, we adhere to the flow speed time-scale, but exceed the remaining time-scales through a variety of methodologies (see Sec. 4). The focus of this paper is comparing methods to exceed the explicit numerical time-step restrictions of the (often very stiff) parabolic operators. To do this, the operators are first isolated from the rest of the terms in their equations of motion through operator-splitting [2] and are written as

$$\frac{\partial u}{\partial t} = F(u), \tag{1}$$

where F is the parabolic operator acting on the variable u.

There exists a number of methodologies which allow for the efficient integration of Eq. (1) including implicit finite difference methods (e.g. backward-Euler, Crank-Nicholson, etc.), Saul'yev unconditionally-stable explicit schemes [3, 4], and super time-stepping schemes (also known as extended-stability Runge-Kutta schemes) [5]. In some cases, smaller time-scale non-parabolic operators yielding the effect of the original operator can be found (e.g. see Ref [6] where a wave formulation of thermal conduction is used). Implicit schemes typically lead to large systems of equations at each time-step that are often solved using Krylov subspace methods (in the linear case), or Newton-Krylov methods (for nonlinear operators) [7].

This paper seeks to compare the basic advantages, disadvantages, performance, and parallel scaling of two classes of methods (implicit and super time-stepping) by using a representative algorithm of each class. The chosen algorithms are 1) The implicit Backwards-Euler scheme solved with the Preconditioned Conjugate Gradient method (BE+PCG) and 2) The explicit second-order Runge-Kutta Legendre (RKL2) super time-stepping method [8, 9]. The two methods are used to integrate the most time-consuming parabolic operators of our model: Spitzer anisotropic thermal conduction and artificial kinematic viscosity. Since the PCG method is only applicable to linear operators, the thermal conduction operator is linearized using lagged diffusivity [10] (i.e. we fix the temperature-dependent part of the diffusion coefficient using the previous time-step's temperature values).

The criteria for the comparisons consists of looking at accuracy validation, the ease/difficulty of the method's formulation and implementation, the features and limitations of each method, their overall performance (e.g. the number of iterations/sub-steps needed for integrating each large time-step, the computational cost of each iteration/sub-step), and the methods' parallel performance including communication/synchronization requirements and their ability to scale to large numbers of processors.

In addition to a reduced time-to-solution, scalability has an important additional advantage, as it allows for much higher-resolution simulations to be performed within reasonable run times. In the context of studying the solar corona, such high-resolution simulations are desired because current three-dimensional thermodynamic MHD coronal simulations are very under-resolved in relation to some important physical phenomena under investigation (e.g. magnetic reconnection and transition-region dynamics). Therefore, efficient scalable algorithms are an essential component in exploring such small-scale physics and their global effects on the corona.

A previous comparison between PCG and the factored RKC STS method [11] for a 1D diffusion test case using the RAMSES code was done in Ref. [12], where the authors found the STS method to be much faster (and more accurate) than their PCG method. Comparisons between RKL2 and a Newton-Krylov method using GMRES and with a multigrid solver was done for several test problems by the authors of the RKL method in Ref. [9], where similar speed improvement results were found. A key element to the current comparison is the use of a large, full production simulation to test the methods.

Since it is not feasible to cover every possible combination of problem size, problem type, HPC environment, etc., our results presented here are not to be taken as generally applicable in all situations, but rather as a single real-world example to help gain some flavor of the advantages/limitations of using these two very different methodologies. The production run used here is a full-scale thermodynamic coronal simulation (described in Sec. 5) computed using two state-of-the-art HPC systems (San Diego Supercomputing Center's Comet and Texas Advanced Computing Center's Stampede machines).

The paper is organized as follows: The two methods are described in Sec. 2 (BE+PCG) and Sec. 3 (RKL2). Sec. 4 describes the MAS thermodynamic MHD model used to simulate the production run described in Sec. 5. The validation of the two methods is discussed in Sec. 6 including important applicability concerns for RKL methods. The timing results and analysis are described in Sec. 7, and we conclude in Sec. 8.

2. Implicit Backward Euler Solved with Preconditioned Conjugate Gradient

There exists a wide range of implicit discretization schemes to integrate Eq. (1) and numerous methods to solve the resulting system of equations. For the present context, we limit ourselves to the backward Euler (BE) scheme solved with a preconditioned conjugate gradient (PCG) sparse solver (we denote the full method as BE+PCG). The BE+PCG method is chosen as it is commonly used and is the current method applied in the MAS code. Although the method is well-known, we review it in this section in order to highlight its advantages and disadvantages, especially in relation to parallel performance.

2.1. Implicit Backward Euler

One of the easiest-to-implement implicit methods for integrating Eq. (1) is the BE scheme given by

$$\frac{u^{n+1}-u^{n}}{\Delta t} = F(u^{n+1}), \tag{2}$$

where $\Delta t = t^{n+1} - t^{n}$ is the chosen time-step and $u^{n} = u(t^{n}, \mathbf{x})$. If one assumes that F is a linear self-adjoint operator, then Eq. (2) can be written as

$$\frac{u^{n+1}-u^{n}}{\Delta t} = \mathbf{M}\, u^{n+1}, \tag{3}$$

where $\mathbf{M}$ is a symmetric positive-definite matrix. Given the solution at step n (u^{n}), Eq. (3) leads to a system of equations,

$$\mathbf{A}\, u^{n+1} = u^{n}, \tag{4}$$

where $\mathbf{A} = 1 - \Delta t\,\mathbf{M}$, which, when solved yields u^{n+1}. The BE method is a low-order method ($O(\Delta t)$ accurate in time), but because it is unconditionally stable, L-stable (i.e. its amplification factor tends to 0 as the wave mode number tends to ∞), and very simple to implement, it remains a useful and wide-spread method.

2.2. Preconditioned Conjugate Gradient

The matrix $\mathbf{A}$ of Eq. (4) is often large and very sparse. A common method for solving such sparse systems are Krylov subspace iterative solvers (see Ref. [13] for a detailed review), and in the case of symmetric operators, the Conjugate Gradient (CG) method. In order to allow the CG method to converge to the solution efficiently, the system is preconditioned with an approximate inverse of the matrix $\mathbf{P}^{-1} \approx \mathbf{A}^{-1}$, leading to the preconditioned-CG method (PCG). The application of the preconditioner ($\mathbf{P}^{-1}x$) is almost always performed without explicitly formulating the inverse matrix $\mathbf{P}^{-1}$. The PCG algorithm as used in MAS is shown in Fig. 1 where we have highlighted the expressions that require inter-process communication with a green box to denote local communication/synchronization (stencil interchange) and a red box to denote global communication/synchronization (global summation for the dot products). While the basics of the PCG method are relatively simple to implement, it is difficult to properly choose optimal convergence criteria/values [14] and, even more so, an efficient preconditioner **P**.

2.3. Preconditioners

A major difficulty with using the PCG method is selecting and computing an inexpensive yet efficient preconditioner (PC). Preconditioning is a very complex topic with many decades of research and a multitude of methods [15]. The use of any method has many weighing factors such as the amount of computational work, complexity of implementation, ability to parallelize efficiently, applicability to the linear solver and matrix representation, as well as problem-specific considerations that can reduce the usefulness of the PC (and in some cases can even cause the PCs to break down [16]). Another difficulty is that certain PCs which work very well to improve convergence, can be very difficult to implement efficiently (or at all) on large HPC systems.

$$
\begin{aligned}
&x_0 = u^n \\
&r_0 = b - \mathbf{A}\, x_0 \\
&z_0 = \mathbf{P}^{-1}\, r_0 \\
&p_0 = z_0 \\
&r_r = r_0 \cdot z_0 \\
&\texttt{do k = 0 : k}_{\texttt{max}} \\
&\quad y_k = \boxed{\mathbf{A}\, p_k} \\
&\quad \alpha_k = r_r / \boxed{(p_k \cdot y_k)} \\
&\quad x_{k+1} = x_k + \alpha_k\, p_k \\
&\quad r_{k+1} = r_k - \alpha_k\, y_k \\
&\quad z_{k+1} = \mathbf{P}^{-1}\, r_{k+1} \\
&\quad r_{\text{old}} = r_r \\
&\quad r_r = \boxed{r_{k+1} \cdot z_{k+1}} \\
&\quad \boxed{\texttt{Check}\, r_r\, \texttt{for convergence}} \\
&\quad \beta_k = r_r / r_{\text{old}} \\
&\quad p_{k+1} = \beta_k\, p_k + z_k, \\
&\texttt{enddo} \\
&u^{n+1} = x_{k+1}
\end{aligned}
$$

Figure 1. The PCG algorithm with inter-processor communication terms boxed (red: global, green: point-to-point)

For the comparisons in this paper, we use two relatively simple communication-free PCs: 1) A Point-Jacobi/diagonal scaling (PC1), which simply uses the inverse of the diagonal of **A**. The PC1 is very inexpensive to formulate and apply, but is limited in its effectiveness. 2) A non-overlapping domain decomposition with zero-fill incomplete LU factorization (PC2), which is more expensive to formulate and apply than PC1, but is also much more effective at reducing iterations. One drawback of PC2 is that, due to the non-overlapping implementation (used to avoid adverse scaling performance), it becomes less effective as the number of processors increases. In fact, in the non-practical upper limit of one processor core per grid point, PC2 becomes equivalent to PC1. However, for realistic uses, the degradation in convergence remains small for reasonable grid points per processor ratios (see Sec. 7). Pseudo-code descriptions of the implementation for the initial formulation and subsequent application of both PCs is described in Appendix A.

Many new methods/preconditioners such as Geometric/Algebraic multigrid [17] and inter-processor overlapping PCs [18] exist, which would yield different performance and scaling results than presented here. However, such methods are often substantially more difficult to implement and parallelize (especially into legacy code) and often have trade-offs in effectiveness and parallel scaling that are hard to predict.

2.4. Advantages and disadvantages of the BE+PCG method

Some advantages of using the BE+PCG approach are that; (a) it is widely used, robust, and has well-known convergence properties; (b) the BE scheme is L-stable, which efficiently damps out high wave modes (see Sec. 6); (c) it can be used for slightly non-symmetric operators [19]; (d) its basic form is fairly straightforward to implement; (d) there exists many available 'black-box' implementations, and (e) it can be efficient given a good preconditioner and implementation.

IOP Conf. Series: Journal of Physics: Conf. Series **837** (2017) 012016 doi:10.1088/1742-6596/837/1/012016

However, the BE+PCG approach does suffer from some important disadvantages. One limitation in using the PCG method (or any Krylov solver) is that it requires a linear operator. Often, (as in the case of thermal conduction) the parabolic terms have non-linear coefficients and/or are desired to be flux-limited. In order to do this, using BE, one would have to employ a nonlinear system solver such as the popular Newton-Krylov methods [7] which can be computationally expensive. Efficiently porting the PCG method to accelerator hardware is a formidable challenge and, as mentioned in the previous section, choosing an optimal stopping criteria and efficient preconditoner is often very difficult.

Running PCG solvers on large HPC systems exposes other disadvantages. The global communication/synchronization requirement of the dot-products in the PCG algorithm can adversely effect the scaling of the method, especially in the presence of (even small) load imbalance (see Sec. 7). There are ongoing efforts to mitigate this problem [20], but they have yet to show substantial improvement. This is problematic for domain decomposed problems (such as ours) because the grid resolution in each dimension does not typically divide evenly into the number of processors assigned to that dimension, leading to some sub-domains having more grid points than others. Therefore, as the problem is spread to more and more processors, the load imbalance will increase (see Fig. 11 in Sec. 7).

3. Explicit Runge-Kutta Legendre Super Time-stepping scheme

A relatively recent class of schemes for integrating parabolic operators explicitly without numerical restriction on the time-step size are 'super time-stepping'/'extended stability' (STS) methods. The main idea behind them is to use a Runge-Kutta multi-stage scheme with stages added for stability rather than accuracy. For a given desired time-step, a number of stages/sub-steps (s) and specialized coefficients for each stage are found such that the overall advance is stable.

$$
\begin{aligned}
&M_0 = \boxed{\mathbf{M}\, u^n}\\
&u_1 = u^n + \tilde{\mu}_1\, \Delta t\, M_u\\
&\texttt{do k = 2 : s}\\
&\quad u_k = \mu_j\, u_{k-1} + \nu_j\, u_{k-2} + (1-\mu_k-\nu_k)\, u^n\\
&\qquad\quad + \tilde{\mu}_k\, \Delta t\, \boxed{\mathbf{M}\, u_{k-1}} + (b_k-1)\, \tilde{\mu}_k\, \Delta t\, M_0\\
&\texttt{enddo}\\
&u^{n+1} = u_s,
\end{aligned}
$$

Figure 2. The RKL2 algorithm with inter-processor communication terms boxed (green: point-to-point)

STS methods were made popular as a practical numerical method in 1996 [5] but remained relatively unknown and unused until recent years where interest in the methods and their development have rapidly increased. Current STS methods are based on either Chebychev polynomials (RKC) [11] or Legendre polynomials (RKL) [8, 9], and are either formulated as factored [21] or recursive [5] methods.

The RKC STS method has been used successfully for parabolic operators in MHD codes [22, 23, 24] and the RKL method is currently being implemented into the Lare3D MHD code [25, 26] and has been shown to be accurate and efficient for some coronal loop simulations [27]. Ref. [9] has shown that for nonuniform/nonlinear diffusion coefficients (such as ours), the RKL STS method has more desirable properties than the RKC. We therefore use the RKL method, and for accuracy and robustness, we use the second-order accurate version (RKL2) [9].

The algorithm for the RKL2 scheme is shown in Fig. 2. The pre-computed coefficients for each stage are defined as

$$b_0 = b_1 = b_2 = \frac{1}{3}, \qquad b_k = \frac{k^2 + k - 2}{2\,k\,(k+1)},$$
$$\tilde{\mu_1} = \frac{4}{3\,(s^2 + s - 2)}, \qquad \mu_k = \frac{2\,k-1}{k}\,\frac{b_k}{b_{k-1}},$$
$$\tilde{\mu_k} = \frac{4\,(2\,k-1)}{k\,(s^2+s-2)}\,\frac{b_k}{b_{k-1}}, \qquad \nu_k = -\frac{k-1}{k}\,\frac{b_k}{b_{k-2}}.$$

The number of stages/sub-steps needed for stability (given a desired Δt) is given by

$$s = \frac{1}{2}\left\lceil\sqrt{9 + 16\,\frac{\Delta t}{\Delta t_{\mathrm{Euler}}}} - 1\right\rceil, \tag{5}$$

where $\Delta t_{\mathrm{Euler}}$ is the explicit Euler time-step stability limit (for a simple and fast way to calculate $\Delta t_{\mathrm{Euler}}$ that avoids too much underestimating, see Appendix B). For better accuracy, the authors of Ref. [9] recommend always ensuring s is odd (although in our experience, when using more sub-steps than Eq. (5), this is unnecessary).

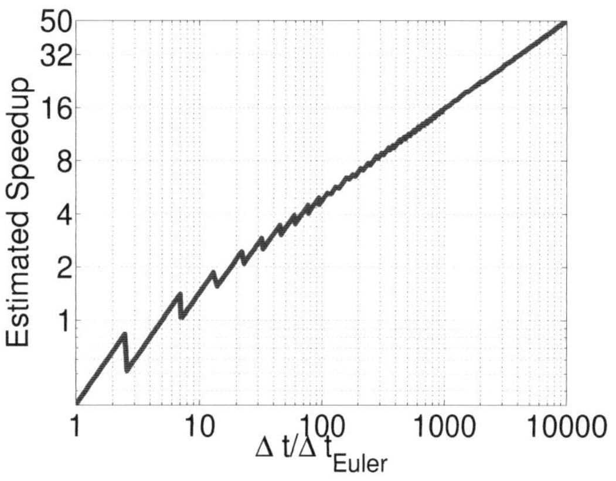

Figure 3. Estimated speedup of using the RKL2 versus the explicit Euler method assuming one Euler step has the same computational cost as one RKL2 sub-step. The jaggedness of the plot is due to the discrete nature of Eq. (5).

Since $\Delta t_{\mathrm{Euler}} \propto \Delta x^2$, the number of sub-steps follows $s \propto 1/\Delta x$ whereas sub-cycling the Euler method would take $\propto 1/\Delta x^2$ steps. Therefore, the RKL2 scheme requires many less steps than the Euler method and the larger the overall time-step is, the more speedup the RKL2 method exhibits (see Fig. 3). As in describing the PCG solver, in Fig. (2) we have highlighted the terms in the RKL2 method that require parallel communication. Unlike the PCG solver, the RKL2 scheme only has local communication/synchronization, so is expected to scale much better in the presence of load imbalance and system performance fluctuations (see Sec. 7 where this is indeed the case).

3.1. Advantages and disadvantages of RKL2

Some of the main advantages of the RKL2 method (and other STS methods) is that it is very easy to implement (both in a matrix and matrix-free manner), lends itself to easy implementation of directive-based parallelism (such as OpenMP/OpenACC), does not require global synchronizations/communications, and can be used with nonlinear operators/flux-limiting. Its performance gains can vary, but in cases where the accuracy of the fast-time scale of the operator is not needed, it can have substantial speedup over standard explicit methods, and, as we will show in Sec. 7, can sometimes outperform implicit methods.

One disadvantage of the RKL2 method is that it is very new with very few real-world implementations to test its robustness in a variety of settings. Also, in some cases, the number of sub-steps/stages can become very large, in which case implicit methods may outperform it. One additional aspect of the RKL2 scheme is that, as will be discussed in more detail in

Sec. 6, its amplification factor does not generally monotonically decrease for increasing wave mode numbers, and is quite high for the highest modes. This can cause solution artifacts when using only a few total time-steps on solutions containing high wave mode structures, and, when using large time-steps in coupled systems, may fail to damp the high wave modes.

4. The MAS Thermodynamic MHD Model

The MHD model we use to test the RKL2 and BE+PCG methods is the Magnetohydrodynamic Algorithm outside a Sphere (MAS) code [28]. MAS integrates the time-dependent resistive thermodynamic MHD equations in spherical coordinates and has been used extensively in models of coronal structure [29, 30, 28, 31], coronal dynamics [32, 33, 34] and coronal mass ejections [35, 36]. It is also the primary MHD model in CORHEL (Corona-Heliosphere), a suite of models [37] for describing the solar corona and inner heliosphere that is available for public use at NASA's CCMC [1]. In this section we describe the model equations and numerical methods used in MAS as they are configured for our comparisons (see Sec. 5).

4.1. Model equations

In MAS, the one-fluid ideal MHD equations [1] are extended to include additional physics with a wide variety of models options and parameters. For the problem used in this paper, the governing equations take the form:

$$\frac{\partial \mathbf{A}}{\partial t} = \mathbf{v}\times(\nabla\times\mathbf{A}) - \boxed{\frac{c^2\,\eta}{4\,\pi}\,\nabla\times\nabla\times\mathbf{A}}\,, \tag{6}$$

$$\frac{\partial \rho}{\partial t} = -\nabla\cdot(\rho\,\mathbf{v}), \tag{7}$$

$$\begin{aligned}\frac{\partial T}{\partial t} = {} & -\nabla\cdot(T\mathbf{v}) - (\gamma-2)\,(T\,\nabla\cdot\mathbf{v}) - \frac{1}{2}\nabla\cdot\left(f_{\rm nc}(r)\,T\,\mathbf{v}\,\hat{\mathbf{b}}\hat{\mathbf{b}}\right) \\ & + \boxed{\frac{(\gamma-1)\,m_p}{2\,k\quad\rho}\left[\nabla\cdot\left(\beta_{\rm Tcut}(T)\,f_{\rm c}(r)\,\kappa_0\,T^{5/2}\,\hat{\mathbf{b}}\hat{\mathbf{b}}\cdot\nabla T\right)\right.} - \left.\frac{\rho^2}{m_p^2}\,\frac{Q(T)}{\beta_{\rm Tcut}(T)} + H\right],\end{aligned} \tag{8}$$

$$\frac{\partial \epsilon_+}{\partial t} = -\nabla\cdot\left(\epsilon_+\left[\mathbf{v}+v_A\,\hat{\mathbf{b}}\right]\right) - \frac{1}{2}\,\epsilon_+\,\nabla\cdot\mathbf{v}, \tag{9}$$

$$\frac{\partial \epsilon_-}{\partial t} = -\nabla\cdot\left(\epsilon_-\left[\mathbf{v}-v_A\,\hat{\mathbf{b}}\right]\right) - \frac{1}{2}\,\epsilon_-\,\nabla\cdot\mathbf{v}, \tag{10}$$

$$\begin{aligned}\frac{\partial \mathbf{v}}{\partial t} = {} & -\mathbf{v}\cdot\nabla\mathbf{v} + \frac{1}{\rho}\left[\frac{1}{c}\mathbf{J}\times\mathbf{B} - \nabla p - \nabla\left(\frac{\epsilon_+ + \epsilon_-}{2}\right) + \rho\,\mathbf{g}\right] \\ & + \boxed{\frac{1}{\rho}\nabla\cdot(\nu\rho\nabla\mathbf{v})} + \frac{1}{\rho}\nabla\cdot\left(S\,\rho\,\nabla\frac{\partial\mathbf{v}}{\partial t}\right),\end{aligned} \tag{11}$$

where $\mathbf{A}$ is the magnetic vector potential, $\mathbf{B} = \nabla\times\mathbf{A}$ is the magnetic field, $\mathbf{J} = \frac{c}{4\pi}\nabla\times\mathbf{B}$ is the current density, ρ is the plasma density, T is the temperature, $p = 2\,k\,T\,\rho/m_p$ is the plasma pressure, $\mathbf{v}$ is the plasma velocity, $\hat{\mathbf{b}} = |\mathbf{B}|/\mathbf{B}$ is the normalized direction of the magnetic field, c is the speed of light in a vacuum, $\gamma = 5/3$ is the adiabatic index, m_p is the proton mass, k is Boltzman's constant, κ_0 is the Spitzer's constant, $f_{\rm c}(r) = 0.5\,(1 - \tanh[(r - 10R_\odot)/0.5\,R_\odot])$ is a profile that limits the radial extend that collisional thermal conduction is active, $f_{\rm nc}(r) = 1 - f_{\rm c}(r)$ is the equivalent profile for collisionless thermal conduction (the last term of Eq. (8) [38]), $Q(T)$ is a radiative loss function (here we select the piece-wise exponential approximation of Ref. [39]), H is a series of combined empirical heating models that are a function of $\mathbf{B}$ [28], $v_A = \sqrt{|\mathbf{B}|^2/4\,\pi\,\rho}$

[1] http://ccmc.gsfc.nasa.gov/

is the Alfvén wave speed, $\mathbf{g} = -g_0 R_\odot^2 \hat{\mathbf{r}}/r^2$ is the gravitational force, and the last expression in Eq. (11) is a semi-implicit term described in the next section. Eqs. (9) and (10) are a WKB approximation for Alfven wave pressure advance [29] where ϵ_+ and ϵ_- are the forward and backward Alfvén wave pressures.

The algorithms used to advance wave-like terms (described in Sec. 4.2) do not guarantee monotonicity, resulting in the potential emergence of oscillatory behavior on the order of the grid size. In order to avoid such behavior (and to prevent physical structures smaller than the grid from trying to form), the dissipative resistivity term in Eq. (6) and the artificial viscosity term of Eq. (11) are added. For the current simulations, we use a constant resistivity coefficient value of $\eta = 4.678 \times 10^{-8}\,$s corresponding to a magnetic diffusion coefficient of $\sim 10^{12}\,\text{cm}^2/\text{s}$. While this is much larger than the typical chromospheric ($\sim 10^7\,\text{cm}^2/\text{s}$) and coronal ($\sim 10^4\,\text{cm}^2/\text{s}$) values [1], the corresponding magnetic diffusion time scale remains much larger than the typical Alfvén travel time (yielding Lundquist numbers of $\sim 10^6$). The kinematic viscosity coefficient is set to $\nu = 6.7 \times 10^{15}\,\text{cm}^2/\text{s}$, which, while also very large, yields a diffusion time-scale for velocity that is still 500 times larger than the typical Alfvén travel time. The large diffusion times of the resistivity and viscosity ensure that the addition of these terms does not substantially affect the overall solution.

The function $\beta_{\text{Tcut}}(T)$ is a cut-off function that sets a minimum temperature that can be used for the thermal conduction coefficient as $\beta_{\text{Tcut}} = (T/T_{\text{cut}})^{5/2}$ for $T < T_{\text{cut}}$ and $\beta_{\text{Tcut}} = 1$ for $T \geq T_{\text{cut}}$, where $T_{\text{cut}} = 5 \times 10^5 K$. Applying this cut-off function serves to broaden the width of the transition region (to make it more numerically resolvable) in such a way that it has a minimal effect on the global coronal solution [40].

In describing the MHD equations above, we have boxed the parabolic terms, where those used to test the two methods are colored blue (the resistivity term uses a very small number of PCG iterations to solve, and is therefore excluded from the method comparisons).

4.2. Numerical methods

The MAS code is written in FORTRAN and parallelized using MPI. It computes the MHD equations using finite difference on a logically rectangular non-uniform spherical grid. The non-uniformity of the grid allows MAS to efficiently resolve small-scale structures such as the transition region and active regions, while allowing for coarser grid points over the global scale. Each component of the field vectors are staggered which, combined with the use of $\mathbf{A}$ as a primitive, ensures that $\nabla \cdot \mathbf{B} = 0$ exactly. Advective terms are differenced using first-order upwinding, while parabolic and gradient terms are centrally differenced. The non-uniformity of the grid is implemented into the discretizations using the standard method 'A' of Ref. [41]. For the anisotropic thermal conduction operator, we use the 'standard' scheme of Ref. [42]. The viscosity operator is written in a self-adjoint form (excluding boundary conditions) that leads to a nearly-symmetric solver matrix.

The code uses an adaptive time-step that conforms to the plasma flow CFL condition. As this is often a much slower time-scale than the magnetosonic waves, a semi-implicit term is added to Eq. (11) that stabilizes the algorithm for time-steps larger than the fast magneto-sonic wave limit. The semi-implicit factor is given as $S = (\Delta t^2\,\tilde{k}^2)^{-1}\,(C_w^2/(1-C_f)^2 - 1)$, where $\tilde{k}^2$ is the upper limit of the combined inverse grid spacing, C_f is the flow CFL, and C_w is the fast magnetosonic wave CFL [43].

The temporal differencing uses a predictor-corrector scheme for the advective and semi-implicit operator, where any reactive terms are included in the corrector step. The parabolic terms are first-order operator split. The WKB approximate Alfvén wave pressure advance of Eqs. (9) and (10) are sub-cycled at their explicit advective CFL condition based on v_A.

All parabolic operators and the semi-implicit solves are currently computed using the BE+PCG method with PC1 for the resistivity, and PC2 for the remaining solves. The operator matrices are stored in a DIA sparse storage format [44] for internal grid points, while the boundary conditions are applied matrix-free. The PC2 preconditioner LU matrix is stored in the CSR sparse storage format optimized for memory access as described in Ref. [45]. In our implementation of the STS method for this comparison, we use the same matrix format used for the solver operators.

5. Real-world Test Case

To provide a useful comparison of the methods, we use a production run of the MAS code. The procedure for producing an MHD model of the solar corona has been described previously [30, 28]. In brief, a full-sun map of the radial magnetic field derived from observations is used to specify the boundary conditions at the solar surface. A potential solution is computed using this boundary condition to supply the initial magnetic field condition (see Fig. 4). A spherically symmetric solar wind solution (see Fig. 5) is used as the initial condition for the other variables (ρ, T, $\vec{v}$). This initially non-equilibrium state is advanced in time until a steady-state is approached. The resulting solutions can be compared against a range of solar and heliospheric observations [37], including EUV and X-ray emission [28], and white-light eclipse images [46].

For the case shown here, the surface radial magnetic field is taken from a flux evolution model (ADAPT) that assimilates observational Earth-based magnetograms to create a full-sun magnetogram at a one particular time [47, 48]. Sometimes active regions (AR) can emerge on the far-side of the Sun that are not assimilated into the ADAPT map until they rotate into Earth's view. In the case being simulated here, such an AR was spotted by the STEREO spacecraft in EUV emission images. In order to better approximate the synchronic state of the surface field, the far-side AR (as observed later in time) is added into the ADAPT map. The map is then interpolated onto the computational grid and smoothed. In Fig. 4, we show the final form of the ADAPT map used, and the initial condition of **B** derived from the resulting potential field solution.

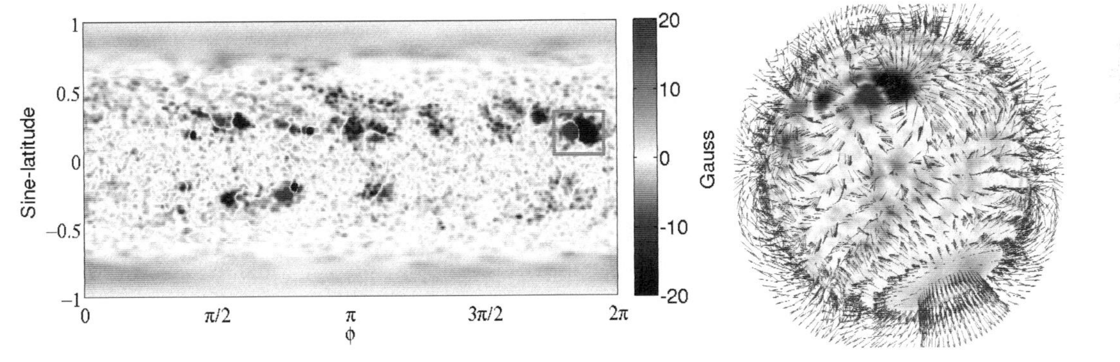

Figure 4. Left: Smoothed initial ADAPT map from 07/07/2010 20:00 with added far-side active region (boxed). Right: Sampled unit vectors of initial condition of magnetic field after solving for potential fields (up to $r = 1.41\,R_\odot$).

The full simulation integrates the initial conditions for 48 simulation-hours. For validation runs (Sec. 6), we integrate from the initial conditions to $t = 8$ simulation-hours as this contains the most dynamical/difficult portion of the total run. For the timing runs, we start with the solution at $t = 8$ simulation-hours (computed using MAS's default algorithms) and then integrate using the two methods for another $t = 6$ simulation-minutes.

The domain is global and covers a radial range from $1\,R_\odot$ to $30\,R_\odot$. It is discretized into a logically-rectangular nonuniform spherical grid (r, θ, ϕ). The grid (summarized in Table 1) is

	N	$\min(\Delta_i)$	$\max(\Delta_i)$	$\max\lvert 1-\Delta_{i+1}/\Delta_i\rvert$
r	181	340km	500,000km	6%
θ	251	0.55°	1.76°	3%
ϕ	602	0.6°	0.6°	0%

Table 1. Grid information for the coronal relaxation run.

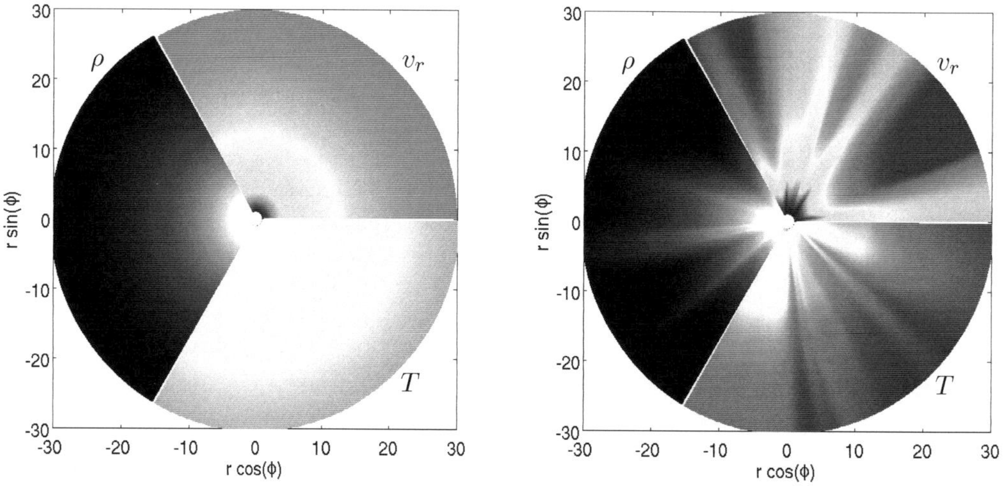

(a) Initial Conditions (b) Relaxed Solution

Figure 5. Initial condition (a) and solution at 48 simulation-hours (b) for the coronal relaxation. Portions within a r–ϕ cut at $\theta = \pi$ are shown for ρ, v_r, and T.

nearly uniform in ϕ, coarsens slightly towards the pole in θ, and is highly non-uniform in r (in order to better resolve the transition from the upper chromosphere to the corona). The time-step changes over the runs to conform to the flow CFL condition and during the 8-simulation-hour validation simulation, ranged from ~ 2.5 seconds to ~ 3.4 seconds with an average step-size of ~ 2.7 seconds.

In Fig. 5, we show some plasma values within an r–ϕ slice at $\theta = \pi/2$ for the initial condition and the final relaxed corona at $t = 48$ simulation-hours.

5.1. HPC environment

To test the parallel scaling and performance of the methods, we run the simulations on state-of-the-art HPC systems. In order not to limit results to one system, we use both the Comet machine at the San Diego Supercomputer Center (SDSC) and the Stampede machine at the Texas Advanced Computing Center (TACC). The hardware and software configuration used for each system are shown in Table 2. We note that although the systems are similar in design, Comet has newer hardware (Haswell AVX2-vectorization processors, more and faster RAM, etc) but on a standard allocation, Stampede allows runs to be performed with more than twice as many processor cores as Comet.

6. Validation

Since our focus is on comparing the two methods (BE+PCG and RKL2) to each other, we do not analyze the solution accuracy of the original BE+PCG method. As mentioned in Sec. 4.1,

	SDSC Comet	**TACC Stampede**
Processor Model	Intel Xeon E5-2660v3	Intel Xeon E5-2680v2
Clock Speed	2.5 GHz	2.7 GHz
CPU Cores/node	24	16
DRAM/node (ECC)	128 GB DDR4-2133MHz	32 GB DDR3-1600MHz
Max # of cores/job	1728	4096
Max Flops/node	960 GFlops	346 GFlops
Network	FDR InfiniBand Hybrid Fat-Tree	FDR InfiniBand 2-level Fat-Tree
MPI Library	MVAPICH 2.1	Intel 2015.2.164
Compiler	Intel 2015.2.164	Intel 2015.2.164
OS	CentOS 2.6.32-573	CentOS 2.6.32-431

Table 2. System hardware and software configuration used for Comet (`http://www.sdsc.edu/support/user_guides/comet.html`) and Stampede (`https://portal.tacc.utexas.edu/user-guides/stampede`).

the MAS code has been used extensively and its accuracy validated [49]. Instead, we focus on comparing the two methods to each other by using a long simulation ($t = 8$ simulation-hours) and qualitatively comparing the results. In Fig. 6, we show r-θ cuts at $\phi \approx 137°$ of various field quantities for each run.

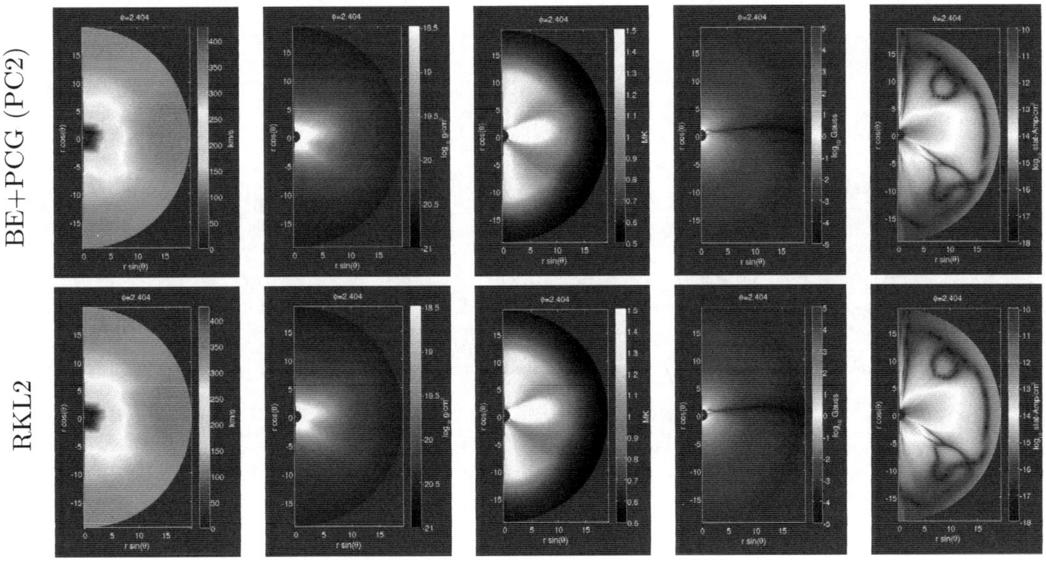

Figure 6. Simulation comparions of selected fields after an 8-simulation-hour relaxation. Left to right: v_r, $\log_{10}(\rho)$, T, $\log_{10}(B_r)$, and $\log_{10}(J_\phi)$. Top: BE+PCG (PC2), Bottom: RKL2.

We see that the two methods produce very similar global results even with complicated structures present. However, as much of the grid is focused near the transition region at roughly $r = 1.02\,R_\odot$, it is difficult to compare the solution in those regions when viewing the full domain. On closer inspection near the solar surface (Fig. 7), we once again see general agreement between the two methods, but also see an alarming difference in some localized regions where the RKL2 method exhibits high wave-mode oscillations in some fields.

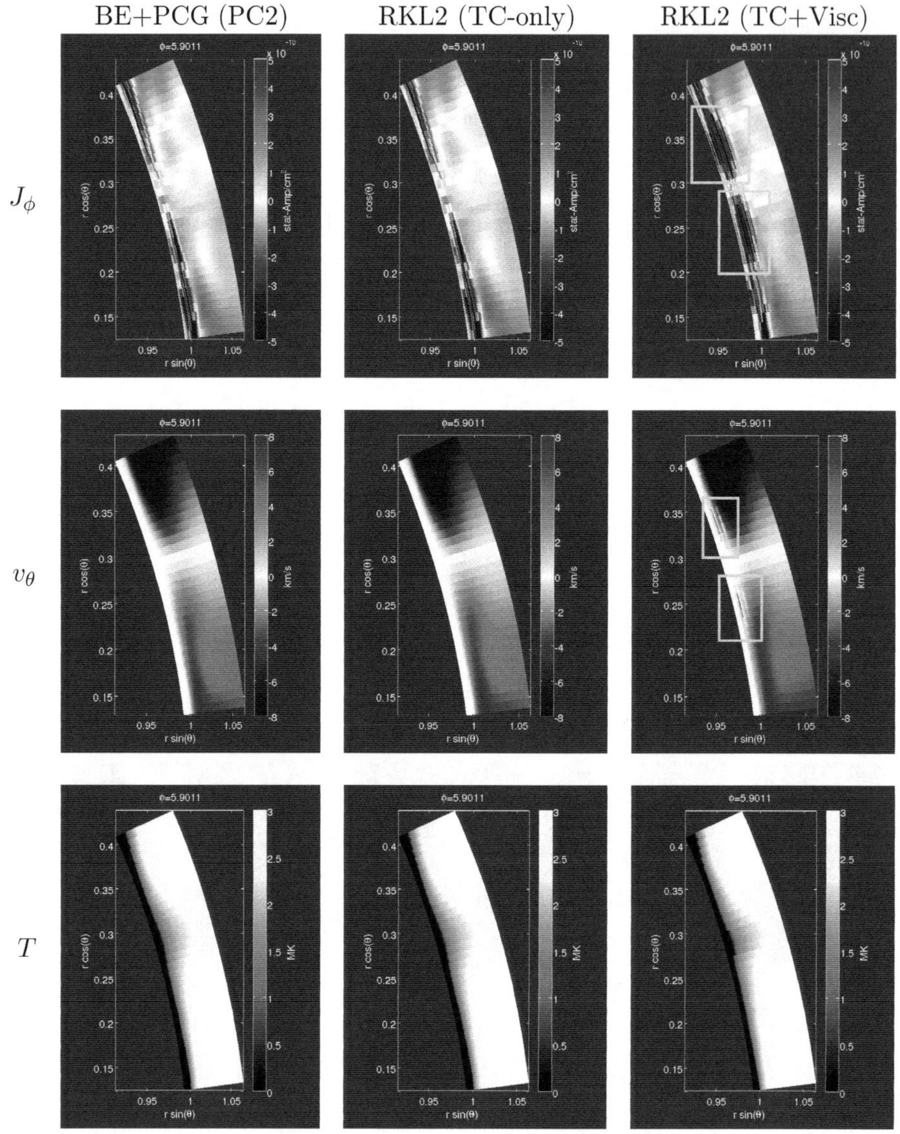

Figure 7. Simulation comparions of J_ϕ, v_θ, and T near the transistion region within an active region. Left-to-right: the result using BE+PCG, RKL2 for thermal conduction only, RKL2 for both thermal conduction and viscosity. Oscillations observed in the RKL2 case are indicated by the green boxes.

A possible explanation of these oscillations is that they are exhibited due to the RKL2's inability to adequately damp high wave modes within one large time-step (as shown below). Due to the extremely difficult conditions in transition-region physics, the standard algorithms in MAS sometimes exhibit oscillatory behavior if the spatial structures are not fully resolved. In order to combat this effect, the artificial viscosity term is added to damp out such oscillations (more than the numerical viscosity due to the algorithm is capable), and smooth out the solution. The true viscosity of the solar corona is so small, this term would not have been included otherwise. In contrast to this, the thermal conduction term is an essential part of the physics of the model. Indeed, if we run the simulations using RKL2 only for the thermal conduction (i.e. using the original BE+PCG for viscosity), the solution becomes visually identical to the BE+PCG case as shown in Fig. 7.

To gain some insight into why the RKL2 method does not damp out the high wave modes, we first note that even though the amplification factor of a convergent scheme is less than 1 (i.e. it is stable and diffusive), it is possible that it is too large to properly damp out a growing wave mode. In Fig. 8, we show the amplification factors for the BE and RKL2 methods for the one-dimensional case of Eq. (1) with a constant diffusion coefficient discretized on a uniform grid with spacing Δx. The factors are derived from the analytic eigenvalues of the system and are shown for integrating at a time-step 1/5, 5, 50, and 500 times the explicit Euler limit $\Delta t_{\mathrm{Euler}}$ (in our simulations, viscosity is being integrated at over 700× the limit in some portions of the grid, and thermal conduction over 10000×). The amplification factor for the exact solution of the semi-discrete system and for the explicit Euler method (in the case where it is stable) are shown for comparison.

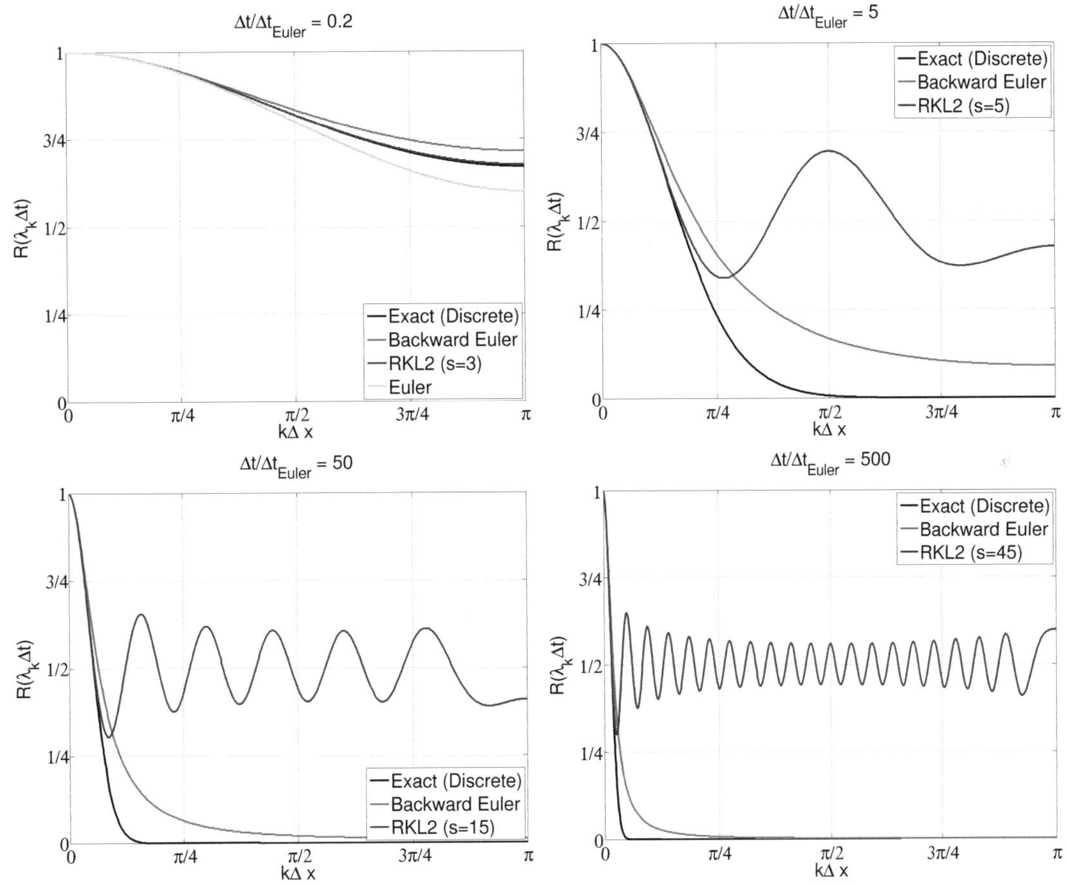

Figure 8. Amplification factors for the BE and RKL2 methods for various values of $\Delta t/\Delta t_{\mathrm{Euler}}$. The exact discrete solution is shown for comparison, and the explicit Euler method is shown for the case where it is stable.

We first notice that for small wave modes, the RKL2 is much more accurate than BE, which is not surprising as it is a second-order accurate method. We also see that when exceeding the Euler limit by a large factor, the amplification factors for both BE and RKL2 are much larger than the true solution, and the RKL2's amplification factor is much larger than the BE's.

Also, the RKL2's amplification factor does not tend towards 0 as $k\Delta x \to \pi$ implying that the RKL2 method is generally not L-stable, although it appears to be strongly A-stable [50] and is positive preserving. The large amplification factor of the RKL2 for high modes will not damp high-frequency solutions efficiently within large time-step because modes which should be virtually eliminated according to Eq. (1) will only be damped by a factor of ~ 0.5. However, with successive application of the method, the wave modes will eventually damp. Therefore, if Eq. (1) were integrated by itself, the RKL2 would be convergent (given enough steps and total simulation time), but when coupled to a larger multi-physics model, the modes may not be damped. We note that BE is not exempt from this issue as its amplification factor is many times larger than the true solution, but since its amplification factor tends towards 0 for high modes, and is generally small, it is expected to be more robust than RKL2.

To support this conclusion, we have tested sub-cycling the RKL2 method within each large time-step for the viscosity operator, and find that it yields a solution visually identical to the BE method given enough sub-cycles (~ 10 in our case). We also found that sub-cycling explicit Euler steps for $\sim 1/4$ of the time-step and completing the remainder of the step with RKL2 also yielded correct results. This is understandable as the explicit Euler method damps the high modes very efficiently such that running it for $\sim 1/4$ of a time-step damps high modes more than even a full step with BE. These considerations may lead to new scheme strategies that combine methods in order to get both an efficient and overall correct solution.

As an aside, we note that the amplification factors shown in Fig. 8 for the RKL2 method are qualitatively different than those shown in Ref. [8] and Ref. [9]. This is because the factors shown there are for the RKL methods where the number of sub-steps s is chosen first, and the time-step factor is then set exactly from the equation for s (such as Eq. (5)). In such an exact case, the RKL2's amplification factor *does* tend towards zero for very high modes, but still remains near $\sim 1/2$ for a large portion of the frequency spectra. In a real-world case where the time-step is set by other factors, and then the minimum stable s is computed (the suggested procedure given in Ref. [9]), the amplification factors become like those shown in Fig. 8. Even one extra s step over the exact amount needed causes the qualitative change. Interestingly, using any extra number of sub-steps for the *RKL1* scheme eliminates the problem shown in Ref. [9] of it approaching magnitude 1 at $k\Delta x = \pi$, possibly making that scheme more useful than otherwise assumed (although it is not positive preserving). These observations imply that when using uniform grids, if one could compute Δt_{Euler} exactly, one could modify the time-step of the system to exactly follow Eq. (5) for the nearest whole value of s and potentially mitigate the problem of the non-decaying amplification factor. However, if computing Δt_{Euler} exactly is not efficiently possible (which is often the case) and/or one is using non-uniform grids, the number of sub-steps s will always be more than exact, at least somewhere on the grid.

The present discussion indicates that the RKL2 (and other STS methods) may be inadequate for use in multi-physics codes when the parabolic operator is being used either for smoothing oscillations and/or shocks (like artificial viscosity) and perhaps even if they are being used in a balanced equation where the diffusion must counter high-mode growth due to reaction or other driving terms.

In the case presented here, we found that the RKL2 works well for thermal conduction (even in the transition region), but fails to damp grid-level oscillations when used for integrating the artificial viscosity. Although these oscillations were limited to the transition region and did not noticeably change the global solution (as seen in Fig. 6), their unchecked continued growth can lead to unphysical solutions that could rapidly propagate into the global result.

It is difficult to generalize from this single case to other coronal simulations as a whole. This is because the case presented here, while being a production run of the code, was in-development and under-resolved, necessitating the need for a large artificial viscosity. It is conceivable that in a refined fully-resolved case, very little artificial viscosity (or none) would be

IOP Conf. Series: Journal of Physics: Conf. Series **837** (2017) 012016 doi:10.1088/1742-6596/837/1/012016

required. Therefore, the results of the RKL2 on thermal conduction are of primary importance when discussing coronal simulations in general. In either case, further research is needed to solidify the applicability of RKL2 (and STS schemes in general), and whether modifications based on the observations made in this Section and/or use-practices can ensure their robustness.

Since in our case, the global solution does not appear to be adversely affected by the limited and localized grid oscillations shown in Fig. 7, we still report the timing results for the runs as a comparison of the scaling of the methods for both thermal conduction and artificial viscosity. Since our main concern in the present context is parallel scaling, we note that any modifications (along the lines described above) to help the RKL2 method damp high wave modes (e.g. sub-cycling Euler steps) may increase the total compute time, but should not effect the scaling, as those steps have the same communication profile as the RKL2 method itself. Since using the RKL2 for thermal conduction alone produces a consistent solution, the timing results for thermal conduction have increased relevance.

7. Timing Results

To time the simulations, we use calls to the MPI library's timing API routines. It is well known that massive parallel codes can have a very wide range of run times when using many processors due to fluctuations of system load and communication occupancy. In order to extract the scaling out of the methods in an as ideal situation as possible, we use multiple runs (between 4 and 6) of each processor count and take the best result. In addition to recording the total time for each portion of the run tested (thermal conduction and viscosity), we also record the 'non-compute' time for each portion which we define as the time the algorithms are involved in inter-processor communication (including buffer loading and unloading, data transfer, and synchronization/waiting).

Since Comet and Stampede each have different number of processor cores per node, the domain decomposition is unique for runs on each system, along with the number of total processors for each run. On Comet, the runs are performed using 1 node up to the maximum of 72 nodes, while on Stampede, due to the smaller amount of system memory per node, the simulations could not be run on fewer than 4 nodes. The domain is split up along processors in each direction yielding 3D block sub-domains. Since the grid size in each dimension will typically not divide evenly into the processor count along that dimension, some cores will have one extra grid point along some dimensions than others. In the worst case, cores may have one extra grid point along all three dimensions. For a fixed problem size, this causes an intrinsic load imbalance that increases as the number of total processors increases. In Fig. 9 we show the processor topologies used on each system and the maximum grid-based load imbalance.

An additional possible source of load imbalance comes from cores containing sub-domains along physical boundaries, which have more or less work to perform based on the boundary conditions. For example, in the case of the viscosity solver, the cores containing sub-domains along the polar boundaries require calculating a polar average involving communication/synchronization with all other cores conditioning the polar boundaries.

The timing results for thermal conduction and viscosity are shown in Fig. 10. Since the BE+PCG simulations with PC1 were found to be much slower than either BE+PCG PC2 or RKL2 during the Comet runs, they were not repeated on Stampede.

We first observe that all methods scale well up to 512 processors. Higher then that, we see that the BE+PCG method starts to loose scaling, and in some cases, the run time starts to increase with larger number of cores. The BE+PCG PC1 seems to scale a bit better than PC2, but is overall a much slower method. Therefore, if the RKL2 method is within the desired accuracy, in this case it is shown to perform far better than BE+PCG PC1 (which is still a commonly used method). The RKL2 is shown to scale better than the BE+PCG PC2 method in all cases. For thermal conduction, the run time for RKL2 is very similar to the BE+PCG

Comet		Stampede	
N_{total}	(N_r, N_θ, N_ϕ)	N_{total}	(N_r, N_θ, N_ϕ)
24	$(2, 3, 4)$	64	$(2, 4, 8)$
48	$(2, 4, 6)$	128	$(4, 4, 8)$
96	$(3, 4, 8)$	256	$(4, 8, 8)$
216	$(4, 6, 9)$	512	$(4, 8, 16)$
432	$(6, 8, 9)$	1024	$(8, 8, 16)$
864	$(6, 8, 18)$	2048	$(8, 16, 16)$
1796	$(6, 12, 24)$	4096	$(8, 16, 32)$

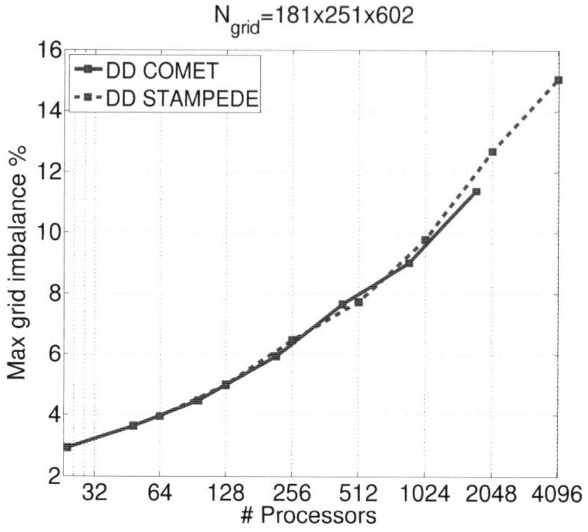

Figure 9. Left: Total cores and processor topology use for domain decomposition (DD) on Comet and Stampede. Right: Maximum possible grid-based load imbalance due to DD.

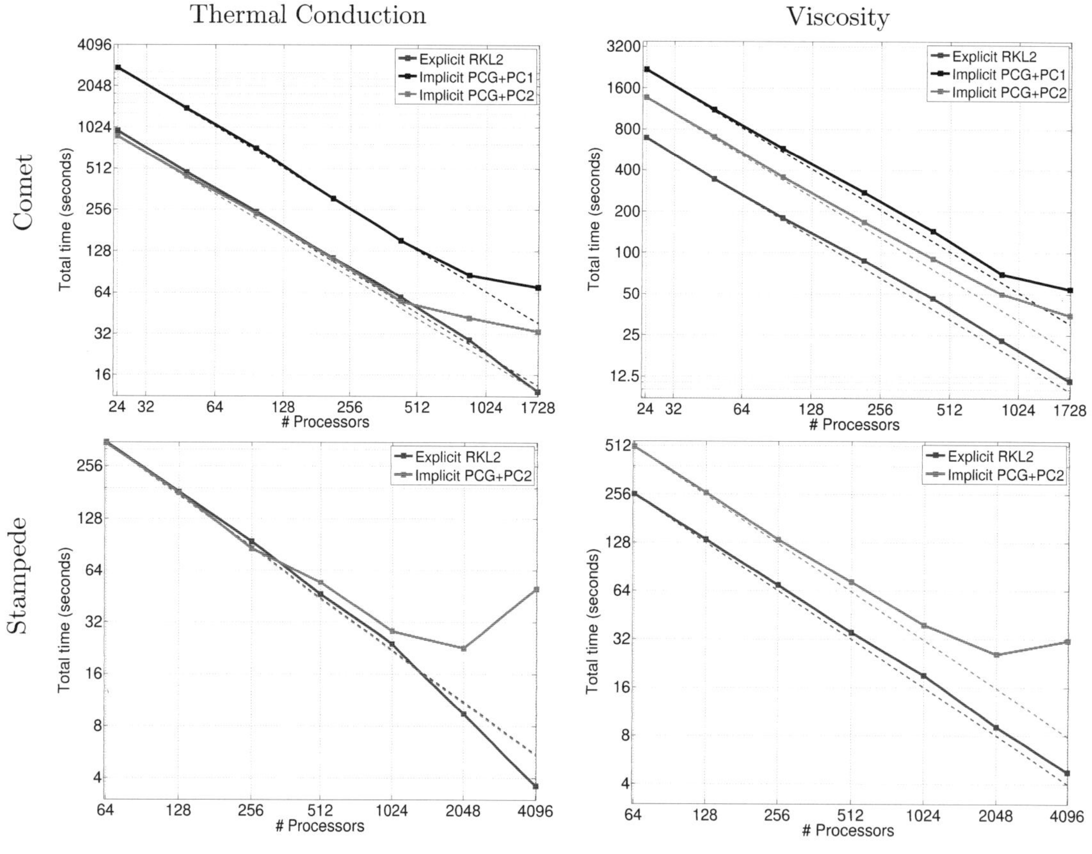

Figure 10. Timing results for the thermal conduction (left) and viscosity (right) portions of the simulations computed on Comet (top) and Sampede (bottom).

PC2 up to the point that the scaling starts to diminish. This is so even though it is using many more sub-steps than the number of iterations of BE+PCG PC2 (see Fig. 11). This shows that one sub-step of the RKL2 method is much faster than one iteration of the PCG PC2.

Using the maximum number of processors on a standard allocation, the thermal conduction using RKL2 exhibited a speedup of over 2.5X for Comet and 13.7X for Stampede compared to the BE+PCG PC2 method. For viscosity, the RKL2 is at least twice as fast as BE+PCG PC2 for all processor counts, and due to its better scaling, reaches a maximum speed increase over BE+PCG PC2 of $\approx$ 3X on Comet and over 6.5X on Stampede.

Another interesting observation is that it appears that the thermal conduction operator does not scale as well as the viscosity operator when using BE+PCG. One possible reason for this is that the number of iterations required for the PCG solver increases as the processor count increases (due to the degradation of the PC2 – see Sec. 2.3) much more so for thermal conduction than it does for viscosity.

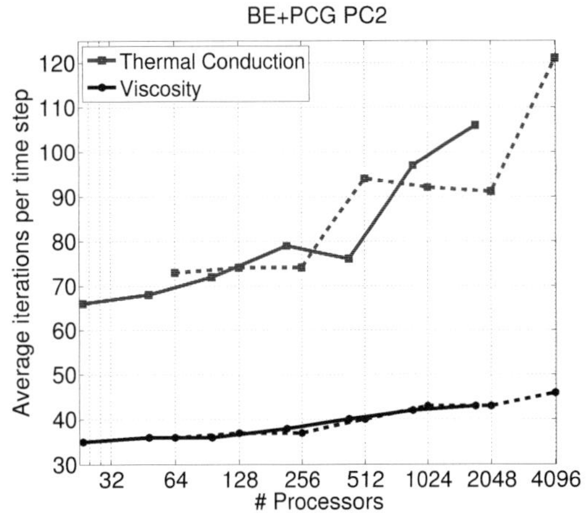

Figure 11. Average number of iterations for the BE+PCG PC2 method applied to thermal conduction and viscosity for runs on Comet (solid line) and Stampede (dashed line). The RKL2 method takes an average of 195 sub-steps for thermal conduction and 49 for viscosity.

It is informative to explore the communication (or 'non-compute') time of the runs, which we display (per iteration/sub-step) in Fig. 12. We have separately recorded the non-compute time for both local (peer-to-peer) and global communications. We see that the peer-to-peer non-compute time for the RKL2 method is somewhat higher than for the PCG methods (which is unexpected as they both use the same matrix-multiply routines). As the processor count increases, this non-compute time decreases (due to the reduction in the number of transferred data points) as expected. However, for the PCG methods, the global non-compute time (which is always larger than the point-to-point time) stays relatively constant for a while, but then starts to increase as the processor count gets high. For Stampede, this effect is the greatest. Since at the time these runs were performed, Stampede was much more oversubscribed then Comet, the network congestion could be a cause of this drastic increase. In either case, we see that the RKL2's lack of global communications is key to its improved scaling. Through testing using MPI barriers, we have found that it is not the AllReduce MPI routine itself in the dot products that is the main cause of the PCG's lack of scaling, but rather it is the routine's global synchronization acting as a barrier, causing all imbalances encountered before the routine to 'pile up'. As the amount of computational work per processor decreases, the effect of these imbalances increases.

Finally, although our focus is on the timing of the thermal conduction and viscosity operators, in Fig. 13 we show the total wall time of the MAS code for the Stampede runs of Fig. 10 and what percentage of the total run time each part of the MAS code contributed to (based on the validation runs of Sec. 6).

It is interesting to note that although the parabolic operators used to test the methods only accounted for $\sim$ 30% of the code's run time when using BE-PCG PC2, the effect of using RKL2 seems to improve scaling to the extent that the overall code exhibits a speedup of over 3X for 4096 cores on Stampede. Why this is the case is still unclear, but may have to do with easing the load on the infiniband buffers, and/or allowing the load imbalance to equalize between operators.

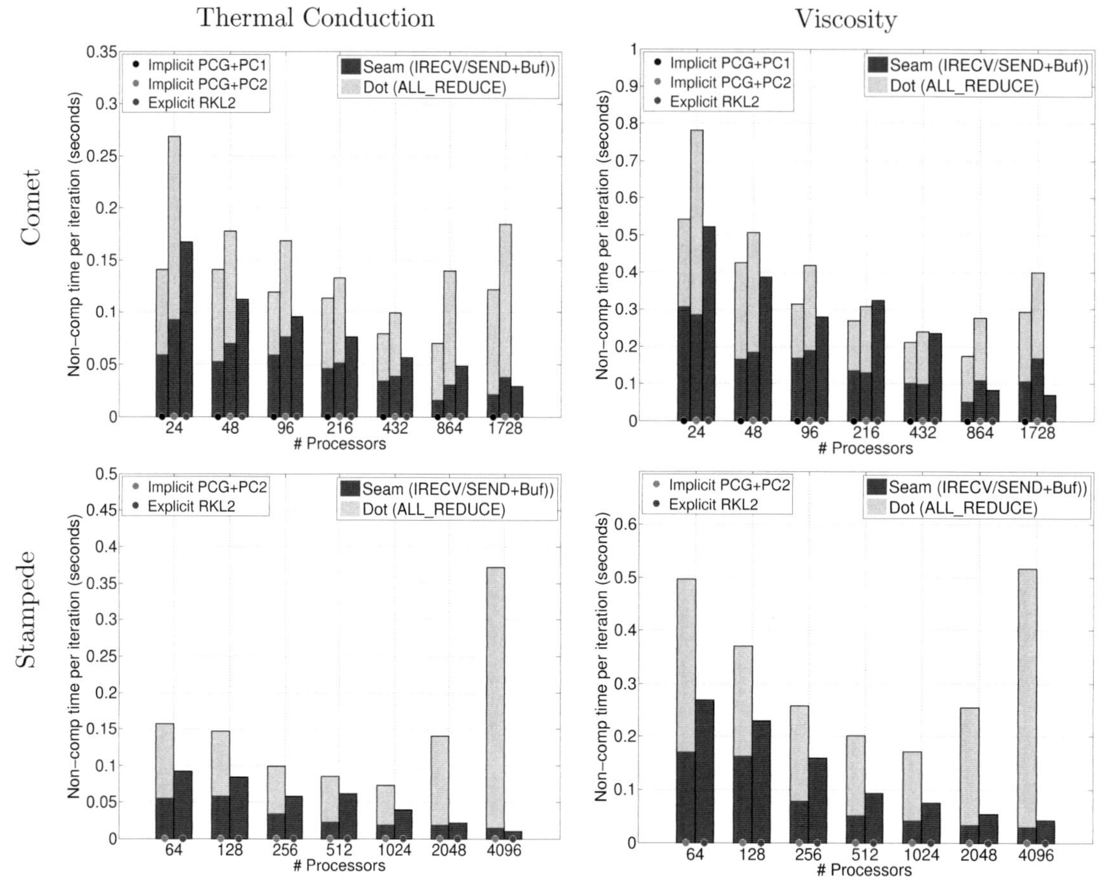

Figure 12. Non-compute time per iteration/sub-step for thermal conduction (left) and viscosity (right) portions of the simulations computed on Comet (top) and Stampede (bottom). The method used for each bar is indicated by the colored dots along the horizontal axis.

Time Breakdown from Relaxation	
Semi-implicit Pred+Corr	46%
Viscosity	16%
Thermal Conduction	14%
Alfvén Waves	19%
Resistivity	1%
Rest of code	4%

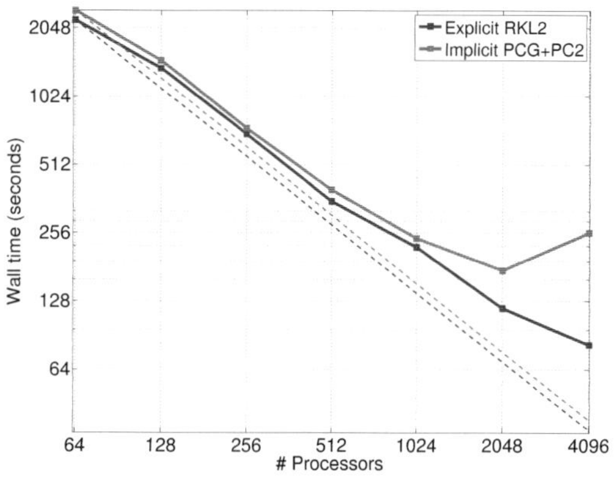

Figure 13. Left: Breakdown of the computational cost of each section of the MAS code for the validation run of Sec. 6 using BE+PCG PC2. Right: Total wall time for the timing runs computed on Stampede for the two methods. Even though the two parabolic operators tested only accounted for 30% of the wall time in the BE+PCG PC2 runs, the wall time when using RKL2 on 4096 cores is reduced by over 3X.

8. Conclusions

In this paper we have compared a representative algorithm from two classes of methods for integrating parabolic operators past the standard explicit time-step limits in the context of thermodynamic MHD simulations of the solar corona. The first method is an implicit scheme solved with a Krylov solver: the backward-Euler scheme with the preconditioned conjugate gradient (BE+PCG) with either a diagonal (PC1) or incomplete LU (PC2) preconditioner. The other is a super time-stepping method: the second-order Runge-Kutta Legendre (RKL2) scheme. The methods were described in detail and tested on a production-level coronal relaxation simulation of the described MAS code. Special attention was given to the performance and scaling of the methods.

We found for our specific case, that the RKL2 scales much better for high core counts than the BE+PCG methods, and matches or exceeds the performance for low numbers of cores. For integrating viscosity, the RKL2 method performed over twice as fast as the BE+PCG. In other tests (not reported here), the speedup of the RKL2 method was even larger (over 3X). The global synchronization of the dot products in the BE+PCG method were found to be the most likely cause of the poor scaling at high core counts, since any load imbalance would be bottle-necked there.

Although the RKL2 method has superior performance and scaling, we found that, when used for viscosity, it exhibited slowly-growing grid-level oscillations in a few localized portions of the domain. An analysis of the methods showed that these oscillations (generated by other parts of the model due to the presence of unresolved structures) are not adequately damped by the RKL2 method due to its high amplification factor for large wave modes. Some methods of avoiding these errors were discussed, but each would slow down the algorithm (although they would not adversely affect the scaling). Using RKL2 for only the thermal conduction operator yielded correct results, implying that it may be an advantageous method, but limited to specific use-cases. Our analysis also showed that the BE+PCG method, while not immune to the same problem, is much more robust due to it being L-stable.

Future work will include searching for ways to modify the RKL methods to alleviate the damping issues discussed above. As for the Krylov solvers, investigation into new non-blocking PCG methods has the potential to improve their scaling. Another interesting direction would be to formulate hybrid schemes, which combine the advantages of the multiple methodologies. Research into the application of STS-type methods to wave equations is of particular interest, as it could yield a scalable alternative to semi-implicit operators.

Acknowledgments

Work at Predictive Science was supported by AFOSR, NASA, and NSF. Computational resources at the Texas Advanced Computing Center (TACC) and the San Diego Supercomputer Center (SDSC) were provided by the NSF-supported XSEDE program. The ADAPT map was produced collaboratively between AFRL and NSO/NISP.

References

[1] Priest E 2014 *Magnetohydrodynamics of the Sun* (Cambridge University Press)
[2] Simpson M J and Landman K A 2008 *Mathematics and Computers in Simulation* **77** 9–21
[3] Karahan H 2007 *Advances in Engineering Software* **38** 80–86
[4] Chin S A 2014 *Numerical Methods for Partial Differential Equations* **30** 1961–1983
[5] Verwer J G 1996 *Applied Numerical Mathematics* **22** 359–379
[6] Jardin S C 2012 *Journal of Computational Physics* **231** 822–838
[7] Kelley C T 2003 *Solving nonlinear equations with Newton's method* vol 1 (Siam)
[8] Meyer C D, Balsara D S and Aslam T D 2012 *Monthly Notices of the Royal Astronomical Society* **422** 2102–2115
[9] Meyer C D, Balsara D S and Aslam T D 2014 *Journal of Computational Physics* **257** 594–626

[10] Galligani E 2012 *Advances in Difference Equations* **2012** 1–21
[11] Alexiades V, Amiez G and Gremaud P A 1996 *Communications in numerical methods in engineering* **12** 31–42
[12] Commercon B, Teyssier R, Audit E, Hennebelle P and Chabrier G 2011 *Astronomy & Astrophysics* **529** A35
[13] Saad Y 2003 *Iterative methods for sparse linear systems* (Siam)
[14] Ashby S F, Holst M J, Manteuffel A and Saylor P E 2001 *BIT Numerical Mathematics* **41** 26–52
[15] Benzi M 2002 *Journal of computational Physics* **182** 418–477
[16] Chow E and Saad Y 1997 *Journal of Computational and Applied Mathematics* **86** 387–414
[17] Briggs W L, McCormick S F *et al.* 2000 *A multigrid tutorial* (Siam)
[18] Shen C and Zhang J 2002 *Parallel Computing* **28** 1451–1475
[19] Shin D H, Kim D H and Song M S 1994 *Communications of the Korean Mathematical Society* **9** 439–448
[20] Eller P Nov 15-20, 2015 *Poster session presented at: Super Computing 2015* (Austin, Texas) URL `http://sc15.supercomputing.org/sites/all/themes/SC15images/src_poster/src_poster_pages/spost116.html`
[21] O'Sullivan S 2015 *Journal of Computational Physics* **300** 665–678
[22] O'Sullivan S and Downes T P 2007 *Monthly Notices of the Royal Astronomical Society* **376** 1648–1658
[23] Lee D, Xia G, Daley C, Dubey A, Gopal S, Graziani C, Lamb D and Weide K 2011 *Astrophysics and Space Science* **336** 157–162
[24] Mignone A, Bodo G, Massaglia S, Matsakos T, Tesileanu O, Zanni C and Ferrari A 2007 *The Astrophysical Journal Supplement Series* **170** 228
[25] Arber T, Longbottom A, Gerrard C and Milne A 2001 *Journal of Computational Physics* **171** 151–181
[26] Arber T D, Longbottom A W, Gerrard C L and Milne A M 2012 Lare3d: Lagrangian-Eulerian remap scheme for MHD Astrophysics Source Code Library (*Preprint* `1208.015`)
[27] Johnston C July 21-23, 2015 *Poster session presented at: Coronal Loop Workshop VII* (University of Cambridge, England) URL `http://www-solar.mcs.st-andrews.ac.uk/\textasciitilde{}cdj3/Talksposters/CJLoopsPoster2015.pdf`
[28] Lionello R, Linker J A and Mikić Z 2009 *The Astrophysical Journal* **690** 902
[29] Mikić Z, Linker J A, Schnack D D, Lionello R and Tarditi A 1999 *Phys. Plasmas* **6** 2217
[30] Linker J A, Mikić Z, Biesecker D A, Forsyth R J, Gibson S E, Lazarus A J, Lecinski A, Riley P, Szabo A and Thompson B J 1999 *Journal of Geophysical Research: Space Physics* **104** 9809–9830
[31] Downs C, Linker J A, Mikić Z, Riley P, Schrijver C J and Saint-Hilaire P 2013 *Science* **340** 1196–1199
[32] Lionello R, Riley P, Linker J A and Mikić Z 2005 *The Astrophysical Journal* **625** 463–473
[33] Lionello R, Linker J A, Mikić Z and Riley P 2006 *The Astrophysical Journal Letters* **642** L69
[34] Linker J A, Lionello R, Mikić Z, Titov V S and Antiochos S K 2011 *The Astrophysical Journal* **731** 110
[35] Linker J A, Mikić Z, Lionello R, Riley P, Amari T and Odstrcil D 2003 *Phys. Plasmas* **10** 1971
[36] Lionello R, Downs C, Linker J A, Török T, Riley P and Mikić Z 2013 *The Astrophysical Journal* **777** 76
[37] Riley P, Linker J A, Lionello R and Mikic Z 2012 *Journal of Atmospheric and Solar-Terrestrial Physics* **83** 1–10
[38] Hollweg J V 1976 *Journal of Geophysical Research* **81** 1649–1658
[39] Athay R G 1986 *Astrophysical Journal* **308** 975–981
[40] Linker J A, Lionello R, Mikić Z and Amari T 2001 *Journal of Geophysical Research: Space Physics (1978–2012)* **106** 25165–25175
[41] Veldman A and Rinzema K 1992 *Journal of engineering mathematics* **26** 119–130
[42] Jahne B, Geissler P and Haussecker H (eds) 1999 *Handbook of Computer Vision and Applications Volume 2* 1st ed (San Francisco, CA, USA: Morgan Kaufmann Publishers Inc.) ISBN 0123797705
[43] Lionello R, Mikić Z and Linker J A 1999 *Journal of Computational Physics* **152** 346–358
[44] Bell N and Garland M 2008 Efficient sparse matrix-vector multiplication on cuda Tech. rep. Nvidia Technical Report NVR-2008-004, Nvidia Corporation
[45] Smith B and Zhang H 2011 *International Journal of High Performance Computing Applications* **25** 386–391
[46] Rušin V, Druckmüller M, Aniol P, Minarovjech M, Saniga M, Mikić Z, Linker J, Lionello R, Riley P and Titov V 2010 *Astronomy & Astrophysics* **513** A45
[47] Hickmann K S, Godinez H C, Henney C J and Arge C N 2015 *Solar Physics* **290** 1105–1118 ISSN 1573-093X URL `http://dx.doi.org/10.1007/s11207-015-0666-3`
[48] Arge C N, Henney C J, Koller J, Compeau C R, Young S, MacKenzie D, Fay A and Harvey J W 2010 Air force data assimilative photospheric flux transport (adapt) model Tech. rep. DTIC Document
[49] Jian L, MacNeice P, Taktakishvili A, Odstrcil D, Jackson B, Yu H S, Riley P, Sokolov I and Evans R 2015 *Space Weather* **13** 316–338
[50] Knabner P and Angermann L 2003 *Texts in Applied Mathematics* **44**
[51] Varga R 2004 Gershgorin and his circles

[52] Caplan R M and Carretero-Gonzlez R 2013 *Applied Numerical Mathematics* **71** 24–40
[53] Scott D S 1985 *Linear algebra and its applications* **65** 147–155

Appendix A

The following table depicts the pseudo-code for formulating and applying the diagonal (PC1) and ILU0 (PC2) preconditioners used for testing the BE+PCG method of Sec. 2.

	Formulation ($\mathbf{P}$)	Application ($z = \mathbf{P}^{-1} r$)

PC1	`do i = 1 : N`
	$P_{ii} = A_{ii}$
	`enddo`
		`do i = 1 : N`
		$z_i = r_i / P_{ii}$
		`enddo`

PC2	$\mathbf{LU} = \mathbf{A}$
	`do i = 2 : N`
	`do k = 1 : i - 1` $(LU_{ik} \neq 0)$
	$LU_{ik} = LU_{ik} / LU_{kk}$
	`do j = k + 1 : N` $(LU_{ij} \neq 0)$
	$LU_{ij} = LU_{ij} - LU_{ik}\, LU_{kj}$
	`enddo`
	`enddo`
	`enddo`
	$\mathbf{P} = \mathbf{LU}$
		`do i = 1 : N`
		$z_i^* = r_i$
		`do j = 1 : i` $(LU_{ij} \neq 0)$
		$z_i^* = z_i^* - LU_{ij}\, z_j^*$
		`enddo`
		`enddo`
		`do i = N : 1`
		$z_i = z_i^*$
		`do j = i + 1 : N` $(LU_{ij} \neq 0)$
		$z_i = z_i - LU_{ij}\, z_j$
		`enddo`
		$z_i = z_i / LU_{ii}$
		`enddo`

Appendix B

In order to compute the minimum required number of sub-steps for the RKL2 method of Sec. 3, one requires a (preferably close) estimate for the Euler stability time-step:

$$\Delta t_{\text{Euler}} = 2/|\lambda|_{\max}, \tag{12}$$

where $|\lambda|_{\max}$ is the largest absolute value eigenvalue of the matrix $\mathbf{M}$ of Eq. (3). For finite difference methods, an upper bound on $|\lambda_{\max}|$ can be given as

$$|\lambda_{\max}| < \max(\alpha\, \tilde{k}^2),$$

where α is the diffusion coefficient, and $\tilde{k}$ is the inverse discrete grid element (e.g. for spherical coordinates, $\tilde{k}^2 \approx 4\,[1/\Delta r^2 + 1/(r\,\Delta\theta)^2 + 1/(r\,\sin(\theta)\,\Delta\phi)^2]$. However, this often leads to a very large under-estimate of Δt_{Euler}, especially when using anisotropic diffusion.

A much better estimate can easily be made by the sometimes overlooked method of using Gershgorin disk theorem [51] as was done in Ref. [52]. A Gershgorin disk is defined as

Definition 1 *Given a square matrix* $\mathbf{A}$, *a Gershgorin disk for every row* j *is defined as a disk in the complex plane, centered at* A_{jj}, *with a radius of:*

$$R_j = \sum_{i \neq j}^{N} |A_{i,j}|.$$

and the following theorem applies to $\mathbf{A}$:

IOP Conf. Series: Journal of Physics: Conf. Series **837** (2017) 012016 doi:10.1088/1742-6596/837/1/012016

Theorem 1 *Every eigenvalue λ of a square matrix $\mathbf{A}$ lies in one of its Gershgorin disks:*

$$|\lambda - A_{jj}| \leq R_j$$

.

Since in our case, $\lambda \in \mathcal{R}$ and $A_{jj} \in \mathcal{R}$, we get a bound on $|\lambda_j|$ for each row j as $|A_{jj}| - R_j \leq |\lambda_j| \leq |A_{jj}| + R_j$. Therefore the maximum value of $|\lambda|$ is bound by:

$$|\lambda|_{\max} \leq \max\left\{\sum_{i=1}^{N} |A_{i,j}|, \forall j \text{ rows}, \right\}. \tag{13}$$

Since the matrix coefficients are already explicitly formed for use in the solvers/integrators, Eq. (13) is trivial to calculate.

The stability bound on Δt_{Euler} found by using Eq. (13) is often much closer to the true bound than the method using $\tilde{k}$ above. The accuracy of the bound for real symmetric matrices was shown in Ref. [53] to be within a factor of $\sqrt{p}$ of the true bound, where p is the maximum number of non-zero values in the rows of $\mathbf{A}$. For the cases tested in this paper, $p = 15$ for viscosity and $p = 19$ for thermal conduction. However, we note that for many cases of finite difference operators, the Gershgorin bounds can be much closer to the true bounds than the maximum error of $\sqrt{p}$ implies [52].

IOP Conf. Series: Journal of Physics: Conf. Series **837** (2017) 012017 doi:10.1088/1742-6596/837/1/012017

AMITIS: A 3D GPU-Based Hybrid-PIC Model for Space and Plasma Physics

Shahab Fatemi[1,3], **Andrew R. Poppe**[1,3], **Gregory T. Delory**[1,3], **William M. Farrell**[2,3]

(1) Space Sciences Laboratory, University of California, Berkeley, California, USA, (2) NASA Goddard Space Flight Center (GSFC), Greenbelt, Maryland, USA, (3) Solar System Exploration Research Virtual Institute (SSERVI), NASA Ames Research Center, Moffett Field, California, USA

E-mail: shahab@ssl.berkeley.edu

Abstract. We have developed, for the first time, an advanced modeling infrastructure in space simulations (AMITIS) with an embedded three-dimensional self-consistent grid-based hybrid model of plasma (kinetic ions and fluid electrons) that runs entirely on graphics processing units (GPUs). The model uses NVIDIA GPUs and their associated parallel computing platform, CUDA, developed for general purpose processing on GPUs. The model uses a single CPU-GPU pair, where the CPU transfers data between the system and GPU memory, executes CUDA kernels, and writes simulation outputs on the disk. All computations, including moving particles, calculating macroscopic properties of particles on a grid, and solving hybrid model equations are processed on a single GPU. We explain various computing kernels within AMITIS and compare their performance with an already existing well-tested hybrid model of plasma that runs in parallel using multi-CPU platforms. We show that AMITIS runs ~10 times faster than the parallel CPU-based hybrid model. We also introduce an implicit solver for computation of Faraday's Equation, resulting in an explicit-implicit scheme for the hybrid model equation. We show that the proposed scheme is stable and accurate. We examine the AMITIS energy conservation and show that the energy is conserved with an error $< 0.2\%$ after 500,000 timesteps, even when a very low number of particles per cell is used.

1. Introduction

Similar to many other fields of science and engineering, the complex nature of the problems encountered in space and plasma physics has demonstrated the need for powerful computer simulations [4, 28]. One computational approach that has been widely used is the kinetic particle-based models. In contrast to magnetohydrodynamics (MHD), kinetic models consider a more detailed analysis of plasma through particle interactions with electromagnetic fields [4]. However, developing a fully kinetic particle-in-cell (PIC) model is computationally intensive, especially for three-dimensional (3D) modeling in space plasma physics and planetary science [57]. In many plasma systems where the ion temporal and spatial scales are of primary interest, hybrid models provide a compromise, and have been successfully applied to a broad range of problems in space and plasma physics [37, 38]. A typical scheme of hybrid models consider ions as kinetic particles, while electrons are a charge neutralizing fluid. Electromagnetic fields are treated in the low-frequency Darwin approximation with the standard assumption of plasma quasi-neutrality ($n_i \approx n_e$, where n is plasma density and sub-index i and e denote ions and electrons, respectively) [26, 38, 58].

In the last two decades, significant effort has been expended on designing and optimizing high performance hybrid models using multi-CPU parallel programming [7, 29, 43]. However, very few hybrid models exist today that are 3D in both spatial and velocity spaces, fully self-consistent, efficiently parallelized, and have genuine impact and sustained productivity in the space and plasma physics community (e.g., Hybrid-FLASH [29], Halfshell [6], and A.I.K.E.F [43]). These models, similar to any other CPU-based parallel model (e.g., iPIC3D in PIC modeling [39] and BATS-R-US in MHD modeling [23]), require the use of large-scale expensive super-computers, which are not generally affordable and/or readily accessible for every desired simulation. Furthermore, slow communication speeds between individual CPU processors, mainly through networks, introduce significant latencies, limiting the performance of any CPU-based model. Thus, CPU-based models face persistent difficulties in tackling new and/or higher fidelity applications, which limits our ability to make fundamental advances.

Over the last few years, graphics processing units (GPUs) have initiated a revolution in heterogeneous parallel computing, and they are exponentially being used for general purpose applications [33, 44]. The large gap in the performance (shown in Figure 1), availability, capital cost, and power consumption between GPUs and CPUs are the main reasons to migrate from CPU-based towards GPU-based models. The advantages in using GPUs compel us to take a step forward in the field of space and plasma simulation by developing a modeling infrastructure in space simulations (AMITIS) with an embedded three-dimensional (3D in space, velocity, and electromagnetic fields) hybrid plasma solver that runs on GPUs.

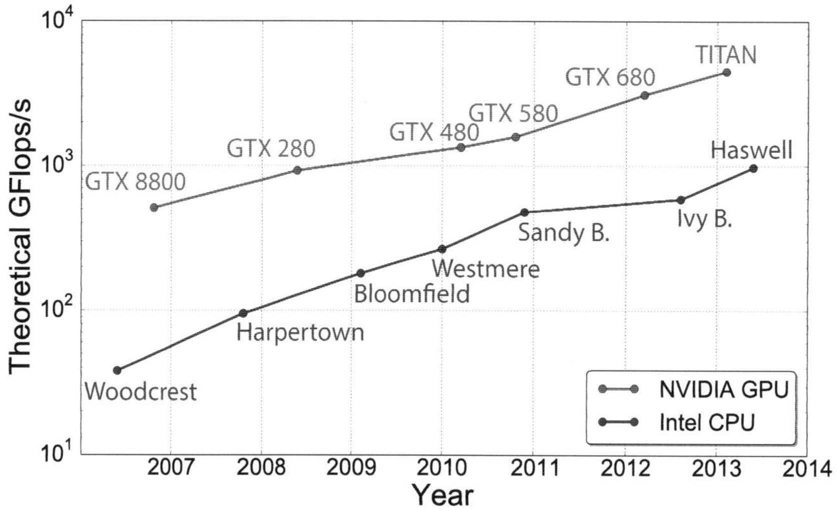

Figure 1. Performance gap between GPUs and CPUs, adopted from www.nvidia.com.

Recently, a few GPU-based models have been developed in plasma physics. However, these models, by the time of this writing, are PIC [1, 8, 12, 15, 16], fluid (MHD) [20, 54], or Vlasov-Boltzmann models [50]. In addition, some of these models are developed in 1D, 2D, and 2.5D [15, 16, 35], and only a few of them are 3D [8, 20]. Here, we present development of the first GPU-based 3D self-consistent hybrid model of plasma for space plasma physics and planetary science, AMITIS, that runs entirely on GPUs. In Section 2 we briefly explain NVIDIA's GPU architecture and compare it with CPUs. Then, in Section 3 we describe hybrid models in detail.

In Section 4 we explain various sections of the AMITIS hybrid model, and in Section 5 provide some of the standard test results for our model and examine its performance, stability, accuracy, and energy conservation.

2. NVIDIA's GPU Architecture

CPUs and GPUs have fundamental differences in their hardware and software architectures. Figure 2 shows a simplified diagram comparing CPU and GPU architectures. A CPU is optimized to enhance the performance of a single thread by using sophisticated control logic while taking advantage of cached memories to reduce any data access latencies [33]. A GPU, in contrast, consists of many cores optimized for parallel tasks, and in total is capable of handling thousands of light-weight threads simultaneously to maximize throughput for parallel programs [33]. Today, a typical CPU with four quad cores can at most run 32 threads concurrently, while a typical NVIDIA GPU runs several thousand active threads concurrently. GPUs also have significantly different memory architecture than CPUs. GPUs contain a hierarchy of progressively smaller, yet faster memory, each of which can serve a particular function in any GPU-based model. The GPU memory types available in NVIDIA devices, listed from the smallest size but the fastest to the largest but slowest, are: registers, shared memory, local memory, constant memory, texture memory, and global memory [11, 55]. In contrast to CPUs, this complex GPU memory hierarchy requires any programmer to engage in sophisticated and intricate memory management in order to achieve high levels of performance in any GPU-based model [9]. Indeed, utilizing memory bandwidth efficiently is critical to algorithm implementation on GPUs [55]. Due to the fundamental differences between GPUs and CPUs, developing a 3D self-consistent GPU-based particle framework and a hybrid plasma field solver requires advanced algorithms and intensive code development, optimization, testing, and diagnostics, completely different from those developed for CPUs.

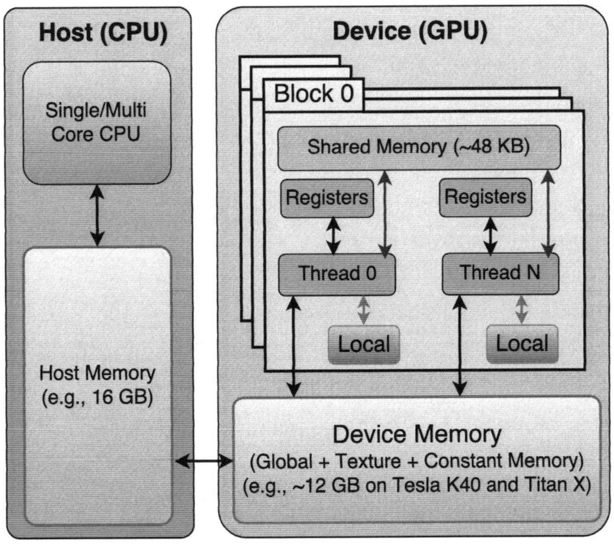

Figure 2. A simplified diagram comparing CPU and GPU architectures.

3. Hybrid Model Description

Hybrid modeling of plasma is a self-consistent kinetic modeling approach that involves solving Maxwell's equations, where a part of plasma is considered as kinetic particles and the rest is a fluid. Depending on the spatial and temporal scales of problems, various types of hybrid models can be defined [56, 58]. One of the most common types that has a broad range of applications in space and plasma physics considers positively charged ions as kinetic particles whereas electrons are a charge-neutralizing fluid. In this type of hybrid model phenomena occurring at ion inertial and gyroradius scales can be resolved. Examples includes ion-driven plasma waves and instabilities, and plasma interaction with the surface and magnetosphere of objects with sizes on the order of a few ion inertial lengths and the incident ion gyroradius. Figure 3 shows some of the solar system objects covering a broad spectrum of ion length scales. The areas covered with green, blue, and red are suitable for fully kinetic PIC, hybrid, and magnetohydrodynamics (MHD) modeling, respectively. Objects that have a radius (R_{obj}) much smaller than proton gyroradius (r_{gyro}) and much smaller than ion inertial length (δ_i) are mainly suitable for fully kinetic PIC modeling (e.g., Phobos, Deimos, Gaspra, and Itokawa). The objects with radius larger or comparable to the ion gyroradius and the ion inertial length are more suitable for hybrid modeling with kinetic ions and fluid electrons (e.g., Moon, Mars, Callisto, and Rhea). However, if the ion gyroradius and inertial length become too small compared to the size of an object, hybrid models are computationally too expensive and MHD models are more applicable, as long as the kinetic behavior of the ions is not of interest and/or the object does not have a substantial exosphere or ionosphere (which can introduce significantly non-Maxwellian populations unsuitable for a MHD description). We note that defining sharp boundaries between the characteristic scales that each plasma model can resolve is not straightforward; however, the limits, advantages, and disadvantages of any model applied to any problem should be accurately and correctly understood and appreciated.

Generally in PIC and/or hybrid plasma solvers there are a number of basic steps in the calculations that have to be made. These steps are illustrated in Figure 4 and they are as follows [57]:

- **Initialization** where the initial plasma particles and field parameters, simulation domain geometry, and boundary conditions are defined;
- **Solver** where the positions and velocities of all particle species are advanced in time with a relatively small timestep (Δt). Then, macroscopic properties of particles are mapped into grid (if the model is grid-based), fields are solved, and finally, the forces acting on particles are calculated. These steps are the core for every particle solver and are repeated until a final time and/or steady-state solution is achieved; and,
- **Output** where the simulation results are stored in files and the results are analyzed through appropriate diagnostics and visualization tools.

3.1. Hybrid Model Approximations and Equations

Several approximations and assumptions are made in different types of hybrid models. Some of which are rather common in all hybrid models, including our model presented here, and they are all explained in detail in previous literature [10, 26, 37, 38]. Here we only list them without further discussion, but interested readers are encouraged to read the aforementioned references.

- Quasi-neutrality: The total plasma charge density (ρ) is zero.

$$\rho = \rho_e + \rho_{\rm I} = 0, \tag{1}$$

where ρ_e denotes electron charge densities, and $\rho_{\rm I}$ is the total ion charge density defined as

$$\rho_{\rm I} = \sum_{i=1}^{N} \rho_i \tag{2}$$

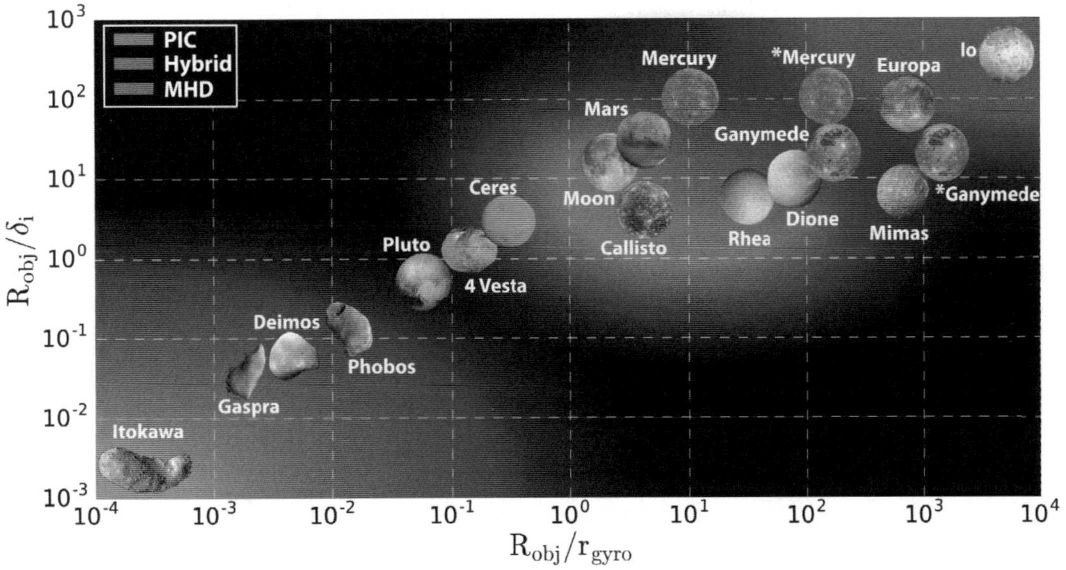

Figure 3. The ratio between the radius of a few solar system objects without any dense atmosphere (R_{obj}) and the typical plasma gyroradius (r_{gyro}) and ion inertial length (δ_i) near those objects. The areas covered with green, blue, and red are suitable for fully kinetic particle-in-cell (PIC), hybrid, and magnetohydrodynamics (MHD) modeling, respectively. The brighter the background, the more feasible those objects are for their associated plasma model. We have included the objects intrinsic magnetic field for those that have an asterisk prior to their name, i.e., Mercury and Ganymede, otherwise only the magnetic field from their surrounding environment has been considered in the ion gyroradius calculation.

where ρ_i denotes ion charge densities, and N is the number of ion species. The quasi-neutrality assumption (Equation 1) removes most electrostatic instabilities and is only valid for grid cells larger than the plasma Debye length λ_D.

- Darwin approximation: The transverse displacement current is neglected from Ampére's law. Thus, Ampére's law is reduced to

$$\mathbf{J} = \frac{\nabla \times \mathbf{B}}{\mu_0}, \tag{3}$$

where $\mathbf{J}$ is the total current density and $\mathbf{B}$ is the magnetic field. Then the electron current density can be written as

$$\mathbf{J}_e = \mathbf{J} - \mathbf{J}_\mathrm{I}, \tag{4}$$

where $\mathbf{J}_\mathrm{I}$ is the total ion current densities defined as

$$\mathbf{J}_\mathrm{I} = \sum_{i=1}^{N} \mathbf{J}_i, \tag{5}$$

where N is the number of ion species.

- Massless electrons ($m_e = 0$): The electron momentum equation with $m_e = 0$ gives an

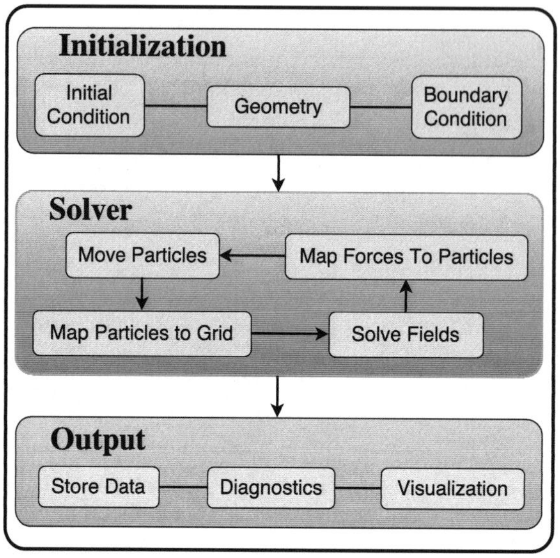

Figure 4. Basic steps in grid-based kinetic plasma simulations inspired from [57].

explicit expression for electric field, $\mathbf{E}$,

$$m_e \frac{d(n_e \mathbf{u}_e)}{dt} = 0 = +\rho_e \mathbf{E} + \mathbf{J}_e \times \mathbf{B} - \nabla \cdot \overline{\mathbf{P}}_e - \rho_e \eta \mathbf{J}, \tag{6}$$

where $\overline{\mathbf{P}}_e$ is the electron pressure tensor and η is resistivity. By replacing $\mathbf{J}$ from Equation 3, $\mathbf{J}_e$ from the expression in Equation 4, and ρ_e from plasma quasi-neutrality (Equation 1), a straightforward solution is obtained for the generalized Ohm's law to calculate the electric field

$$\mathbf{E} = \underbrace{-\frac{\mathbf{J}_\mathrm{I} \times \mathbf{B}}{\rho_\mathrm{I}}}_{\text{convective term}} + \overbrace{(\frac{\nabla \times \mathbf{B}}{\mu_0}) \times \frac{\mathbf{B}}{\rho_\mathrm{I}}}^{\text{Hall term}} - \underbrace{\frac{\nabla \cdot \overline{\mathbf{P}}_e}{\rho_\mathrm{I}}}_{\text{ambipolar term}} + \underbrace{\frac{\eta}{\mu_0} \nabla \times \mathbf{B}}_{\text{Ohmic term}}. \tag{7}$$

The commonly used names for the various electric field terms in Equation 7 are indicated for future reference and ρ_I is defined in Equation 2.

- Electron pressure: Various approaches are taken to treat the electron pressure in Equation 7. As reviewed in [38] some hybrid models assume the electron pressure, and consequently ambipolar electric field term, to be constant or entirely ignored. In some models, the electron pressure is assumed to be isotropic ($\nabla \cdot \overline{\mathbf{P}}_e = \nabla p_e$), where p_e is the scalar electron pressure. Some models determine the electron pressure tensor from the gyrotropic lowest order distributions and reduce it to a diagonal form $\overline{\mathbf{P}}_e = p_\perp (I - \hat{b}\hat{b}) + p_{||} \hat{b}\hat{b}$, where $\hat{b} = \mathbf{B}/|\mathbf{B}|$. A few models also consider the full evolution of the electron pressure tensor derived from Vlasov's equation.
  In the current version of AMITIS, we assume the electron pressure $p_e = n_e k_b T_e$ to be adiabatic, as explained by [29, 19]. The electron pressure is then related to the electron charge density by $p_e \propto |\rho_e|^\gamma$, where γ is the adiabatic index.
- Faraday's law is used to advance the magnetic field in time,

$$\frac{\partial \mathbf{B}}{\partial t} = -\nabla \times \mathbf{E}, \tag{8}$$

where the electric field, $\mathbf{E}$, is obtained from Equation 7 and contains no unknowns, thus can be solved straightforwardly. After the electric field is calculated, Faraday's law (Equation 8) is solved to advance the magnetic field, $\mathbf{B}$, in time.

The equation of motion for a charged particle with mass m_i and charge q_i at position $\mathbf{r}_i$ and velocity $\mathbf{v}_i$ is

$$\frac{d\mathbf{r}_i}{dt} = \mathbf{v}_i, \quad \frac{d\mathbf{v}_i}{dt} = \frac{q_i}{m_i}(\mathbf{E} + \mathbf{v}_i \times \mathbf{B}), \tag{9}$$

which is an ordinary differential equation.

For this type of hybrid model where the ions are kinetic particles and electrons are a massless charge-neutralizing fluid, modelers are often asked about the treatment of the electron fluid. It is worth mentioning that no flowing "fluid" exists in this type of hybrid models in a way that the fluids are treated in the MHD models, and no continuity equation is required to be explicitly solved for the electron fluid. Hybrid models take advantage of the massless electron fluid concept to calculate the electric field, as shown in Equation 6. However, if so desired, one can estimate the electron current density from Equation 4, and find the electron bulk flow velocity, $\mathbf{u}_e$, through the quasi-neutrality assumption as

$$\mathbf{u}_e = \frac{\mathbf{J}_e}{\rho_e} = \frac{\mathbf{J} - \mathbf{J}_\mathrm{I}}{-\rho_\mathrm{I}}. \tag{10}$$

However, since quasi-neutrality is one of the core assumptions in hybrid models, wherever a positively charged ion exists throughout the simulation domain, a cloud of electrons are assumed to exist around them to maintain quasi-neutrality.

3.2. Hybrid Model Numerical Schemes

Several numerical schemes exist to discretize the hybrid model equations (Equations 7, 8, and 9) and are mostly described in [38], and reviewed by [36, 37, 56, 58]. Two of the most commonly used schemes are Predictor-Corrector (PC) and Current Advance Method and Cyclic Leapfrog (CAM-CL). A detailed description with numerical implementation of the PC scheme can be found in [26, 29], and [56], and CAM-CL is described in [3, 41]. An overview on comparing these two schemes, their advantages and disadvantages, as well as their limitations are explained by [36] and [37].

4. AMITIS Hybrid Model for GPUs

4.1. Model Infrastructure

We developed AMITIS as a 3D, self-consistent hybrid model of plasma that runs on a single CPU-GPU pair. We use the CUDA parallel computing platform and the C/C++ programming language throughout our model. We use NVIDIA GPUs because of their high processing power, high memory bandwidth, low cost and power consumption, and industry support. We use a single CPU as a "host" (shown in Figure 2) to initialize our simulations, call the GPU kernels[1], and store the results into files on the hard drive (initialization and output sections shown in Figure 4). We use a single GPU, termed the "device", for our simulations to perform all computational steps, shown inside the green box in Figure 4, to maximize the model performance.

The NVIDIA GPUs we use for our simulations provide very high throughput and memory bandwidth using single precision operations. Thus, we use single precision floating-point format throughout our model. To minimize round-off errors for single precision operations and to reduce the number of operations, we normalize input parameters to a unit system used in our model.

[1] The parallel portion of an application that is executed on a GPU is called kernel.

All particles and fields are initialized on the host and transferred to the device's global memory. To improve performance during calculations we store all the constant values to the device's constant memory, and we take the advantage of on-chip and fast registers and shared memories, when necessary (see Figure 2 for GPU's memory architecture) [11].

We call the kernels in a sequential order to move the particles and solve the hybrid equations. Since the sequence of every operation per timestep in a plasma solver is crucial, we use synchronization at any level if necessary, and only use asynchronous kernel streams to update grid boundaries and to calculate 3D current densities and fields. We transfer our simulation results from the device to the host memory after every T number of timesteps (often $T \geq 100$), and store them in a Numpy (a Python library) compressed file format. We use Python scripts to visualize our results using Matplotlib library.

Our model is grid-based and we use regular-spaced cell-centered Cartesian grids to solve our equations using finite difference approximations. We do not use adaptive (irregular) grids because of the non-physical forces and instabilities they can introduce into models [22, 53]. We stochastically inject particles into phase space, and we use inversion of cumulative density functions to form distribution functions [48], including Maxwellian, bi-Maxwellian, and Kappa for velocity distributions [40, 46] and uniform, beam-like, and Gaussian for spatial distributions [48].

4.2. Model Sections and Individual Performance Results

Due to the differences between CPU and GPU architectures, several areas of the AMITIS model required intensive research, development, and optimization. In this section, we explain the methods we used in AMITIS to numerically solve every section of the particle model shown in the green box in Figure 4, and benchmark its performance against a highly optimized similar code developed for CPUs. The CPU version of the code used in this section is not a parallel code and only uses a single CPU core. It has been intentionally developed for benchmarking and is highly optimized and well-tested to provide a fair comparison between CPU and GPU performance results. In Section 5, as noted later, we compare AMITIS performance against a well-tested parallel hybrid model than runs on multi-CPU platforms.

For all the simulation results presented in this section we used single precision operations and ran on a machine equipped with an Intel Core i7-4790 3.60GHz CPU, 16 GB DDR3 synchronous 1600 MHz system memory, and a single GeForce GTX TITAN X GPU (from now on called TITAN X) with compute capability 5.0 and 12 GB global memory. We used a single CPU core to run our CPU-based simulations and used a single CPU-GPU pair for AMITIS simulations. We ran only one simulation at a time on our benchmarking machine and made sure that during run-time for each simulation no unnecessary processes were running either on the CPU or on the GPU. We used Ubuntu 14.04 operating system, Linux hybrid 3.13.0-88-generic kernel. We used Intel compiler (icc version 16.0.3) to compile our CPU-based model, and NVIDIA's CUDA compiler (nvcc) to compile AMITIS, unless noted otherwise. Our NVIDIA GPU driver version was 352.68, and our CUDA version was 7.5. Table 1 compares some of the main specifications of Intel Core i7-4790 CPU with TITAN X GPU.

For benchmarking we loaded a 3D simulation box with $64\times64\times64$ cubic-cells of size 200 km with $\sim$17 million protons as kinetic particles (64 particles per cell), advanced their trajectories, and solved the electromagnetic fields for over 1000 timesteps. All the particles (protons) are singly charged particles with mass 1.0 amu, and they are all uniformly distributed in configuration space with a drifting Maxwellian velocity distribution of bulk speed 350 km/s, thermal speed $\sim$50 km/s, and background magnetic field 5 nT (typical solar wind parameters near the Earth). We used profilers to estimate runtime simulations and averaged them over 1000 timesteps.

	Intel Core i7-4790	**GeForce TITAN X**
Cores	4	3072
Active threads	8	49152
Core frequency (GHz)	3.60	1.0
Peak performance (GFLOPS/s)	460.8	6144
Peak mem. bandwidth (GB/s)	25.6	336.5
Power consumption TDP (W)	84	250
Power cons. per operation (W/GFLOPS)	0.18	0.04
Current price ($)	300	1000
Price per operation ($/GFLOPS)	0.65	0.16

Table 1. Comparison between Intel Core i7-4790 CPU and GeForce GTX TITAN X GPU. Presented data are taken from http://ark.intel.com and http://www.geforce.com.

4.2.1. Moving Particles Numerous methods exist to numerically solve the charged particles equation of motion (Equation 9) with various levels of accuracy and stability. The most commonly used method to integrate charged particle velocities is the Boris-Buneman method [5]. This method is an explicit scheme with second order accuracy $\mathcal{O}(\Delta t^2)$, has bounded energy error for all timesteps, and conserves phase space volume [34, 49].

We used the Boris-Buneman method to integrate particle velocities and the midpoint method to integrate particle positions with global error of order $\mathcal{O}(\Delta t^2)$. We utilized two different layouts to organize particle data: structure of arrays (SoAs) and array of aligned structures (AoaSs). Generally, many different parallel computer architectures, particularly Single Instruction Multiple Data (SIMD) style like GPUs, prefer SoAs because of its efficient coalesced access to global memory [11, 52]. However, Mei and Tian, 2016 [42] have shown that under some circumstances AoaSs can also provide acceptably high performance compared to SoAs. Figure 5 shows the SoA and AoaS data layouts we have examined in our models. In the AoaS structure we have added two padding elements (foo1 and foo2) to make a 32 bytes data structure and take advantages of aligned memory access. For consistency and to provide a fair comparison between the CPU and GPU versions of our code, we implemented and examined both of these data layouts using different compilers and algorithms. Our benchmarking results are shown in Table 2. We see the impact of compilers (gcc 4.8.4 and intel icc 2016.0.3) as well as the data layouts in the CPU version of our code. We also see that the CPU version of our code runs faster if we advance particle velocities and positions in two "separated" function calls rather than having both "together" in one function call. In contrast, the GPU version of our code runs faster when we advance particles velocity and position "together" in one single function call. This is mainly due to the fundamental differences between the process scheduler on CPUs and GPUs, indicating the importance of finding the most optimized algorithm on each device. Table 2 shows that in general, SoAs provide higher performance compared to AoaSs for both CPU and GPU versions of the particle pushing code, as expected. We also highlighted the fastest performance we achieved running our code on a single CPU core (5.43 ns/particle) and on GPU (0.27 ns/particle) using 256 threads per block. We see that the GPU version of our code provides a speed up ~20X to solve the equation of motion for charged particles.

The maximum number of threads that can be launched per block on the TITAN X is 1024. We examined various configurations of threads per block including 32, 64, 128, 256, 512, and 1024 threads per block. We did not observe noticeable changes in our performance results using any of these configurations to move the charged particles. We also found these results independent from the initial velocity and thermal speed of the particles. It is also worth noting that if we did a)

```
#define N 10000

 struct SoA{
    float   rx[N];
    float   ry[N];
    float   rz[N];
    float   vx[N];
    float   vy[N];
    float   vz[N];
};
struct SoA ps;
```

b)

```
 struct AoaS_32{
    float   rx;
    float   ry;
    float   rz;
    float   vx;
    float   vy;
    float   vz;
    float   foo1;
    float   foo2;
};
struct AoaS_32 pa[N];
```

Figure 5. C/C++ data layouts for particles. (a) an example of structure of arrays (SoA), (b) an example of an arrays of aligned structure of size 32 bytes.

Device	Compiler	SoAs		AoSs (Aligned)	
		Separated [ns/particle]	Together [ns/particle]	Separated [ns/particle]	Together [ns/particle]
CPU	gcc/g++	34.47	39.09	33.29	33.28
[Intel Core i7-4790]	intel icc	**5.43**	18.53	6.77	7.74
GPU [TITAN X]	nvcc	0.31	**0.27**	0.52	0.42

Table 2. Performance results for pushing particles using CPUs and GPUs for various data structures, compilers, and algorithm developments. Marked numbers in bold are the fastest runtime achieved using CPU and GPU.

not make a fair comparison, for example by comparing our GPU code performance results with those obtained from CPU code compiled with gcc, a speed up over ~100X could be achieved. Such large speed ups are questionable, if a fair comparison is not provided.

4.2.2. Mapping Particles to Grid In particle-based kinetic models the particles distribution in space and velocity is defined by a distribution function $f_s(\mathbf{x}, \mathbf{v}, t)$ for each species s, where

$$f_s(\mathbf{x}, \mathbf{v}, t) = \sum_{p=1}^{P} S(\mathbf{x}, \mathbf{x}_p(t))\ \delta(\mathbf{v}, \mathbf{v}_p(t)), \tag{11}$$

where P is the number of particles, S is the shape function, and δ is a Dirac delta function. Macroscopic particle quantities (e.g., charge density, ρ_i, and current density $\mathbf{J}_i$) are mapped onto a grid using shape functions, where

$$\rho_i(\mathbf{x}, t) = \sum_{p=1}^{P} q_p S(\mathbf{x}, \mathbf{x}_p(t)), \qquad \mathbf{J}_i(\mathbf{x}, t) = \sum_{p=1}^{P} q_p \mathbf{v}_p S(\mathbf{x}, \mathbf{x}_p(t)), \tag{12}$$

where q_p is the charge of a particle. Two commonly used shape functions are the cloud-in-cell (CIC) and triangular-shaped cloud (TSC), which are shown in Figure 6 [28]. The macroscopic quantities are used to calculate the electric field (Equation 7) on a grid in hybrid models, and thus must be accurately calculated.

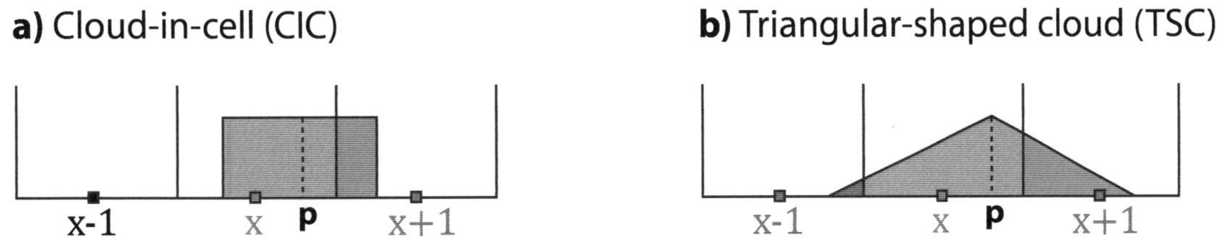

Figure 6. The cloud shape interpolation of charge and current density assignment, taken from [28]. The fraction of charge assignment for a single particle at position p to its neighboring cells using (a) could-in-cell and (b) triangular-shaped cloud shape functions are color-coded here.

Calculating macroscopic quantities associated with particles, and storing these quantities on grid-based models represents one of the largest performance bottleneck operations on GPUs [51]. This is mainly due to race conditions that may arise when several threads attempt to access the same memory address on global or shared memory to store calculated quantities from several particles. In the last decade, several sophisticated and complex algorithms have been proposed to efficiently solve particle mapping and field interpolation for particle algorithms. Examples include clustering a group of grid cells, known as tiling, and particle reordering/sorting [16, 17, 35, 51], and hierarchical Barnes-Hut tree [9]. However, most of these methods have only been developed, optimized, and tested for 1D, 2D and at most 2.5D models [15, 17, 35, 51]. In addition, NVIDIA has included several scalable synchronization mechanisms to its GPUs such as atomic operations. Generally, atomic operations perform mathematical instructions without interference from any other threads. Traditionally, these operations are slow because they may cause serialization for threads in a warp [11]. This has prompted the development of advanced algorithms in the past to avoid race conditions and provide high performance particle mapping to a grid. However, NVIDIA has significantly improved its atomic instructions, especially after the release of CUDA v7.5. From now on,"the atomic functions do not act as memory fences and do not imply synchronization or ordering constraints for memory operations", as stated in the CUDA v7.5 programming guide [14].

The majority of the non-atomic-based algorithms developed to calculate particle macroscopic quantities on a grid ([15, 17, 35]) require particle sorting/reordering and allocations of large memory buffers for particles in order to handle their motion from one grid tile to another. Although this can be minimized, particle sorting adds an extra overhead to model performance. Memory buffers also enforce the use of lower number of particles in a simulation. Since our model is 3D and accommodating more particles per grid cell in the limited global memory of a GPU is crucial, we prefer to use CUDA's atomic operations. Thus, we take the advantage of these operations in AMITIS to calculate particles macroscopic quantities and store them on a grid (mapping particles to grid). Extensive research comparing methods based on atomic and non-atomic operations is required, which is outside the scope of this study.

In AMITIS we have implemented both the CIC and TSC shape functions, and selection of the applied shape function is a free parameter in our model. Due to the larger memory access and complexity of TSC compared to CIC, a lower speed up for TSC is expected compared to CIC. As previously explained, we perform a fair comparison between optimized CPU and GPU versions of our code. Table 3 compares model performance to calculate particle charge density on a regularly spaced 3D grid using various number of grid cells and particles per cell (ppc) using the CIC shape function (averaged over 1000 timesteps). We see a minimum speed up of 25X with over 30X for large number of particles. This is mainly due to the higher kernel occupancy and cached memory access for larger sized problems on GPUs. Unlike solving the equation of motion, the results here depend on the number of threads per block. For these benchmark experiments we used 256 threads per block. A lower speed up (~8X-16X) would be obtained using more than 256 threads per block, because of limited resources available on each streaming multiprocessor (SM) on GPUs to perform atomic operations and using registers to calculate particle shape functions on a grid.

Run	Num grid cell	ppc	Total particles	CPU [ms]	GPU [ms]	Speed up
1	32×32×32	32	1048576	14.28	0.57	25.05
2	32×32×32	64	2097152	28.61	1.07	26.73
3	32×32×32	128	4194304	57.13	2.10	27.20
4	64×64×64	32	8388608	132.32	4.23	31.28
5	64×64×64	64	16777216	263.72	8.41	31.35
6	64×64×64	128	33554432	527.03	16.74	31.48

Table 3. Performance results for mapping particles to grids, averaged over 1000 timesteps and presented in milliseconds (ms).

4.2.3. Explicit-Implicit Scheme for Hybrid Field Solver We use the hybrid model solver explained by Holmström, 2010 [29] and solve the hybrid equations using finite differencing scheme in Cartesian coordinates. If the Ohmic term ($\eta\nabla \times \mathbf{B}/\mu_0$) is not included in Equation 7, an explicit finite difference scheme to solve the Faraday's law (Equation 8) will be stable and accurate as long as the Courant-Friedrichs-Lewy (CFL) condition is satisfied for time stepping [29, 36, 41]

$$\Delta t < \frac{\Omega_g}{\sqrt{3}\pi}\left(\frac{\Delta h}{v_A}\right)^2, \tag{13}$$

where Ω_g is the ion gyrofrequency, and v_A is the Alfvén speed. Δh is the length of a grid cell, and theoretically should satisfy $\Delta h \gg \delta_e$, where δ_e is the electron inertial length. For typical solar wind conditions near the Earth with solar wind bulk flow speed $v_{\text{sw}} \simeq 400$ km/s, proton gyrofrequency $\Omega_g \simeq 0.5$ rad/s, and Alfvén speed $v_A \simeq 50$ km/s, the simulation cell size should be $\Delta h \gg 2.5$ km. If Δh=100 km, then the CFL condition is satisfied for $\Delta t < 0.35$ s. The timestep also needs to satisfy $\Delta t < \Delta h/v_{\text{sw}}$ to ensure that a particle does not cross a grid cell in a single timestep. Thus, for the solar wind condition mentioned above, $\Delta t < 0.25$ s.

Vacuum regions naturally occur in plasmas and present difficulties for hybrid and kinetic models as ρ_I becomes zero in Equation 7, in the absence of the Ohmic term. One proposed solution to address this challenge requires including the Ohmic term from Equation 7 [30]. Another method is including the electron inertial term in Equation 7, which requires considering non-zero electron mass ($m_e \neq 0$) [2]. Omelchenko and Karimabadi, 2012 [45] also proposed an asynchronous framework with inhomogeneous timescales to deal with this problem. We follow Holmström, 2013 [30] by using the Ohmic term in Equation 7, which provides the advantage of handling any low-density and vacuum regions in plasma, as well as allowing the definition of conductive/resistive interior structures of objects (e.g., the Moon [31], Europa [32]). The main challenge in using this term in Equation 7 is that in regions where ρ_I=0 (e.g., plasma wakes and objects interior) Faraday's law (Equation 8) becomes,

$$\frac{\partial \mathbf{B}}{\partial t} = -\nabla \times \left(\frac{\eta}{\mu_0}\nabla \times \mathbf{B}\right), \tag{14}$$

where $\eta=\eta(\mathbf{x},t)$ is resistivity in space, $\mathbf{x}$, and time, t. For a spatially constant resistivity, Equation 14 can be given as

$$\frac{\partial \mathbf{B}}{\partial t} = \frac{\eta}{\mu_0}\nabla^2\mathbf{B}, \tag{15}$$

which is a diffusion equation, and is numerically stable for timesteps

$$\Delta t_d < \frac{\mu_0 \Delta h^2}{2\eta}, \tag{16}$$

which is usually much smaller than the CFL condition (Equation 13). For typical solar wind conditions described previously, and for Δh=100 km, the CFL condition is satisfied for $\Delta t < 0.25$ s. However, if $\eta = 10^7\ \Omega\cdot\mathrm{m}$, then the stability for magnetic diffusion (Equation 16) requires $\Delta t_d < 6.0\times10^{-4}$ s, which is nearly 400X smaller than the CFL timestep. In order to physically model the interior structure of various solar system objects that are feasible for hybrid modeling, shown in Figure 3, a much larger resistivity is required. For example, lunar crustal resistivity is $\eta \gg 10^7\ \Omega\cdot\mathrm{m}$ [24, 31]. Thus, for Δh=100 km, $\Delta t_d \ll 6.0\times10^{-4}$ s. Hybrid modeling of solar wind plasma interaction with the Moon using explicit finite difference schemes with timesteps smaller than 0.01 s is computationally expensive, and for timesteps smaller than 0.001 s is computationally impossible, especially if the electromagnetic response of the interior to a magnetic discontinuity needs to be fully resolved.

To overcome the diffusive restriction on the timestep, we propose an explicit-implicit scheme. We break down the electric field from Equation 7 into two terms: the electric field without the Ohmic term ($\mathbf{E}_{\mathrm{nonOhmic}}$), and the electric field with the Ohmic term ($\mathbf{E}_{\mathrm{Ohmic}}$),

$$\mathbf{E}_{\mathrm{nonOhmic}} = -\frac{\mathbf{J}_i\times\mathbf{B}}{\rho_{\mathrm{I}}} + (\frac{\nabla\times\mathbf{B}}{\mu_0})\times\frac{\mathbf{B}}{\rho_{\mathrm{I}}} - \frac{\nabla\cdot\overline{\mathbf{P}}_e}{\rho_{\mathrm{I}}}, \tag{17}$$

$$\mathbf{E}_{\mathrm{Ohmic}} = \frac{\eta}{\mu_0}\nabla\times\mathbf{B}, \tag{18}$$

such that

$$\mathbf{E} = \mathbf{E}_{\mathrm{nonOhmic}} + \mathbf{E}_{\mathrm{Ohmic}}. \tag{19}$$

From Faraday's law (Equation 8), we have

$$\frac{\partial \mathbf{B}_{\mathrm{nonOhmic}}}{\partial t} = -\nabla\times\mathbf{E}_{\mathrm{nonOhmic}} \tag{20}$$

and

$$\frac{\partial \mathbf{B}_{\mathrm{Ohmic}}}{\partial t} = -\nabla\times\mathbf{E}_{\mathrm{Ohmic}}. \tag{21}$$

By the distributive properties of vector operations, this also yields

$$\mathbf{B} = \mathbf{B}_{\mathrm{nonOhmic}} + \mathbf{B}_{\mathrm{Ohmic}}. \tag{22}$$

Now we can solve the non-Ohmic term explicitly (at the CFL timestep) while the Ohmic term is solved *implicitly* using an implicit solver (e.g., Crank-Nicholson [13]). The implicit solver allows us to take a single timestep for the Ohmic term as large as the CFL constraint, bypassing the restrictive diffusive timestep. The non-Ohmic and Ohmic solutions are then re-combined (Equations 19 and 22) for the full solution to Maxwell's equations.

In the AMITIS code, we have implemented the scheme presented above to solve our hybrid equations. We explicitly solve Equation 20 using finite difference scheme presented by [29], and use the well-known Crank-Nicholson method [13] to solve Equation 21. The stability and accuracy of this explicit-implicit method is discussed in Section 5.3 and the results are compared with a fully explicit method.

4.2.4. Finite Difference Stencil Calculation of the total current density (Equation 3), electric field (Equation 7), and magnetic field (Equation 8) requires the use of derivatives and vector operations. In AMITIS we implemented two schemes: a three-point and a five-point stencil along each dimension to numerically solve the derivatives in their finite-differenced form and allow the user to select which stencil schema to choose as an input to simulations. The three-point stencil requires one layer of guard cells (also known as ghost cells or halos) to be added in all directions outside the user-defined simulation grids, as shown in Figure 7a. These guard cells are used to establish the boundary conditions and to allow use of the stencil for the outer most cells of the user-defined simulation grids. The five-point stencil requires two layers of guard cells, shown in Figure 7b. The three-point stencil runs faster but at reduced accuracy, while the five-point stencil has improved accuracy but reduced performance. Examples of five-point stencils of two adjacent grid cells in a 3D domain are shown in Figure 7c. In a five-point stencil, the first derivative of a quantity Q at a point x can be approximated as

$$\frac{dQ(x)}{dx} = \frac{Q(x-2h) - 8Q(x-h) + 8Q(x+h) - Q(x+2h)}{12h} + \mathcal{O}(h^4), \tag{23}$$

where h is the distance between two adjacent cells along x.

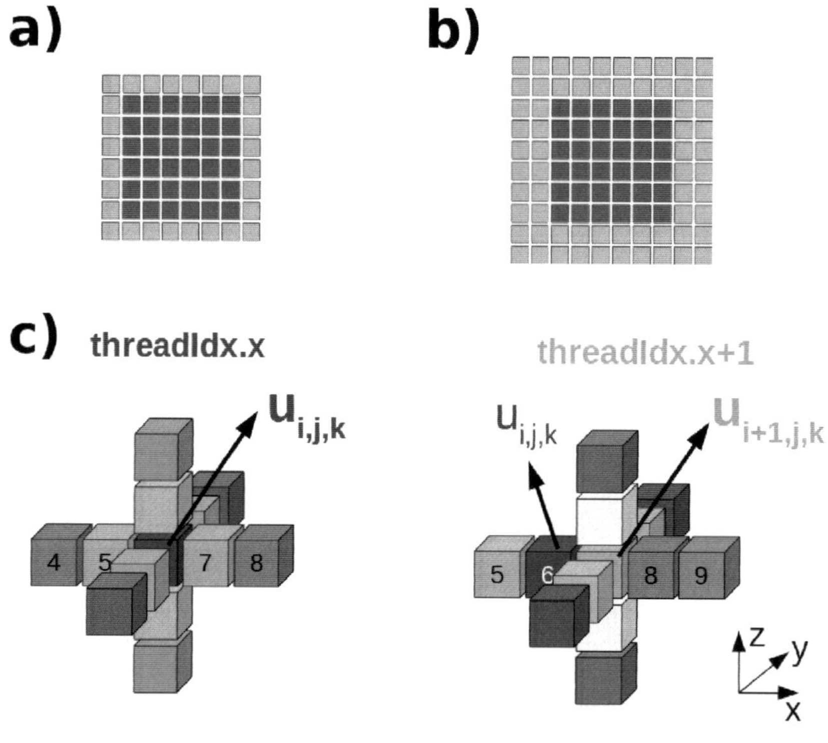

Figure 7. (a) An example of a 6 × 6 data tile with one layer of guard cells, shown in green, for three-point stencils, (b) an example of a 6 × 6 data tile with two layers of guard cells for five-point stencils, (c) high order approximations for central based finite difference of a quantity u assigned to a thread of a block.

Since all the grid values are stored in the global memory, reducing the number of reads and writes from global memory and preventing race conditions while calculating derivatives can improve the model performance. Different algorithms have been proposed to minimize global memory access, which are mainly based on shared memory [11, 55]. Again, the majority of these algorithms have been developed for 1D and 2D domains. Based on our experience in 3D modeling using Maxwell GPU generations and later, we found that the shared memories not only fail to provide considerable improvements in the model performance, but also add a high level of complexity in the model development and programming. In AMITIS we use the global memory to calculate derivatives, but to prevent race conditions we calculate derivatives of every third grid cell in the three-point stencil and every fifth grid cell in the five-point stencil at every call of our computation kernel. Then we repeat the calculation procedure until the derivatives for all grid cells are calculated. Table 4 compares AMITIS performance to calculate a 3D vector curl operation with the CPU version of our code. For a small simulation domain, the GPU code does not perform very well. This is, again, mainly due to the low device occupancy to solve small size problems. However, Table 4 shows a speed up of over 17X can be achieved in calculating three-point stencil curl operations using GPUs compared to CPUs.

Run	Num grid cells	CPU [ms]	GPU [ms]	Speed up
1	16×16×16	0.009	0.019	0.47
2	32×32×32	0.053	0.015	3.53
3	64×64×64	0.436	0.041	10.63
4	128×128×128	4.166	0.234	17.80

Table 4. Performance results to calculate 3D curl operation, averaged over 1000 timesteps.

4.2.5. Mapping Forces to Particles Similar to mapping particles to a grid, shape functions are used for interpolating electromagnetic fields and other forces (e.g., gravity) to the particles position [4, 28]. These forces are used to advance particles in velocity space by solving the equation of motion (Equation 9). Given a single particle p located at position $\mathbf{x}_p$, the force, $\mathbf{F}$, applied to the particle is given by

$$\mathbf{F}(\mathbf{x}_p) = \sum_c \mathbf{F}(\mathbf{x}_c)\; W(\mathbf{x}_c, \mathbf{x}_p), \tag{24}$$

where c are all the cell vertices sounding the particle, and W is the weight function given by the shape function, S, integrated over the cell volume [27]

$$W(\mathbf{x}_c, \mathbf{x}_p) = \int_{\mathbf{x}_c-\Delta x/2}^{\mathbf{x}_c+\Delta x/2} S(\mathbf{x}_c, \mathbf{x}_p)\Delta x, \tag{25}$$

where Δx is the grid cell size. Consistency in the shape function used for mapping particles to grid and interpolating fields to particles is necessary [28]. Therefore, we have included two weight functions based on the CIC and TSC in our model, which can be selected as a free input parameter for a simulation. Since the particles are moving together, if we select a subset of particles, there is a high chance that most of the particles need to access the same grid cells for force interpolation. Therefore, in AMITIS we assign every 32 particles to a single thread of a block. Each thread reads its own dedicated particles, reads the electromagnetic fields from the global memory for the neighboring cells, and interpolates forces to the particles position. Each thread needs to access the global memory for all the neighboring cells only once, unless the particles position have a widely spread spatial distribution. Table 5 compares the AMITIS performance with a highly optimized CPU-based code. With our clustering method on GPU, and without using any shared memory, we achieved >150X speed up in force interpolation to the particles. However, these results may vary by changing the particles thermal speed. We have chosen an ion thermal speed of ∼50 km/s for the results presented in Table 5. A speed up of ∼25 X was achieved for ions with thermal speed ∼200 km/s. A part of this large speed up in because we do not need to store the interpolated fields into the global memory for every particle. Thus, we do not write into the global memory for this procedure. The method proposed above for particle clustering and assigning each cluster to a thread gives rise to load imbalance and thread divergence on GPUs, which has direct impact on the model performance. This problem can be resolved by assigning every single particle to a single thread per block. This requires every thread reads several grid points from global memory, which again impacts model performance. We did not experience noticeable performance changes using either of these methods.

Run	Num grid cell	ppc	Total particles	CPU [ms]	GPU [ms]	Speed up
1	32×32×32	32	1048576	9.42	0.059	159.6
2	32×32×32	64	2097152	16.33	0.108	151.2
3	32×32×32	128	4194304	32.64	0.202	161.5
4	64×64×64	32	8388608	67.82	0.379	178.9
5	64×64×64	64	16777216	133.21	0.729	182.7
6	64×64×64	128	33554432	256.28	1.441	177.8

Table 5. Performance results for mapping electromagnetic forces to particles, averaged over 1000 timesteps.

5. Test Results

In order to examine the accuracy and stability of our GPU based hybrid model (AMITIS), we analyzed a few standard diagnostics which are explained here. We also compare our model performance with a parallel CPU-based hybrid model implemented in the FLASH code (from now on called hybrid-FLASH). FLASH is a modular, parallel multiphysics simulation code developed by the FLASH Center for Computational Science at the University of Chicago [21]. It is mainly written in Fortran90, and it uses message passing interface (MPI) for interprocessor communications. Although FLASH particle and grid modules have fundamental differences compared to AMITIS, it is the only parallel CPU-based hybrid model of plasma that we have access to. Thus, it is worth noting that we cannot provide a fair comparison between AMITIS and hybrid-FLASH because of their fundamental differences.

5.1. Periodic Plasma Box

We first tested a simple periodic plasma box without any obstacle. We made three simulation runs with different configurations, listed in Table 6. We loaded the simulation box with typical solar wind plasma at 1 AU (near the Earth and the Moon) with ion density $n_0 \simeq 7$ cm^{-3}, ion temperature $T_i \simeq 12$ eV, electron temperature $T_e \simeq 15$ eV, plasma bulk flow speed $|\mathrm{u}_0| \simeq 350$ km/s, and magnitude of the magnetic field $|\mathrm{B}_0| \simeq 5$ nT. Consequently, the ion gyrofrequency $\Omega_g \simeq 0.5$ rad/s, ion inertial length $\delta_i \simeq 86$ km, electron inertial length $\delta_e \simeq 2$ km, Alfvén speed $v_A \simeq 41$ km/s, and ion sound speed $c_s \simeq 46$ km/s.

We only included solar wind protons in the simulations ($m_i \sim 1.0$ amu and $q_i \sim 1.0e$). We assumed the solar wind flows along the -x axis and the magnetic field is only along the +y axis, perpendicular to the plasma flow direction. Thus, the convective electric field $\mathbf{E} = -\mathbf{v} \times \mathbf{B}$ is along the +z axis. The simulation domain is entirely filled in with solar wind protons at initialization (t=0 s), and the domain is assumed to be periodic along all directions. Therefore, when a particle moves out of the simulation domain, it comes back into the box through the opposite boundary. No sources for particle injection and/or loss are considered in the plasma box. The simulation results are expected to be stable and should only show small statistical

runs	grid cells	cell size (Δh)	timestep (Δt)	ppc	total particles
#1	48×24×24	$1.16\delta_i$ = 100 km	$2\times10^{-3}\Omega_c = 1\times10^{-3}$ s	32	884736
#2	48×24×24	$1.16\delta_i$ = 100 km	$2\times10^{-3}\Omega_c = 1\times10^{-3}$ s	4	110592
#3	48×24×24	$0.58\delta_i$ = 50 km	$2\times10^{-4}\Omega_c = 1\times10^{-4}$ s	4	110592

Table 6. List of simulation runs to study the performance cost of the periodic plasma box problem.

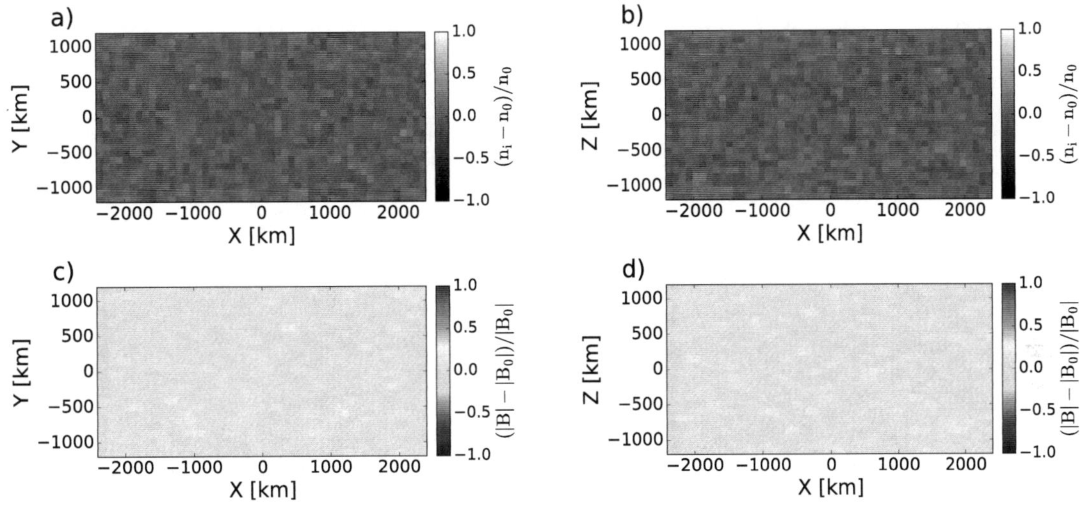

Figure 8. Periodic plasma box simulation results obtained from AMITIS for run #1 in Table 6. (a,b) Plasma number density perturbations relative to the background (initial) ion number density, n_0, (c,d) magnitude of the magnetic field perturbations relative to the background (initial) magnetic field $\mathbf{B}_0$. Panels a and c are cuts in the XY plane at $Z = 0$, viewed from the $+Z$ axis. Panels b and d are cuts in the XZ plane at $Y = 0$, viewed from the $-Y$ axis. In all panels the solar wind plasma flows from right to the left, and the background magnetic field is along the $+Y$ axis and is perpendicular to the plasma flow direction.

fluctuations, depending on hybrid solver scheme, plasma thermal speed, simulation timestep, number of particles per cell, and system round-off/truncation error. We ran each of these experiments for 5×10^5 timesteps, equivalent for solar wind plasma to transit the entire domain for ~35 times for runs #1 and #2, and for ~7 times for run #3. We also set η=0 in the ohmic term (Equation 7), thus no implicit solver was used in these experiments.

Figure 8 shows the simulation results for run #1 after 5×10^5 timesteps. The top panels show the variation of the ion number density from the initial solar wind density (n_0) and the bottom panels show the deviation of the magnetic field magnitude ($|\mathbf{B}|$) from the initial magnetic field ($|\mathbf{B}_0|$). We see that the model is stable and does not contain any large fluctuations after 5×10^5 timesteps.

We also calculated the total particle kinetic energy ($\sum_i m_i v_i^2/N_i$, where N_i is the total number of particles), magnetic field energy ($\sum_g B^2/N_g$, where N_g is the total number of grid cells), and electric field energy ($\sum_g E^2/N_g$) at every 200 timesteps, and the results for our three experiments are shown in Figure 9. The energy is conserved throughout all timesteps for our simulation runs with an error of $< 0.2\%$. As shown in Figure 9a, the energy error for run #1 where the simulation cell sizes are slightly larger than ion inertial length ($\sim1.2\delta_i$) and the grid is loaded with moderate number of particles (32 particles per cell) is $< 0.05\%$. In run #2, we intentionally used a smaller

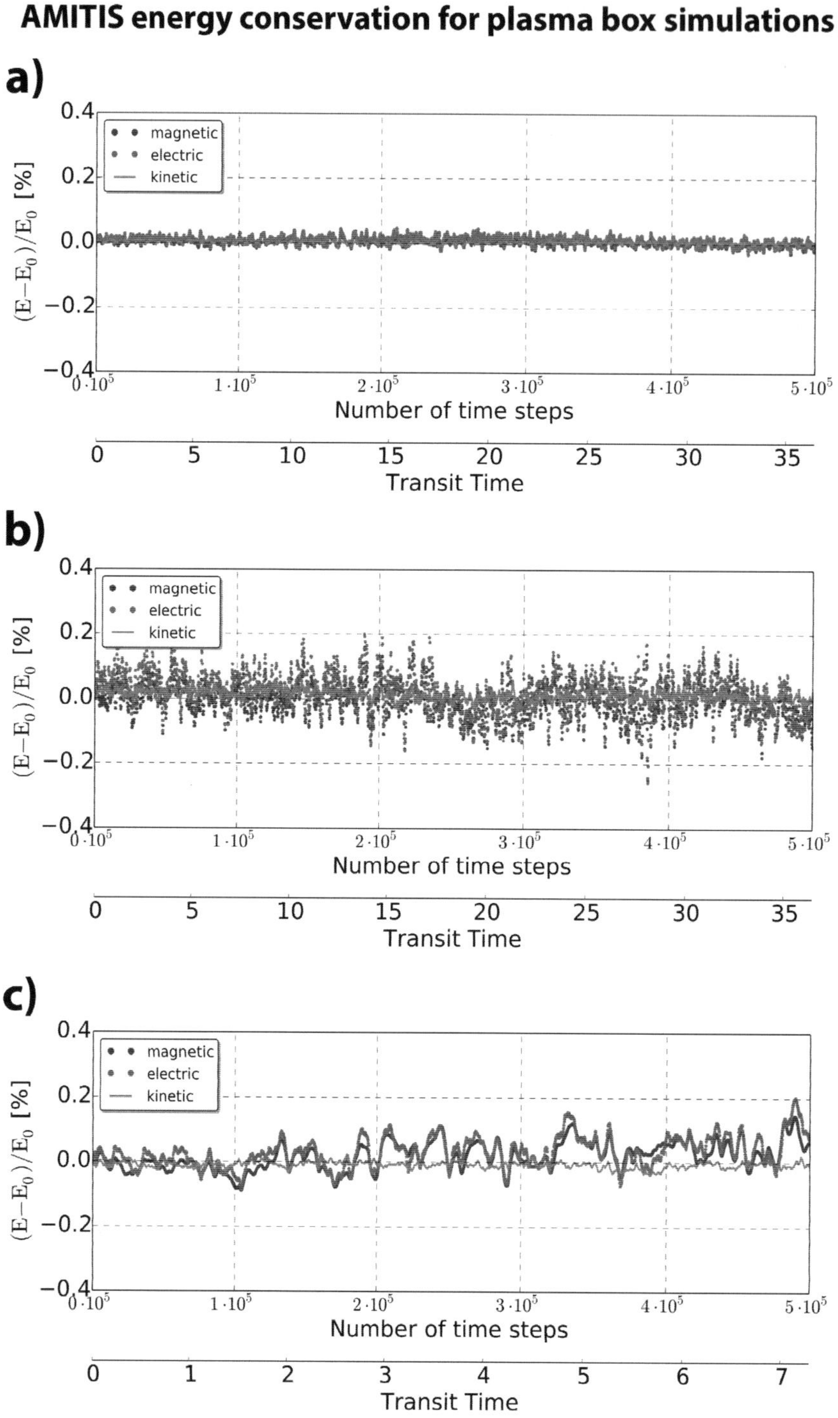

Figure 9. AMITIS energy error $(E - E_0)/E_0$ in percentage, where E_0 is the initial energy, for (a) run #1, (b) run #2, and (c) run #3 listed in Table 6. In all panels, the solid red line indicates particle kinetic energy, blue dots are magnetic field energy, and green dots are electric field energy. The horizontal axes are binned for the number of simulation timesteps and their associated number of transits that the plasma have gone through the entire simulation domain.

number of particles (4 particles per cell) to examine our model stability and energy conservation for fewer number of particles. Figure 9b shows the energy error for run #2 is $\sim 0.2\%$. In run #3, we reduced the simulation cell sizes to $\sim 0.6\delta_i$ and loaded the system with 4 particles per cell, and we see that the energy is conserved with $< 0.2\%$ error, as shown in Figure 9c.

5.2. Scaling Experiment
In order to examine the AMITIS performance and scaling, we provided 9 simulation runs for various number of grid cells and number of particles per cell (ppc), listed in Table 7. We also made identical simulations in FLASH to compare the AMITIS performance against hybrid-FLASH. Again, our simulation domain is periodic along all directions and was loaded with plasma conditions exactly identical to those explained in Section 5.1 (η=0, and no implicit solver is used in AMITIS). We ran the AMITIS code using a single CPU and a single TITAN X GPU. We ran hybrid-FLASH simulations on the Abisko supercomputer[2] using various number of processors, and no GPUs.

run	grid cells	cell size	ppc	total particles
#1			16	432,000
#2	30×30×30	100 km	32	864,000
#3			64	1,728,000
#4			16	3,456,000
#5	60×60×60	100 km	32	6,912,000
#6			64	13,824,000
#7			16	27,648,000
#8	120×120×120	100 km	32	55,296,000
#9			64	110,592,000

Table 7. List of simulation runs for the model scaling experiment.

Figure 10a shows AMITIS execution times for the simulation runs listed in Table 7. As shown in Table 7, the total number of particles for each simulation is increased by a factor of 2 from its previous run. Figure 10a also shows the execution run time for each run takes nearly twice as long compared to the previous run, indicating that AMITIS scales very well as a function of the domain size and the total number of particles. This scaling is very well pronounced for runs #5 to #9, while it is not as good for runs #1 to #4. This is because the total number of particles used in runs #1 to #4 are relatively low for TITAN X to achieve its maximum kernel occupancy, maximum throughput, and maximum memory bandwidth. As soon as the number of particles reach a moderately high value for the TITAN X, a linear scaling as a function of total particle number is achieved.

We ran similar experiments using FLASH by dedicating 48, 96, 384, and 768 CPUs (one processing node has 48 cores), and compared the execution run time per timestep per particle with AMITIS. Results are shown in Figure 10b. We see that depending on the total number of particles and the number of processors, AMITIS runs about 6–40 times faster than hybrid-FLASH.

[2] The Abisko supercomputer is provided by the Swedish National Infrastructure for Computing (SNIC) at the High Performance Computing Center North (HPC2N), Umea University, Sweden. As stated in https://www.hpc2n.umu.se/resources/abisko "the Abisko is comprised of 332 nodes with a total of 15936 CPU cores of which 328 nodes are available to the users. Each node is equipped with 4 AMD Opteron 6238 (Interlagos) 12 core (48 cores per node) 2.6 GHz processors, and 128 GB of DDR3-1600 RAM. The interconnects are Mellanox 4X QSFP 40 Gb/s InfiniBand. The Abisko compute nodes are to be considered complex Non-Uniform Memory Access (NUMA) machines, where the layout of memory in relation to the computing cores used can make a huge difference in real-life performance."

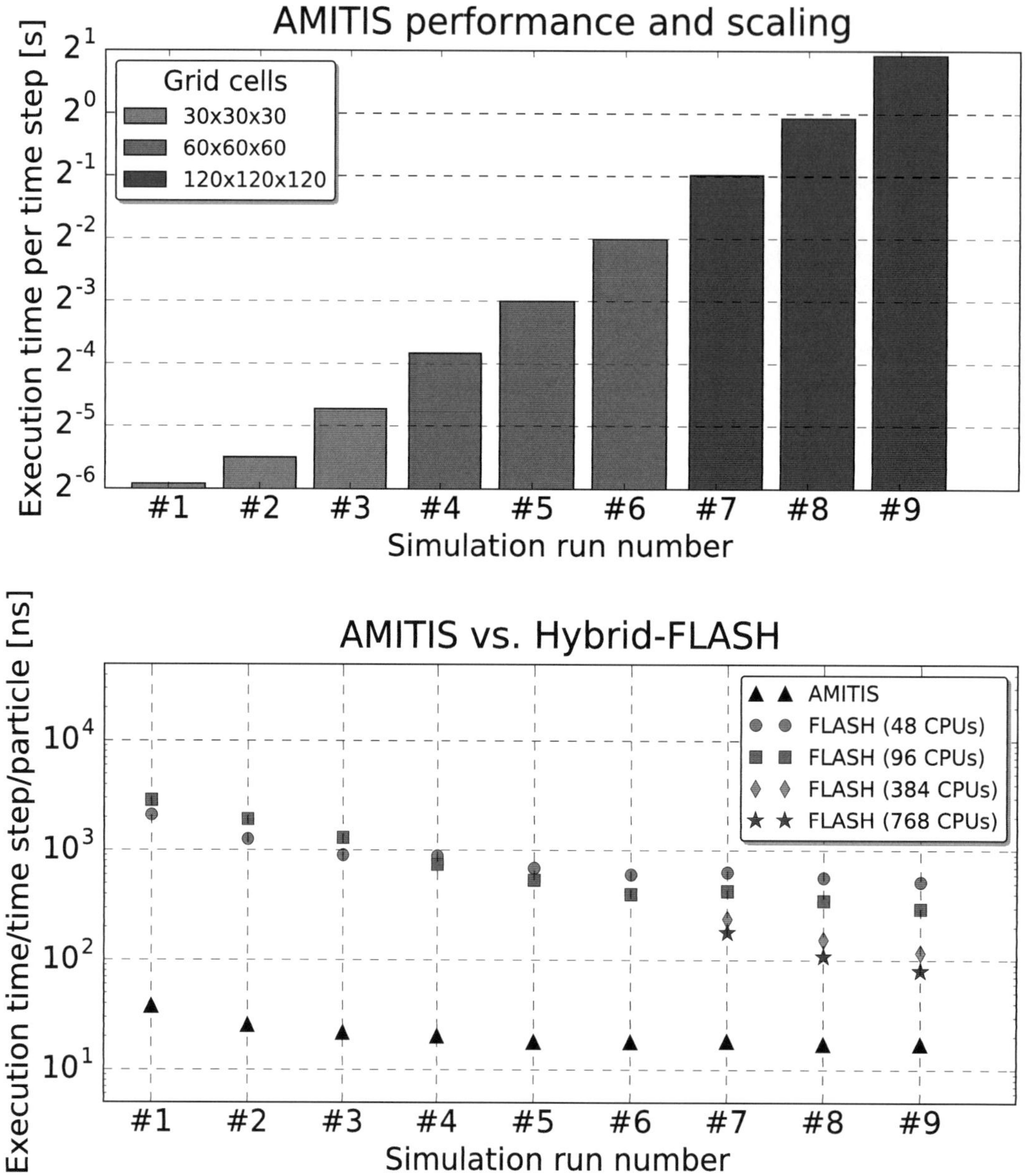

Figure 10. (a) AMITIS performance and scaling for the simulation runs listed in Table 7, (b) Comparison between the AMITIS and hybrid-FLASH execution time. In AMITIS we have used a single CPU-GPU pair, and in hybrid-FLASH we have used different number of CPUs.

5.3. Explicit-Implicit Solver

In order to examine the stability and accuracy of the explicit-implicit scheme discussed in Section 4.2.3, we provided a simulation test for the solar wind plasma interaction with the Moon. Recently, Poppe et al., 2014 [47] compared magnetic field and plasma density observations from the ARTEMIS spacecraft with hybrid simulation results using an explicit solver implemented in the hybrid-FLASH. We took the Poppe et al., 2014 [47] hybrid simulation configurations, listed in Table 8, and applied them into AMITIS. We used a uniform internal resistivity inside the Moon $\eta = 10^7\ \Omega \cdot \text{m}$, identical to that chosen by Poppe et al., 2014 [47], and used the explicit-implicit solver in the AMITIS code.

Parameters	**Values**	**Units**
Solar wind velocity	[-307.0, 0.0, 0.0]	km/s
Solar wind density	10.5	cm^{-3}
Magnetic field	[3.68, 2.23, 0.0]	nT
Ion temperature	43.0	eV
Electron temperature	10.0	eV
Alfvén speed	~29	km/s
Sound speed	~69	km/s
Plasma β	~9	

Table 8. Solar wind plasma parameters during an ARTEMIS spacecraft lunar wake crossing taken from [47].

The Moon, as a first order approximation, can be considered as a non-conducting, atmosphereless body without an intrinsic magnetic field. Thus, the solar wind plasma directly impacts the lunar surface and is absorbed by the Moon. This forms a wake structure downstream and leaves a plasma cavity (vacuum) behind the Moon [25]. Figure 11 provides a global overview on the solar wind plasma interaction with the Moon and compares AMITIS hybrid simulation results using the explicit-implicit solver (Figures 11a and 11b) with hybrid-FLASH explicit solver results presented by [47] (Figures 11c and 11d). Figures 11a and 11c show a density cavity (vacuum) forms downstream behind the Moon, which is due to plasma absorption on the lunar dayside ($X^2 + Y^2 = R_L^2$, where $R_L = 1730$ km is the radius of the Moon). Then, plasma expands into the vacuum region to maintain pressure balance. As the expansion front moves into the vacuum, the density of the ions near the front decreases with distance, forming a rarefied region around the vacuum region. Figures 11b and 11d show the magnitude of the magnetic field in the central lunar wake increases to nearly 4 times higher than the ambient solar wind magnetic field, surrounded by magnetic depressions in the rarefaction region. Magnetic field enhancement and plasma density reduction in the central lunar wake, and magnetic field and plasma density reduction in the rarefaction region are the typical features of the solar wind plasma interaction with the Moon [25]. Since it is impossible to distinguish the differences between the two model results we show the residuals from subtraction of the hybrid-FLASH results from the AMITIS simulations in Figures 11e and 11f. There are minor differences between the two model results which are mainly due to the differences between the explicit-implicit scheme used in the AMITIS and the explicit solver used in the hybrid-FLASH. There are also differences associated with the particle statistics, mapping particles to grids, and mapping forces to particles.

To examine the accuracy of our explicit-implicit solver, in Figure 12 we compared the AMITIS simulation results with hybrid-FLASH simulations and ARTEMIS spacecraft observations presented by [47]. The left panels in Figure 12 compare ARTEMIS spacecraft observations (black lines) with Poppe et al., 2014 [47] explicit hybrid-FLASH simulations (blue lines), and AMITIS explicit-implicit solver results (red lines). We see that both hybrid model simulations have good

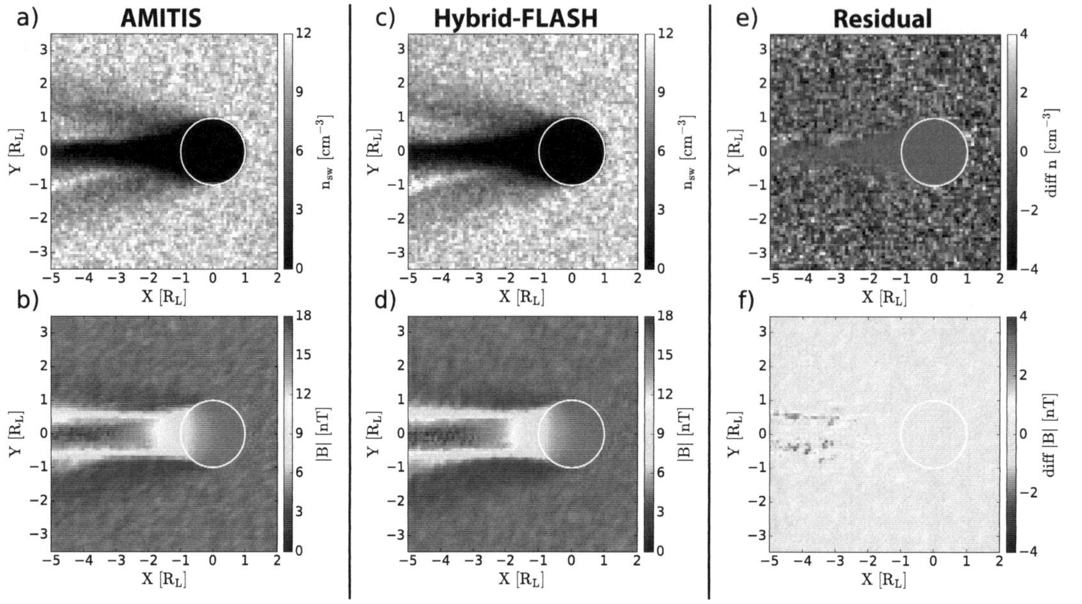

Figure 11. Hybrid simulation results for the moon-solar wind plasma interaction obtained from (a,b) the explicit-implicit solver developed in AMITIS, (c,d) explicit solver developed in hybrid-FLASH, (e,f) residuals from subtraction of the hybrid-FLASH results from AMITIS simulation results. (a,c) Total solar wind plasma density, (b,d) magnitude of the magnetic field. The Moon is located at the center of the coordinate system and is shown by a white circle. The geometry of the cuts is the same as those in Figure 8, and the upstream plasma parameters are listed in Table 8.

agreement with observations. Similar to the results presented in Figure 11 it is impossible to distinguish the differences between the two model results. Thus, in the left panels in Figure 12 we show the residuals from subtraction of the hybrid-FLASH results from the AMITIS simulations on the right panels in Figure 12. The two model results are very similar, but some disagreement between them is expected due to the fundamental differences between the two solvers. This test shows that our explicit-implicit scheme is accurate, and agrees with observations and simulation results obtained from the explicit solver implemented in hybrid-FLASH.

To examine the stability and energy conservation for our explicit-implicit solver, we calculated the particle kinetic energy, magnetic field energy, and electric field energy similar to that in Section 5.1, and the results are shown in Figure 13. At the early stage of the lunar wake development (timesteps $< 2 \times 10^4$), the particle kinetic energy decreases. This is due to the particle absorption on the lunar dayside. As the lunar wake evolves, an ambipolar electric field forms around the lunar wake [18, 25]. This electric field together with particles thermal motion accelerates particles into the wake to fill in the vacuum region shown in Figure 11a. This particle acceleration at later times results in increasing the kinetic energy, shown in Figure 13. After the lunar plasma wake has evolved and reached a steady-state solution, the kinetic energy and electromagnetic energy remain constant with relatively small fluctuations throughout the rest of the simulation. We see from Figure 13 that the explicit-implicit scheme developed for the AMITIS hybrid model is fairly stable and conserves energy with an error $< 0.2\%$.

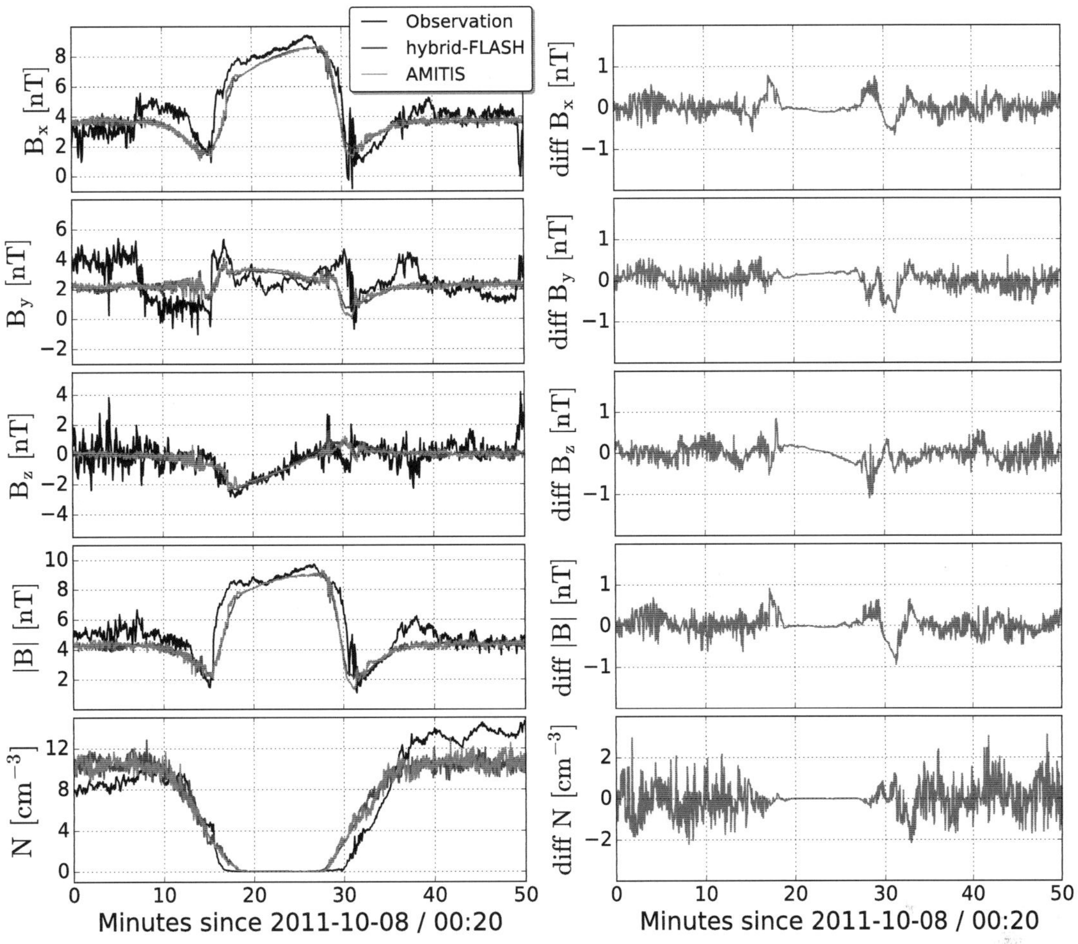

Figure 12. Left panels are a comparison of (black) ARTEMIS data, (blue) hybrid-FLASH using explicit solver, and (red) AMITIS using explicit-implicit solver. Right panels are the magnetic field residuals from subtraction of the hybrid-FLASH results from AMITIS simulation results. Shown from top to the bottom are the three components of the magnetic field, magnetic field magnitude, and plasma density, respectively.

6. Discussion

AMITIS is a GPU-based high performance particle model infrastructure for space simulations. The current version includes a three-dimensional self-consistent hybrid model of plasma and it has the capability to be converted to a full PIC solver by changing its electromagnetic field solver and adding an electron species. The current version of AMITIS runs on a single CPU-GPU pair and on average runs nearly 30X-50X faster than a serial CPU-based model, and ∼5X-10X faster than a parallel CPU-based hybrid model in FLASH. The average cost to buy and maintain a workstation/desktop with a single GPU and a reasonable CPU at the time of this writing is less than ∼3000 US dollar, while a supercomputer with only 128 processors costs over 50,000 USD (see Table 1 for a single CPU vs. single GPU cost comparison). Thus, AMITIS and any other GPU-based models that entirely run on GPUs are more economical compared to parallel CPU-based models. We also showed that AMITIS scales very well with the size of a problem

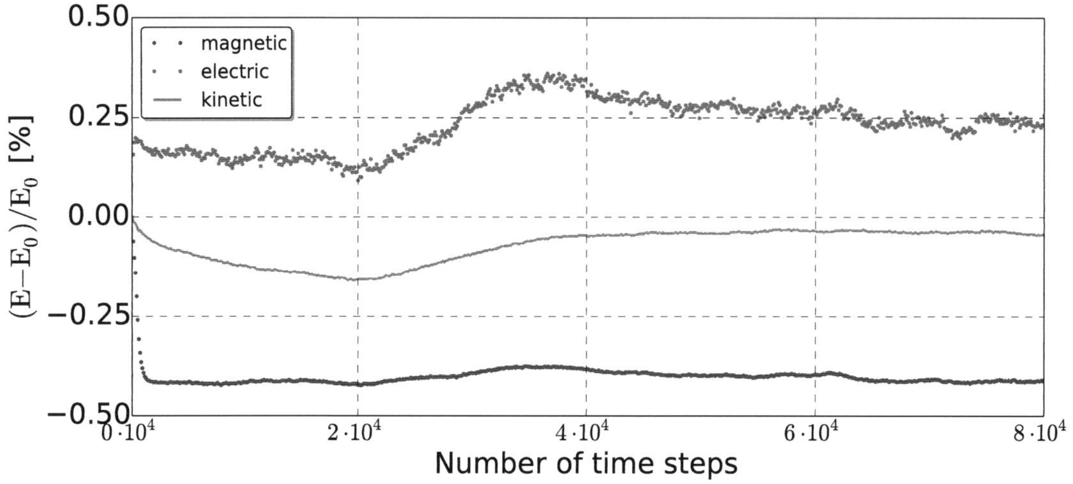

Figure 13. AMITIS explicit-implicit scheme energy error for hybrid simulation of plasma interaction with the Moon, shown in Figure 11. The solid red line indicates particle kinetic energy, blue dots are magnetic field energy, and green dots are electric field energy.

(Figure 10), and conserves energy with an energy error < 0.2% (Figures 9 and 13).

In the current version of AMITIS the maximum domain size that fits into a 12 GB global memory of the TITAN X GPU is 256×256×256 with 16 particles per cell, or any configuration with similar number of grid cells and particles. This domain size is sufficient to study plasma interactions with different solar system objects that fit within the scales of hybrid modeling shown in Figure 3. However, some plasma physics applications require larger domains (e.g., 1024^3) and/or larger number of particles per cell, which provides motivation to move towards a multi-GPU version of AMITIS in the future.

7. Acknowledgements

Shahab Fatemi is grateful to Mats Holmström for developing a hybrid model of plasma based on FLASH, full and stimulating discussions on hybrid modeling, and sharing his valuable knowledge to the corresponding author. The authors are also grateful to Martin Burtscher at Texas State University for enlightening and fruitful discussions on GPU's memory structure, and CUDA programming. The authors gratefully acknowledge support from NASA's SSERVI institute, grant #NNX14AG16A. We thank NVIDIA's GPU Research Center Program and GPU Technology Conference at San Jose, California for their great support and providing GPUs used in this research. We also acknowledge the DOE NNSA ASC-supported and DOE Office of Science ASCR-supported FLASH Center for Computational Science at the University of Chicago. This research was mainly conducted using GPUs provided by NVIDIA Research Center on computers at Space Sciences Laboratory, University of California at Berkeley, and partly by the Swedish National Infrastructure for Computing (SNIC) at the High Performance Computing Center North (HPC2N), Umeå University, Sweden.

References

[1] Abreu, P., R. A. Fonseca, J. M. Pereira, and L. O. Silva. PIC codes in new processors: A full relativistic PIC code in CUDA-enabled hardware with direct visualization. *IEEE Trans. Plasma Sci.*, 39(2):675–685, 2011.

[2] Amano, T., K. Higashimori, and K. Shirakawa. A robust method for handling low density regions in hybrid simulations for collisionless plasmas. *J. Comput. Phys.*, 275:197–212, 2014.

[3] Bagdonat, T. and U. Motschmann. 3D Hybrid Simulation Code Using Curvilinear Coordinates. *J. Comput. Phys.*, 183(2):470–485, 2002.

[4] Birdsall, C. K. and B. Longdon. *Plasma physics via computer simulation.* McGraw Hill, New York, 1984.

[5] Boris, J. P. Relativistic plasma simulation-optimization of a hybrid code. *Proc, of 4th Conf. on Numerical Simulations of Plasmas*, (Nov.), 1970.

[6] Brecht, S. H. and S. A. Ledvina. The solar wind interaction with the martian ionosphere/atmosphere. *Space Sci. Rev.*, 126(1-4):15–38, 2006.

[7] Brecht, S. H. and S. A. Ledvina. Control of ion loss from mars during solar minimum. *Earth, Planets and Space*, 64(2):165–178, 2012.

[8] Burau, H., R. Widera, W. Honig, G. Juckeland, A. Debus, T. Kluge, U. Schramm, T. E. Cowan, R. Sauerbrey, and M. Bussmann. PIConGPU: A Fully Relativistic Particle-in-Cell Code for a GPU Cluster. *IEEE Trans. Plasma Sci.*, 38(10):2831–2839, 2010.

[9] Burtscher, R. and K. Pingali. An Efficient CUDA Implementation of the Tree-based Barnes Hut N-Body Algorithm. In Wen-Mei, W. H., editor, *GPU Computing Gems Emerald Edition*, chapter 6. Elsevier, 2011.

[10] Byers, J.A, B.I Cohen, W.C Condit, and J.D Hanson. Hybrid simulations of quasineutral phenomena in magnetized plasma. *J. Comput. Phys.*, 27(3):363–396, 1978.

[11] Cheng, J., M. Grossman, and T. McKercher. *Professional CUDA C Programming.* John Wiley & Sons, Inc., 2014.

[12] Claustre, J., B. Chaudhury, G. Fubiani, M. Paulin, and J. P. Boeuf. Particle-In-Cell Monte Carlo Collision Model on GPU-Application to a Low-Temperature Magnetized Plasma. *IEEE Trans. Plasma Sci.*, 41(2): 391–399, 2013.

[13] Crank, J. and P. Nicholson. A practical method for numerical evaluation of solutions of partial differential equations of the heat conduction type. *Proc. Camb. Phil. Soc.*, 43(1):50–67, 1947.

[14] CUDA C Programming Guide v7.5, NVIDIA CUDA Toolkit Documentation, . `http://docs.nvidia.com/cuda/cuda-c-programming-guide/index.html`, 2015. Last accessed: Nov 18, 2015.

[15] Decyk, V. K. and T. V. Singh. Adaptable Particle-in-Cell algorithms for graphical processing units. *Comp. Phys. Comm.*, 182(3):641–648, 2011.

[16] Decyk, V. K. and T. V. Singh. Particle-in-Cell algorithms for emerging computer architectures. *Comp. Phys. Comm.*, 185(3):708–719, 2014.

[17] Decyk, V. K., T. V. Singh, and S. A. Friedman. Graphical Processing Unit-based Particle-in-Cell Simulations. In *Proc. Int. Computational Accelerator Physics Conf.*, 2009.

[18] Fatemi, S., M. Holmström, and Y. Futaana. The effects of Lunar surface plasma absorption and solar wind temperature anisotropies on the solar wind proton velocity space distributions in the low-altitude Lunar plasma wake. *J. Geophys. Res.*, 117(A10105):A10105, 2012.

[19] Fatemi, S., M. Holmström, Y. Futaana, S. Barabash, and C. Lue. The lunar wake current systems. *Geophys. Res. Lett.*, 40(1):17–21, 2013.

[20] Feng, X. S., D. K. Zhong, C. Q. Xiang, and Y. Zhang. GPU-accelerated computing of three-dimensional solar wind background. *Sci. China, Earth Sci.*, 56(11):1864–1880, 2013.

[21] FLASH, . Flash Center for Computational Science. University of Chicago and Collaborators. `http://flash.uchicago.edu/site/index.shtml`.

[22] Friedman, A. A second-order implicit particle mover with adjustable damping. *J. Comput. Phys.*, 90(2): 292–312, 1990.

[23] Gombosi, T. I., D. L. Dezeeuw, C. P. T. Groth, K. G. Powell, C. Robert Clauer, and P. Song. *From Sun to Earth: Multiscale MHD Simulations of Space Weather*, pages 169–176. American Geophys. Union, 2003.

[24] Grimm, R. E. and G. T. Delory. Next-generation electromagnetic sounding of the Moon. *Adv. Space Res.*, 50(12):1687–1701, 2012.

[25] Halekas, J. S., D. A. Brain, and M. Holmström. *Moon's Plasma Wake*, pages 149–167. in Magnetotails in the Solar System (eds A. Keiling, C. M. Jackman and P. A. Delamere) John Wiley & Sons, Inc, 2015.

[26] Harned, D. S. Quasineutral hybrid simulation of macroscopic plasma phenomena. *J. Comput. Phys.*, 47 (452):452–462, 1982.

[27] Haugbølle, T., J. T. Frederiksen, and Å Nordlund. photon-plasma: A modern high-order particle-in-cell code. *Physics of Plasmas*, 20(6):062904, 2013.

[28] Hockney, R. W. and J. W. Eastwood. *Computer Simulation Using Particles.* Adam Hilger, Bristol and New

York, 1988.

[29] Holmström, M. Hybrid modeling of plasmas. In *Proceedings of ENUMATH, the 8th European Conference on Numerical Mathematics and Advanced Applications.* Springer, 2010.

[30] Holmström, M. Handling vacuum regions in a hybrid plasma solver. *ASTRONUM-2012, ASP conference series*, (474):202–207, 2013.

[31] Hood, L. L., F. Herbert, and C. P. Sonett. The deep lunar electrical conductivity profile: Structural and thermal inferences. *J. Geophys. Res.*, 87(B7):5311–5326, 1982.

[32] Khurana, K. K., M. G. Kivelson, D. J. Stevenson, G. Schubert, C. T. Russell, R. J. Walker, and C. Polanskey. Induced magnetic fields as evidence for subsurface oceans in Europa and Callisto. *Nature*, 395(6704):777–780, 1998.

[33] Kirk, D. B. and W. W. Hwu. *Programming Massively Parallel Processors: A Hands-on Approach.* Morgan Kaufmann Publishers, a imprint of Elsevier, 2012.

[34] Knapp, C., A. Kendl, A. Koskela, and A. Ostermann. Splitting methods for time integration of trajectories in combined electric and magnetic fields. *Phys. Rev.*, 92:063310, 2015.

[35] Kong, X., M. C. Huang, C. Ren, and V. K. Decyk. Particle-in-cell simulations with charge-conserving current deposition on graphic processing units. *J. Comput. Phys.*, 230(4):1676–1685, 2011.

[36] Krauss-Varban, D. From Theoretical Foundation to Invaluable Research Tool: Modern Hybrid Simulations. *Proceedings of the 7th International Symposium for Space Simulations (ISSI 7)*, Kyoto University:3, 2005.

[37] Ledvina, S. A., Y.-J. Ma, and E. Kallio. Modeling and Simulating Flowing Plasmas and Related Phenomena. *Space Sci. Rev.*, 139(1-4):143–189, 2008.

[38] Lipatov, A. S., . *The hybrid multiscale simulation technology: an introduction with application to astrophysical and laboratory plasmas.* Springer Science & Business Media, 2002.

[39] Markidis, S., G. Lapenta, and Rizwan-uddin. Multi-scale simulations of plasma with iPIC3D. *Math. and Comp. in Simulation*, 80(7):1509 – 1519, 2010.

[40] Marsch, E. Kinetic physics of the solar wind plasma. In *Physics of the Inner Heliosphere II*, volume 1, chapter 8, pages 45–133. Springer Berlin Heidelberg, 1991.

[41] Matthews, A. Current Advance Method and Cyclic Leapfrog for 2D Multispecies Hybrid Plasma Simulations. *J. Comput. Phys.*, 112(1):102–116, 1994.

[42] Mei, G. and H. Tian. Impact of data layouts on the efficiency of GPU-accelerated IDW interpolation. *SpringerPlus*, 5(1):104, 2016.

[43] Müller, J., S. Simon, U. Motschmann, J. Schüle, K.-H. Glassmeier, and G. J. Pringle. A.I.K.E.F.: Adaptive hybrid model for space plasma simulations. *Comp. Phys. Comm.*, 182(4):946–966, 2011.

[44] Navarro, C. A., N. Hitschfeld-Kahler, and L. Mateu. A Survey on Parallel Computing and its Applications in Data-Parallel Problems Using GPU Architectures. *Comm. Comp. Phys.*, 15, 2014.

[45] Omelchenko, Y. A. and H. Karimabadi. HYPERS: A unidimensional asynchronous framework for multiscale hybrid simulations. *J. Comp. Phys.*, 231(4), 2012.

[46] Pierrard, V. and M. Lazar. Kappa distributions: Theory and applications in space plasmas. *Solar Physics*, 267(1), 2010.

[47] Poppe, A. R., S. Fatemi, J. S. Halekas, M. Holmström, and G. T. Delory. ARTEMIS observations of extreme diamagnetic fields in the lunar wake. *Geophys. Res. Lett.*, 41(11):3766–3773, 2014.

[48] Press, W. H., S. A. Teukolsky, W. T. Vetterling, and B. P. Flannery. *Numerical Recipes, The Art of Scientific Computing.* Cambridge University Press, UK, 3 edition, 2007.

[49] Qin, H., S. Zhang, J. Xiao, J. Liu, Y. Sun, and W. M. Tang. Why is Boris algorithm so good? *Phys. Plasmas*, 20(8), 2013.

[50] Rieke, M., T. Trost, and R. Grauer. Coupled Vlasov and two-fluid codes on GPUs. *J. Comp. Phys.*, 283: 436–452, 2015.

[51] Stantchev, G., W. Dorland, and N. Gumerov. Fast parallel Particle-To-Grid interpolation for plasma PIC simulations on the GPU. *J. Parallel and Distributed Computing*, 68(10):1339–1349, 2008.

[52] Strzodka, R. Abstraction for AoS and SoA Layout in C++. In *GPU Computing Gems Jade Edition*, chapter 31. Addison-Wesley Professional, 2012.

[53] Vay, J.-L., P. Colella, J. W. Kwan, P. McCorquodale, D. B. Serafini, a. Friedman, D. P. Grote, G. Westenskow, J.-C. Adam, a. Heron, and I. Haber. Application of adaptive mesh refinement to particle-in-cell simulations of plasmas and beams. *Phys. Plasmas*, 11(5), 2004.

[54] Wang, P., T. Abel, and R. Kaehler. Adaptive mesh fluid simulations on GPU. *New Astronomy*, 15(7): 581–589, 2010.

[55] Wilt, N. *The CUDA handbook; A comprehensive guide to GPU programming.* Addison Wesley, 2013.

[56] Winske, D. Hybrid Simulation Codes with Application to Shocks and Upstream Waves. *Space Plasma Simulations*, 42:53–66, 1985.

[57] Winske, D. and N. Omidi. A nonspecialist's guide to kinetic simulations of space plasmas. *J. Geophys. Res.*,

101(A8):17287, 1996.
[58] Winske, D. and L. Yin. Hybrid codes: Past, present and future. *Proceedings of ISSS-6*, 6, 2001.

New High-order Methods using Gaussian Processes for Computational Fluid Dynamics Simulations

Dongwook Lee[1], Adam Reyes[2], Carlo Graziani[3], Petros Tzeferacos[3,4]

[1] Applied Mathematics and Statistics, University of California, Santa Cruz, CA, U.S.A, [2] Department of Physics, University of California, Santa Cruz, CA, U.S.A, [3] Flash Center for Computational Science, Department of Astronomy & Astrophysics, University of Chicago, IL, U.S.A, [4] Physics, Oxford University, U.K

Abstract. We present an entirely new class of high-order numerical algorithms for computational fluid dynamics simulations. The new method is based on the Gaussian Processes (GP) modeling that generalizes the Gaussian probability distribution. Our approach is to adapt the idea of the GP prediction technique that utilizes the covariance kernel functions, and use the GP prediction to interpolate/reconstruct a high-order approximations for solving hyperbolic PDEs. We propose the GP high-order method as a new class of numerical high-order formulations, alternative to the conventional polynomial-based approaches.

1. Introduction

Cutting edge simulations of gas dynamics and magnetohydrodynamics (MHD) have been among the headliner applications of scientific high-performance computing (HPC) [1, 2, 3, 4]. One notable HPC development is that the newer architectures are expected to feature a dramatical change in the current balance between computation and memory resources, with memory per compute core declining drastically from current levels [2, 3, 5]. This trend tells us that there need to be algorithmic strategies to save memory at the cost of increased computation. Recently, this paradigm shift has become a high priority for HPC applications. In the context of numerical methods for computational fluid dynamics, an important approach to accomplish such strategies is the design of very high-order of accuracy methods [3]. As compared to low-order methods, high-order numerical methods achieve an increased target solution accuracy by computing increased higher-order floating-point approximations on a given grid resolution [6, 7, 8], thus embodying, in a concrete manner, the desired tradeoff between memory and computation by exercising more computation per memory (or equivalently, the equal amount of computation with less memory).

Within the broad framework of FD and FV discretization methods, discrete algorithms of data interpolation and reconstruction play a key role in numerical methods for PDE integration [6, 7, 9]. They are frequently a limiting factor in a numerical scheme's overall convergence rate, efficiency, and algorithmic complexity. Interpolation and reconstruction are not only essential for estimating high-order accurate approximations of discrete solutions; but also of interface tracking; of prolonging states from coarse zones to corresponding refined zones in adaptive-mesh refinement (AMR) schemes; and of various other contexts associated with high-order solutions as well. In CFD simulations, these interpolation and reconstruction algorithms therefore must be carried out as accurately as possible, because, by and large, their accuracy is one of the key main factors that determines the overall accuracy of each simulation.

Polynomial based approach has been most successful and popular among interpolation/reconstruction methods in this field. It is easy to understand why this should be the case. In the first place, it is easily relatable to Taylor expansion, the most familiar of function approximations. Secondly, it is well-studied that the nominal N-th order accuracy of polynomial interpolation/reconstruction is derived from using polynomials of degree $N-1$, bearing a leading term of the error that scales with $\mathcal{O}(\Delta^N)$ as the local grid spacing Δ approaches to zero [6, 7, 9]. However, the simplicity of polynomial interpolation/reconstruction has costs. In the first instance, the polynomial approach is notoriously prone to oscillations in data fitting, especially with discontinuous data [16]. Further, in many practical situations the high-order interpolation/reconstruction is carried out on a series of 1D stencils in each of the normal directions. In this sense, such 1D based polynomial interpolation/reconstruction works at the cost of overlooking available information from the neighborhood in transverse directions, which could otherwise be used to improve the accuracy and stability of the interpolation/reconstruction. Moreover, the algorithmic complexity of such schemes typically grows with order of accuracy [17]. Genuinely multidimensional polynomial based schemes can be constructed at the cost of greater computational effort and with more restrictions on mesh geometry [**?**].

We present a new high-order approach that readily overcomes the aforementioned issues in the polynomial approaches by developing the broader outlook afforded by the GP modeling perspective. Our new GP method is a class of high-order schemes designed for numerical evolution of hyperbolic PDEs, $u_t + \nabla \cdot F(u) = 0$. In this paper we describe new high-order GP approximation strategies in two steps focusing on one-dimensional formulation: firstly, Gaussian Process (GP) *interpolation* that works on pointwise values of $u(x_i)$, and secondly, GP *reconstruction* for finite volume methods (FVM) where volume-averaged values, $\bar{u}_i = \frac{1}{\Delta \mathcal{V}_i}\int_{\Delta \mathcal{V}_i} u(x)d\mathcal{V}$, are considered. The first will provide a baseline formulation of using GP as a new high-order interpolator, while the latter will serve as a high-order reconstructor by applying the volume averaged integration to the mathematical components derived from the first. The second approach will provide a scheme that is better suited for compressible fluid flows including shocks and discontinuities.

The theory of Gaussian Processes, and more generally of stochastic functions, goes back to the work of Wiener [18] and Kolmogorov [19]. Modern-day applications are numerous. Just in the physical sciences, GP prediction is in common use in meteorology, geology, and time-series analysis [20, 21], and in cosmology, where GP models furnish the standard description of the Cosmic Microwave Background [22]. Applications abound in many other fields, in particular wherever spatial or time-series data (or data otherwise scattered on some manifold) requires "nonparametric" modeling [21, 23]. Our goal in this study is the use of the ideas of predictive GP modeling that is processed by training observed data, e.g., cell-centered initial conditions of fluid variables. In what follows, we will stick to a relatively informal level of technical detail for the sake of brevity. Readers who wish to pursue the subject in greater detail are referred to [21, 23].

2. Gaussian Process Modelings

Gaussian Processes are a class of stochastic processes, that is, processes that yield sampling data (or functions) from an infinite dimensional function space. The function space is probabilistically constrained by specifying *a prior probability* to every function in the function space, where the functions are not exactly known. Random functions (or samples) drawn from of the prior are regarded as data, and the model is "trained or learned" on the data by means of Bayes' theorem, producing an updated probabilistic constraint on the unknown functions. In this way GP produces a data-informed *posterior* model (an updated posterior mean function and a posterior covariance kernel function) from an "agnostic" *prior* model (a prior mean and a covariance kernel functions). The updated posterior mean function from the posterior is a data-informed mean that is to be used for a data prediction in a GP sense. This GP prediction, by means of the updated mean function, is our target interpolation/reconstruction for FDM and FVM.

A GP is fully defined by two functions: (i) a mean function $\bar{f}(\mathbf{x}) = \mathbb{E}[f(\mathbf{x})]$ over $\mathbb{R}^N$, and (ii) a covariance function which is a symmetric, positive-definite integral kernel $K(\mathbf{x}, \mathbf{y})$ over $\mathbb{R}^N \times \mathbb{R}^N$. For instance, a function $f(\mathbf{x})$ represents a mathematical object describing a fluid density ρ. Such functions f drawn randomly from this distribution are said to be sampled from a Gaussian Process with mean function $\bar{f}(\mathbf{x})$ and covariance function $K(\mathbf{x}, \mathbf{y})$, and we write $f \sim \mathcal{GP}(\bar{f}, K)$. As with the case of finite-dimensional Gaussian distributions, the significance of the covariance is

$$K(\mathbf{x}, \mathbf{y}) = \mathbb{E}[\left(f(\mathbf{x}) - \bar{f}(\mathbf{x})\right)\left(f(\mathbf{y}) - \bar{f}(\mathbf{y})\right)], \tag{1}$$

where the averaging is over the GP distribution.

In standard statistical modeling practice, both $\bar{f}(\mathbf{x})$ and $K(\mathbf{x}, \mathbf{y})$ are typically parametrized functions, with parameters controlling the character (e.g. scale lengths, differentiability, oscillation strength) of "likely" functions. Given a GP, and given n "training" points $\mathbf{x}_i$, $i = 1, \ldots, n$ at which the function values $f(\mathbf{x}_i)$ are known, we may calculate the likelihood $\mathcal{L}$ (the probability of $\mathbf{f}$ given the GP model) of the data vector $\mathbf{f} \equiv [f(\mathbf{x}_1), \ldots, f(\mathbf{x}_n)]^T$ (e.g., n many point-wise values of density ρ at $\mathbf{x}_i$, $i = 1, \ldots, n$) by

$$\mathcal{L} \equiv P(\mathbf{f}) = (2\pi)^{-N/2} \det |\mathbf{K}|^{-1/2} \exp\left[-\frac{1}{2}\left(\mathbf{f} - \bar{\mathbf{f}}\right)^T \mathbf{K}^{-1}\left(\mathbf{f} - \bar{\mathbf{f}}\right)\right], \tag{2}$$

where $\mathbf{K} = [K_{ij}]_{i,j=1,\ldots,n}$ with $K_{ij} \equiv K(\mathbf{x}_i, \mathbf{x}_j)$.

Next, given the function samples $\mathbf{f} \equiv [f(\mathbf{x}_1), \ldots, f(\mathbf{x}_n)]^T$ obtained at spatial points $\mathbf{x}_i$, $i = 1, \ldots, n$, we wish to make probabilistic statements (or GP predictions) about the value $f_* \equiv f(\mathbf{x}_*)$ of the unknown function $f \sim \mathcal{GP}(\bar{f}, K)$ at a new spatial point $\mathbf{x}_*$. In other words, we are interested in making a new prediction of GP for f at any randomly chosen point $\mathbf{x}_*$. This is particularly of interest to us from the perspectives of FDM and FVM, because in this way, we can use GP to predict a function value (e.g., density flux) at cell interfaces $\mathbf{x}_* = x_{i+\frac{1}{2}}$, at which both FDM and FVM seek for the best possible high-order approximations of flux functions in designing conservative discrete hyperbolic PDEs. This can be done in GP as follow. We look at the augmented likelihood function $\mathcal{L}_*$ by considering the joint distribution of the currently available training outputs, $\mathbf{f}$, and the new test output f_*,

$$\mathcal{L}_* \equiv P\left(\mathbf{f}, f_*\right) = (2\pi)^{-(n+1)/2} \det |\mathbf{M}|^{-1/2} \exp\left[-\frac{1}{2}\left(\mathbf{g} - \bar{\mathbf{g}}\right)^T \mathbf{M}^{-1}\left(\mathbf{g} - \bar{\mathbf{g}}\right)\right], \tag{3}$$

where $\mathbf{g}$ and $\bar{\mathbf{g}}$ are the $(n+1)$-dimensional vectors whose components, in partitioned form, are

$$\mathbf{g}^T \equiv [f_*, \mathbf{f}], \quad \bar{\mathbf{g}}^T \equiv \left[\bar{f}(\mathbf{x}_*), \bar{\mathbf{f}}\right], \tag{4}$$

and $\mathbf{M}$ is the $(n+1) \times (n+1)$ augmented covariance matrix, given in partitioned form by

$$\mathbf{M} = \begin{pmatrix} k_{**} & \mathbf{k}_*^T \\ \mathbf{k}_* & \mathbf{K} \end{pmatrix}. \tag{5}$$

In Eq. (5), we've defined a scalar k_{**} and an n-dimensional vector $\mathbf{k}_* = [\mathbf{k}_{*,i}]_{i=1,\ldots,n}$ given by

$$k_{**} \equiv K(\mathbf{x}_*, \mathbf{x}_*), \quad \mathbf{k}_{*,i} \equiv K(\mathbf{x}_*, \mathbf{x}_i). \tag{6}$$

Using Bayes' Theorem, the conditioning property applied to the joint Gaussian prior distribution on the observation $\mathbf{f}$ yields the Gaussian posterior distribution of f_* given $\mathbf{f}$. As a result, the main outcome of the GP prediction [21] can be derived:

$$P(f_*|\mathbf{f}) = \left(2\pi U^2\right)^{-1/2} \exp\left[-\frac{1}{2}\frac{\left(f_* - \bar{f}_*\right)^2}{U^2}\right], \tag{7}$$

where the newly *updated posterior mean function*

$$f_* \equiv \bar{f}(\mathbf{x}_*) + \mathbf{k}_*^T \cdot \mathbf{K}^{-1}\left(\mathbf{f} - \bar{\mathbf{f}}\right), \tag{8}$$

and the newly *updated posterior covariance kernel function*

$$U^2 \equiv k_{**} - \mathbf{k}_*^T \cdot \mathbf{K}^{-1} \cdot \mathbf{k}_*. \tag{9}$$

We emphasize here that, in Eq. (8), the GP on the unknown function f has resulted in a Gaussian probability distribution on the unknown function value f_*, providing a probabilistic mean prediction of f_* at a new desired location $\mathbf{x}_*$ as given in Eq. (8), whose uncertainty is given by the posterior covariance in Eq. (9).

2.1. GP Interpolation

Although in this paper we are more interested in developing a high-order reconstruction method based on the volume averaged data, we consider first an interpolation method based on the pointwise data. We will see that an algorithmic design for GP interpolation of pointwise data will provide a good mathematical foundation for FVM which reconstructs pointwise values from volume averaged data.

The mean f_* of the distribution given in Eq. (8) is our interpolation of the function f at the point $\mathbf{x}_*$, where f is any given fluid variable such as density, pressure, velocity, flux functions, etc. For an exposition purpose, let us use u to denote a fluid variable (e.g., density). The GP interpolation is then to operate on $f = u$, predicting an output value $\hat{u}$ with a highly accurate approximation, say at $x_{i+\frac{1}{2}}$, $\{\hat{u}_{i+\frac{1}{2}}\} = \mathcal{I}_{GP}(u_{i-p}, \dots, u_{i+q})$, where $\mathcal{I}_{GP}(\cdot)$ is the GP interpolation given in Eq. (8). As shown in Eq. (8), the interpolant f_* is a simple linear combination of the observed data $\mathbf{f}$ and the covariance kernels $\mathbf{k}_*, \mathbf{K}$, anchored by one of its arguments to one of the data points, $\mathbf{x}_*, \mathbf{x}_1, \dots, \mathbf{x}_n$. The term added to the mean function $\bar{f}(\mathbf{x}_*)$ in Eq. (8) may also be cast as an inner product between a vector of weights $\mathbf{w} \equiv \mathbf{K}^{-1}\mathbf{k}_*$ and a vector of data residuals $(\mathbf{f} - \bar{\mathbf{f}})$. The noteworthy fact about this is that the weights $\mathbf{w}$ are absolutely independent of the data values $\mathbf{f}$; and they depend only on the n locations of the data points $\mathbf{x}_i$ and on the desired interpolation point $\mathbf{x}_*$. This means that, for an interpolation scheme in which the training point locations (viz. the stencil) and the interpolation point (viz. a cell interface) are known in advance — which is often the case as initial grid configurations are set in static grid setups — the weight vector $\mathbf{w}$ may be computed and stored in advance, as an initialization, and can remain constant throughout the simulation. In case that an adaptive mesh refinement (AMR) configuration is under consideration, $\mathbf{w}$ may be computed at all possible grid refinement levels and stored a priori and used later. Another more efficient way is, a new set of weight vectors $\mathbf{w}$ can be newly computed only once and stored whenever there is a new change in grid resolution on any local AMR region during simulations. The GP interpolations then come at the cost of an inner product of a vector whose size is the number of points in the stencil of the interpolation $[x_{i-p}, x_{i+q}]$. Typically, the sizes of the stencil are three for 2nd order piecewise linear methods [24, 9]; five for both the 3rd order piecewise parabolic [25] and 5th order WENO methods [26]. This fact assures that the sizes of the linear system in Eq. (8) are not large at all, e.g., $\mathbf{K}$ is going to be a 5×5 matrix when using the stencil of five.

An important feature of the GP interpolation is that it naturally supports multidimensional stencil configuration. This can be easily seen in the arguments of the covariance kernels. This multidimensional stencil is a qualitative advantage of the GP interpolation over the usual 1D-based polynomial interpolations, in that, the GP approach does not incur a great cost in algorithmic design due to the inherent complexity that arises in multidimensional polynomial approaches especially when designing high-order algorithms [27, 13, 15]. The inversion of $\mathbf{K}$ – whose size is not big – is also computationally efficient with Cholesky decomposition because the matrix is symmetric positive definite, about a factor two faster than the usual LU decomposition for inversion.

There is an additional piece of information beyond the point estimate f_* – we also have an uncertainty in the estimate, given by U in Eq. (9). This posterior uncertainty is of crucial importance in many GP modeling applications, but it is of limited interest for the purpose of this paper. We will overlook posterior uncertainty in the current study, and focus on the posterior mean formula given in Eq. (8).

2.2. GP Reconstruction

In FVM the fluid variables to be evolved are not point-wise values. Rather, they are volume-averaged quantities, $\bar{u}_i = \frac{1}{\Delta\mathcal{V}_i}\int_{\Delta\mathcal{V}_i} u(x)d\mathcal{V}$. The GP interpolation strategy discussed so far should therefore be modified in order to fit well with the data type in FVM. Note that the integral averages over cell (and face) constitute linear operations on a function $f(\mathbf{x})$. As with ordinary finite-dimensional multivariate Gaussian distributions, where linear operations on Gaussian random variables result in new Gaussian random variables with linearly transformed means and covariances, a set of m linear functionals operating on a GP-distributed function f has an m-dimensional Gaussian distribution with mean and covariance that are linear functionals of the GP mean function and covariance function.

Suppose in a FVM sense, for example, that we have m measures $dg_i(\mathbf{x})$, $i = 1, \ldots, m$, defining m linear functionals,

$$G_i \equiv \int dg_\alpha(\mathbf{x})\, f(\mathbf{x}),\ i = 1, \ldots, m. \tag{10}$$

Then, the vector $\mathbf{G} = [G_1, \ldots, G_m]^T$ is normally distributed with mean $\bar{\mathbf{G}} = [\bar{G}_1, \ldots, \bar{G}_m]^T$ and covariance matrix $\mathbf{C} = [\mathbf{C}_{i,j}]_{i,j=1,\ldots,m}$, where

$$\bar{G}_i = \mathbb{E}[G_i] = \int dg_i(\mathbf{x})\, \mathbb{E}[f(\mathbf{x})] = \int dg_i(\mathbf{x})\, \bar{f}(\mathbf{x}), \tag{11}$$

and

$$\begin{aligned} \mathbf{C}_{i,j} &= \mathbb{E}[(G_i - \bar{G}_i)\,(G_j - \bar{G}_j)] \\ &= \int dg_i(\mathbf{x})\, dg_j(\mathbf{y})\, \mathbb{E}[(f(\mathbf{x}) - \bar{f}(\mathbf{x}))\,(f(\mathbf{y}) - \bar{f}(\mathbf{y}))] \\ &= \int dg_i(\mathbf{x})\, dg_j(\mathbf{y})\, K(\mathbf{x}, \mathbf{y}). \end{aligned} \tag{12}$$

Thus, we see that the GP distribution on the function f leads to a multivariate Gaussian distribution on any m-dimensional vector $\mathbf{G}$ of linear functionals of f. This Gaussian distribution can be used for likelihood maximization in a manner completely analogous to the more standard case where training is performed using point-wise data in the case of FDM, rather than averages in FVM.

For reconstruction, it is also not hard to generalize Eq. (8) – define the m-dimensional *prediction vector* $\mathbf{T}_* = [\mathbf{T}_{*,i}]_{i=1,\ldots,m}$ at $\mathbf{x}_*$ by

$$\mathbf{T}_{*,i} \equiv \int dg_i(\mathbf{x})\, K(\mathbf{x}, \mathbf{x}_*). \tag{13}$$

Then the point-wise function value $\tilde{f}_*$ at the point $\mathbf{x}_*$, reconstructed from the volume-averaged data $\mathbf{G}$, is given by

$$\tilde{f}_* = \bar{f}(\mathbf{x}_*) + \mathbf{T}_*^T \mathbf{C}^{-1}(\mathbf{G} - \bar{\mathbf{G}}), \tag{14}$$

a straightforward generalization of Eq. (8).

For the sake of a reconstruction scheme on a mesh of control volumes, it is clear that the measures $dg_i(\mathbf{x})$ should be chosen as the cell volume-average measures,

$$dg_i(\mathbf{x}) = \begin{cases} d^N\mathbf{x}/\Delta^N, & [\mathbf{x} \in I_i], \\ 0, & [\mathbf{x} \notin I_i], \end{cases} \tag{15}$$

where Δ is the grid spacing, and I_i is the i-th cell, $I_i = \Pi_{d=1}^N I_i^{(d)}$ with 1D cell interval $I_i^{(d)} = [x_i^{(d)} - \Delta^{(d)}/2, x_i^{(d)} + \Delta^{(d)}/2]$ for each d direction (note $N = 3$ for 3D). For the purpose of the current discussion, we will assume a locally-uniform rectilinear grid of cubical cells I_i of unifom size $\Delta = \Delta^{(1)} = \Delta^{(2)} = \Delta^{(3)}$.

2.3. GP Implementation with Squared Exponential Kernel

To illustrate a single interpolation/reconstruction procedure we consider a stencil S_R of radius R points centered on the cell I_i containing $2R + 1$ points:

$$S_R = \{I_{i-R}, \ldots, I_i, \ldots, I_{i+R}\} \tag{16}$$

The fluid variables on this stencil then act as the input data (G) to the GP model in Eq. (14). In the current study, we adopt a GP model consisting of an isotropic "Squared Exponential (SE)" covariance [21],

$$K(\mathbf{x}, \mathbf{y}) \equiv \Sigma^2 \exp\left[-\frac{(\mathbf{x} - \mathbf{y})^2}{2\ell^2}\right], \tag{17}$$

and a constant mean function, $\bar{f}(\mathbf{x}) = f_0\mathbf{u}$, where $\mathbf{u} = [1, \ldots, 1]^T$. The model features three free parameters, f_0, Σ^2 , and ℓ. The later two are called "hyperparameters" which are the parameters that are built into the kernel. The parameter f_0 is determined from the data as part of the interpolation process, by maximizing the likelihood function, Eq. (2). This can easily be done analytically, and it turns out that f_0 is a weighted mean of the function data values,

$$f_0 = \begin{cases} \frac{\mathbf{u}^T\mathbf{C}^{-1}\mathbf{f}}{\mathbf{u}^T\mathbf{K}^{-1}\mathbf{u}} & \text{for interpolation,} \\ \\ \frac{\mathbf{u}^T\mathbf{C}^{-1}\mathbf{G}}{\mathbf{u}^T\mathbf{C}^{-1}\mathbf{u}} & \text{for reconstruction.} \end{cases} \tag{18}$$

The derivation is omitted here for the sake of brevity. The hyperparameter Σ^2 has no effect on the posterior mean function, hence we set $\Sigma^2 = 1$ for simplicity. The hyperparameter ℓ is the correlation scale length of the model. It determines the length scale of variation preferred by the GP model. Our GP predictions of interpolation/reconstruction, which necessarily agree with the observed values of the function at the training points $\mathbf{x}_i$, may wiggle on this scale between training points. In this sense, ℓ is a "rigidity", controlling the curvature scales of the prediction. Since we want function interpolations/reconstructions that are smooth on the scale of the grid, we certainly want $\ell > \Delta$, and would also prefer $\ell \geq R$. The optimal values of ℓ should be determined by experience with the algorithm, and can further be re-estimated by considering relevant physical length scales of the data [28].

This covariance has the property of having a native space of C^∞ functions, and one of the practical advantages of the SE covariance model stems from its dimensional factorization. Therefore, the volume averages in Eqs. (11) and (12) simplify to iterated integrals, and in fact they can be expressed *analytically* in terms of a pre-computed list of error functions of arguments proportional to one-dimensional cell center differences. Additionaly the SE kernel only depends on the spatial distance between data points, belonging to the class of isotropic kernels. This means that from the perspective of the underlying interpolating/reconstructing functions there is no additional complexity or restrictions to using genuinely multidimensional stencils. Extension of the full GP method for finite volume schemes remains a topic of future work.

Other choices of kernel functions are also available, such as the Màtern, rational quadratic and Wendland classes of covariances [23, 29]. Such functions contain additional smoothness hyperparameters that fix the smoothness of the assumed GP model. We observe that use of suitably unsmooth kernel functions can negatively impact the error convergence and overall accuracy of the GP method. In addition such choices of kernels would require numerical approximations for Eqs. (12) and (13), rather than an analytical form that is afforded in the SE case, which can further negatively affect the accuracy. Further, in the limiting case of large smoothness these functions approach the SE kernel function.

3. Numerical Results

3.1. 1D Smooth Advection

We present numerical results using GP reconstruction applied to the 1D compressible Euler equations, using a fourth-order Runge-Kutta (RK4) method for temporal advancement. The test considered here involves the passive advection of a Gaussian density profile. We initialize a computational box on [0,1] with periodic boundary conditions. The initial density profile is defined by $\rho(x) = 1 + e^{-100(x-x_0)^2}$, with $x_0 = 0.5$, with constant velocity, $u = 1$, and pressure, $P = 1/\gamma$. The specific heat ratio is chosen to be $\gamma = 5/3$. The resulting profile is propagated for one period through the boundaries. At $t = 1$, the profile returns to its initial position at $x = x_0$. Any deformation of the initial profile is due to either phase errors or numerical diffusion. We perform this test using a length hyperparameter of $\ell = 0.1$ for stencil raddii $R = 1, 2, 3, 4$ and 5, with a fixed Courant number, $C_{\text{cfl}} = 0.8$ and vary the resolution of computational box, with $N = 32, 64, 128, 256$ and 512.

The results of this study are shown in Fig. 1(a). From these numerical experiments, GP reconstruction shows convergence that goes as the size of the stencil, $2R+1$. The error plateaus at an L_1 error $\sim 10^{-12}$, a few orders of magnitude greater than double-precision. This happens because at high resolution the length hyperparameter, ℓ, becomes very large relative to the grid spacing, Δ. The covariance matrix, $\mathbf{C}$ given in Eq. (12) becomes nearly singular in the regime $\ell/\Delta \gg 1$, yielding very large condition numbers for $\mathbf{C}$. We find the plateau in the L_1 error occurs for condition numbers, $\kappa \sim 10^{18}$, corresponding to the point where the errors in inverting $\mathbf{C}$ in Eq. (14) begin to dominate.

The choice of floating point precision then has an immense impact on the possible ℓ/Δ so that the condition number errors do not dominate. To produce the results in Fig. 1 we use quadruple-precision for the calculation of $\mathbf{z}^T = \mathbf{T}_*^T\mathbf{C}^{-1}$ in Eq. (13). Since $\mathbf{z}^T$ needs to be calculated only once, before starting the simulation, it can then be truncated to double-precision for use in the actual reconstruction procedure. It should be noted that this is necessary for the purposes of a grid convergence study, which is a very unusual situation from the statistical point of view. In a practical simulation it would be an overkill to do so, because in this case the GP algorithm tries to over resolve a feature with the characteristic length ℓ on a relatively very small grid spacing Δ, yielding $\ell/\Delta \gg 1$.

The length hyperparameter, ℓ, provides an additional avenue to tune solution accuracy that is not present in polynomial based methods. Fig. 1(b) shows how the errors in the smooth-advection problem changes with the choice of ℓ, compared with the error from a fifth-order WENO5 + RK4 solution (denoted in dotted lines) [10]. At large ℓ the errors become roughly the same as in WENO5, and at small values becomes worse. The error finds a minimum at a value of ℓ near the half-width of the Gaussian density profile. This would suggest that the optimal choice of ℓ should match the physical length scale of the feature being resolved. It should be noted that, for a given value of R, varying the value of ℓ does not affect the computational efficiency of the scheme.

The result in Fig. 1(b) may give a false impression that a larger value of ℓ would just work appropriately for a wide range of general problems, thus removing the need to tune this parameter from the model consideration. However, such a case is to be considered as an over-simplified case (e.g., Gaussian advection) and only valid if a simulation under study has one dominant physical length scale that is almost invariant during the simulation. This is not true in practice because there could be a range of dynamically important length scales that can vary over time. In this case a method that allows adaptively changing ℓ locally will be ideal. We currently investigate on such an algorithm, "Adaptive ℓ Re-estimate (ALR)" method, to tune the GP model sensitivity on ℓ, the approach of which is logically similar to the standard Adaptive Mesh Refinement (AMR) algorithms.

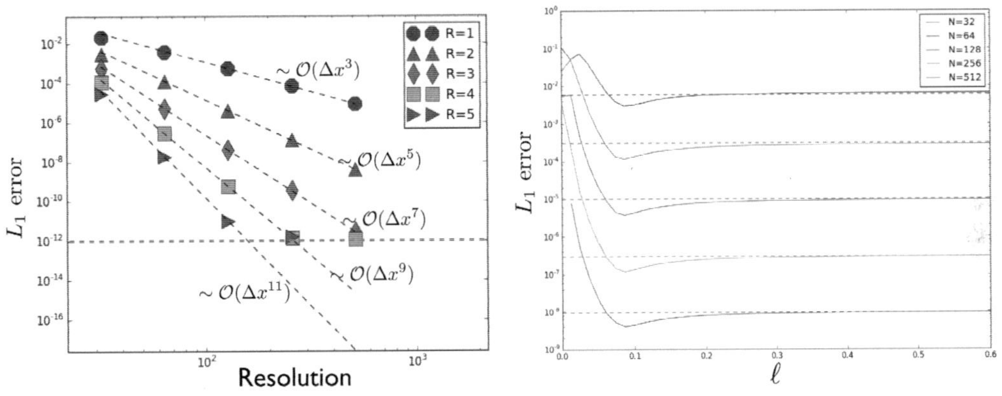

(a) (b)

Figure 1. (a) Convergence for smooth Gaussian density advection using different GP stencil radii, all using the length scale $\ell = 0.1$. Black dotted lines show $2R+1$ convergence rates, ranging from 3rd-order ($R = 1$) to 11th-order ($R = 5$). Red dotted line represents a plateau where a large condition number of the covariance matrix is obtained and no further accuracy is achievable. (b) L_1 errors as a function of the hyperparameter ℓ, using $R = 2$. Dotted lines are the error for fifth-order WENO5 + RK4 method on the same stencil.

Shown in Scheme A of Table 1 is a performance comparison of the GP method with $R = 2$ relative to three most popular piecewise polynomial methods, including first-order Godunov (FOG) [30], PPM [25] with 2nd-order characteristic tracing, and WENO5 + RK4 [10]. As demonstrated, the performance of GP + RK4 is about 40% more expensive than the usual performance of PPM with characteristic tracing, while it is more than twice faster than WENO5 + RK4. Notice that we temporally integrated GP with $R = 2$ using RK4 in order to provide a fair comparison with WENO5+RK4 in terms of delivering the equivalent order of numerical solution accuracies in both space and time. The temporal schemes used in the other two, FOG and PPM, are much cheaper and lighter, delivering the overall solution accuracies first-order and second-order respectively at most.

In Scheme B, we now compare performances of the four spatial methods, all combined with RK4. This reveals us the computational expenses only due to the four different spatial solvers. As seen, GP is only twice expensive than FOG which does not even involve computing any high-order interpolation/reconstruction at all. The relative gain of GP compared to PPM and WENO5 is very significant, in that it only requires less than half of the computing times than those of the two methods.

Scheme A	Speedup	Scheme B	Speedup
GP + RK4	1.0	GP + RK4	1.0
FOG	0.17	FOG + RK4	0.5
PPM + charTr	0.83	PPM + RK4	2.8
WENO5 + RK4	2.4	WENO5 + RK4	2.4

Table 1. (Scheme A) Relative speedup of the GP + RK4 compared with piecewise polynomial schemes of first-order Godunov (FOG), PPM, and WENO5 + RK4 for the 1D Gaussian advection problem. (Scheme B) The same relative speedup except that all spatial schemes are now integrated with RK4. The comparisons have been obtained from a serial calculation on a single CPU.

3.2. 1D Shu-Osher Shock Tube Problem

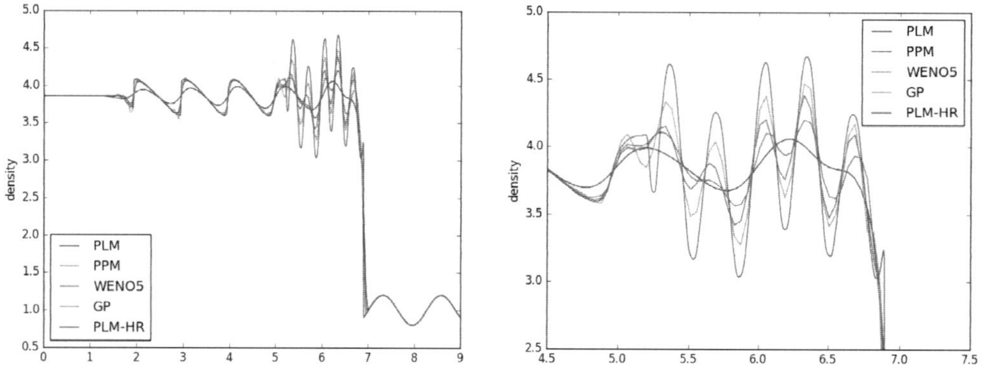

(a) (b)

Figure 2. The Shu-Osher problem. Density profiles at $t = 1.8$ computed using three different polynomial based reconstructions (PLM, PPM, & WENO5 + RK4) as well as GP + RK4 on a 200 grid resolution. The reference solution is obtained using PLM on a 1600 grid resolution (PLM-HR). 2(b) shows a closeup view of the post shock features.

The second test is the Shu-Osher problem [26] to test a shock-capturing scheme's ability to also resolve small-scale flow features in addition to the shock. The test gives a good indication of the method's numerical diffusivity, and is a popular bencmark to demonstrate numerical errors of a given method. In this problem, a (nominally) Mach 3 shock wave propogates into a constant density field with sinusoidal perturbations. As the shock advances, two sets of density features appear behind the shock. One set has the same spatial frequency as the unshocked perturbations, while in the second set the frequency is doubled and follows more closely behind the shock. The test of the numerical method then is to accurately resolve the dynamics and strengths of the oscillations behind the shock.

The results of this test are shown for various reconstruction methods in Fig. 2. The solutions are calculated at $t = 1.8$ using a resolution of $N = 200$ and are compared to a reference solution resolved on $N = 1600$. The GP solution uses $\ell/\Delta = 6$ and the shock detector on (see [31]). It is evident that the GP solution exhibits the least diffusive solution of the shown methods, producing a very-high order accurate solution that is more quickly approaching the high resolution reference solution.

4. Conclusion

We have introduced a new Gaussian Process approach in designing a family of high-order numerical algorithms to numerically solve the hyperbolic system of conservation laws. The GP methods presented here show an extremely fast rate of solution accuracy in smooth advection problems by controlling one single parameter R, strongly suggesting the new promising method is more effective than the conventional polynomial based high-order methods.

References

[1] Dongarra J J, Meuer H W, Simon H D and Strohmaier E 2010 *The Birth of Numerical Analysis* 93
[2] Dongarra J 2012 *On the Future of High Performance Computing: How to Think for Peta and Exascale Computing* (Hong Kong University of Science and Technology)
[3] Subcommittee A 2014 *US Department Of Energy Report*
[4] Keyes D E, McInnes L C, Woodward C, Gropp W, Myra E, Pernice M, Bell J, Brown J, Clo A, Connors J *et al.* 2013 *International Journal of High Performance Computing Applications* **27** 4–83
[5] Attig N, Gibbon P and Lippert T 2011 *Computer Physics Communications* **182** 2041–2046
[6] LeVeque R J 2002 *Finite volume methods for hyperbolic problems* vol 31 (Cambridge university press)
[7] LeVeque R J 2007 *Finite difference methods for ordinary and partial differential equations: steady-state and time-dependent problems* vol 98 (Siam)
[8] Hesthaven J S, Gottlieb S and Gottlieb D 2007 *Spectral methods for time-dependent problems* vol 21 (Cambridge University Press)
[9] Toro E 2009 *Riemann Solvers and Numerical Methods for Fluid Dynamics: A Practical Introduction* (Springer) ISBN 9783540252023 URL http://books.google.com/books?id=SqEjX0um8o0C
[10] Jiang G S and Shu C W 1996 *Journal of Computational Physics* **126** 202 – 228 ISSN 0021-9991 URL http://www.sciencedirect.com/science/article/pii/S0021999196901308
[11] Mignone A, Tzeferacos P and Bodo G 2010 *Journal of Computational Physics* **229** 5896–5920
[12] Lee D 2013 *Journal of Computational Physics* **243** 269–292
[13] McCorquodale P and Colella P 2011 *Communications in Applied Mathematics and Computational Science* **6** 1–25
[14] Shu C W 2009 *SIAM review* **51** 82–126
[15] Buchmüller P and Helzel C 2014 *Journal of Scientific Computing* **61** 343–368
[16] Gottlieb D and Shu C W 1997 *SIAM review* **39** 644–668
[17] Gerolymos G, Sénéchal D and Vallet I 2009 *Journal of Computational Physics* **228** 8481 – 8524 ISSN 0021-9991 URL http://www.sciencedirect.com/science/article/pii/S0021999109003908
[18] Wiener N 1949 *Extrapolation, interpolation, and smoothing of stationary time series, with engineering applications* (Cambridge: Technology Press of the Massachusetts Institute of Technology) "First published during the war as a classified report to Section D 2, National Defense Research Committee."; Stationary time series
[19] Kolmogorov A 1941 *Izv. Akad. Nauk. SSSR* **5** 3–14
[20] Wahba G, Johnson D, Gao F and Gong J 1995 *Monthly Weather Review* **123** 3358–3369
[21] Rasmussen C and Williams C 2005 *Gaussian Processes for Machine Learning* Adaptive Computation And Machine Learning (MIT Press) ISBN 9780262182539 URL http://books.google.com/books?id=vWtwQgAACAAJ
[22] Bond J, Crittenden R, Jaffe A and Knox L 1999 *Computing in science & engineering* **1** 21–35
[23] Stein M 1999 *Interpolation of Spatial Data: Some Theory for Kriging* Springer Series in Statistics Series (Springer New York) ISBN 9780387986296 URL http://books.google.com/books?id=5n_XuL2Wx1EC
[24] Van Leer B 1979 *Journal of computational Physics* **32** 101–136
[25] Colella P and Woodward P R 1984 *Journal of computational physics* **54** 174–201
[26] Shu C W and Osher S 1989 *Journal of Computational Physics* **83** 32–78
[27] Zhang R, Zhang M and Shu C W 2011 *Communications in Computational Physics* **9** 807–827
[28] Ambikasaran S, Foreman-Mackey D, Greengard L, Hogg D W and O'Neil M 2016 *IEEE transactions on pattern analysis and machine intelligence* **38** 252–265
[29] Cressie N 2015 *Statistics for spatial data* (John Wiley & Sons)
[30] Godunov S K 1959 *Matematicheskii Sbornik* **89** 271–306
[31] Balsara D and Spicer D 1999 *Journal of Computational Physics* **149** 270–292

Skew-Symmetric Splitting and Stability of High Order Central Schemes

Björn Sjögreen[1], **H.C.Yee**[2] **and Dmitry Kotov**[3]

[1] Multid Analyses AB, Odinsgatan 28, Göteborg, Sweden, Bjorn.Sjogreen@multid.se
[2] NASA Ames Research Center, Moffett Field CA94035, Helen.M.Yee@nasa.gov
[3] kotov-d-v@yandex.com

Abstract. Skew-symmetric splittings of the inviscid flux derivative for high order central schemes are studied and developed to improve their numerical stability without added high order numerical dissipation for long time wave propagations and long time integration of compressible turbulent flows. For flows containing discontinuities and multiscale turbulence fluctuations the Yee & Sjogreen [33] and Kotov et al. [15, 14] high order nonlinear filter approach is utilized in conjunction with the skew-symmetric form of high order central schemes. Due to the incomplete hyperbolic nature of the conservative ideal magnetohydrodynamics (MHD) governing equations, not all of the skew-symmetric splittings for gas dynamics can be extended to the ideal MHD. For the MHD the Ducros et al. [6] variants are constructed. In addition, four formulations of the MHD are considered: (a) the conservative MHD, (b) the Godunov/Powell non-conservative form, (c) the Janhunen MHD with magnetic field source terms [13], and (d) a MHD with source terms of [3]. The different formulation of the equations in conjunction with the variants of Ducros et al. type skew-symmetric splitting will be shown to have a strong effect on the stability of non-dissipative approximations. Representative test cases for both smooth flows and problems containing discontinuities for the ideal MHD are included. The results illustrate the improved stability by using the skew-symmetric splitting as part of the central base scheme instead of the pure high order central scheme.

1. Introduction

Starting the early 80s skew-symmetric splitting of certain component of the inviscid flux derivatives in conjunction with central schemes was shown to help with numerical stability for long time integration. For a certain splitting it can provide a stable energy norm estimate for the Euler equations with smooth flows. For other skew-symmetric formulations they can provide a discrete momentum conservation or a discrete kinetic preservation property. See Arakawa, Blaisdell et al., Yee et al., Yee & Sjögreen, Sjögreen & Yee and Kotov et al. [1, 2, 35, 6, 32, 33, 21, 15, 14] for some discussions and performance of the combined approach for DNS and LES applications. Some of the skew-symmetric splitting for the gas dynamics flux derivatives are not applicable and/or cannot straightforwardly be extended to the ideal MHD [35]. Their construction is also dependent on the MHD governing equation formulation.

Entropy conservative schemes [27, 5, 31] are another class of methods that might have better stability properties than straightforward pure centered discretizations. In view of the fact that methods proposed in [27, 5, 31] are low order and their linear numerical dissipation approaches for shock-capturing require further improvement, Sjögreen & Yee combined some of these ideas to construct a form of the high order entropy conservative numerical fluxes. Starting with the high order entropy conservative development of Sjögreen & Yee [21] for gas dynamics smooth flows, construction of efficient high order conservative numerical fluxes for problems containing discontinuities and for the ideal MHD are reported in Sjögreen & Yee [23, 24]. Again, extension of high order entropy conservative numerical fluxes that were developed for gas dynamics to the MHD is not straightforward due to the non-strictly hyperbolic nature of the conservative ideal MHD equations. See [23, 24]. Some high order comparisons between skew-symmetric splittings with high order entropy conservative centered schemes can be found in [23, 24, 22].

Objective and Outline: The objective of this paper is to develop and test four variants of Ducros et al. [6] type skew-symmetric splitting methods for equations of compressible ideal MHD, including problems with discontinuities. Their stability for problems with smooth solution will be investigated. Putting the discrete div **B** error issue aside, strictly speaking, there are several slightly different, but equivalent, ways to formulate the equations of MHD. We will show that the exact formulation of the equations will have a strong effect on the stability of non-dissipative approximations. Specifically, four formulations of the MHD are considered: (a) the conservative MHD, (b) the Godunov/Powell non-conservative form, (c) the Janhunen MHD with magnetic field source terms [13], and (d) the MHD source term of [3]. Another complication is that not all skew-symmetric splitting for the perfect gas dynamics can be extended to the MHD. Furthermore, extension of some of these split schemes for the MHD consist of many variants due to the incomplete hyperbolic nature of the conservative MHD equations. For the Ducros et al. type of splitting, the inviscid flux derivatives consist of triple product functions of conservative variables terms. Note that the perfect gas dynamics conservation laws are straightly hyperbolic equations and the inviscid flux derivatives consist of double product functions of conservative variables terms. Among many variants of the Ducros et al. type skew-symmetric splitting for the MHD, only four formulations are considered for the numerical experiment. For flows containing discontinuities and multiscale turbulence fluctuations the Yee & Sjogreen [33] and Kotov et al. [15, 14] high order nonlinear filter approach is extended to include the four skew-symmetric splittings for the MHD as part (pre-processing step) of the base scheme. Different splitting formulations and comparison among the aforementioned four forms of the MHD will be illustrated.

2. An overview of skew-symmetric split approximations for gas dynamics

Standard centered difference approximations of nonlinear conservation laws normally encounter nonlinear instabilities after a short time integration without added numerical dissipation. It is well known that the appearance of these instabilities can be delayed if the convective flux derivatives are written in an equivalent desired split form before the pure central approximation is employed. Hereafter this is referred to as a split approximation.

For example, a split approximation starts from rewriting the derivative of the product $(ab)_x$ as

$$(ab)_x = \alpha(ab)_x + \gamma a b_x + \beta a_x b \tag{1}$$

before discretization. Here a and b are functions of x and α, γ and β are parameters so chosen to be still equivalent to the original $(ab)_x$ before discretization. A common split derivative is by setting $\alpha = \gamma = \beta = 1/2$ resulting in the form

$$(ab)_x = \frac{1}{2}(ab)_x + \frac{1}{2}a b_x + \frac{1}{2}a_x b \tag{2}$$

These methods have a long history in finite difference approximations; see, .e.g., [1, 16]. See also a generalized conservative split convective derivative operators study by Pirozzoli [19]. The key mathematical idea is that formulas of type (2) can be used to estimate the L^2 norm or the energy norm of the computed solution. From physical considerations some of the splittings provide the discrete conservation of momentum or preservation of discrete kinetic energy. A well-known example is the linear system of conservation laws

$$\mathbf{u}_t + A(x)\mathbf{u}_x = \mathbf{0} \quad 0 < x < L, \tag{3}$$

where $A(x)$ is a symmetric matrix, and we solve for the unknown vector $\mathbf{u} = \mathbf{u}(x,t)$ from given initial data $\mathbf{u}(x,0) = \mathbf{u}_0(x)$. Boundary data are given at $x = 0$ and $x = L$. To show how this is done, e.g., we rewrite (3) in an mathematically equivalent form:

$$\mathbf{u}_t + \frac{1}{2}(A(x)\mathbf{u})_x + \frac{1}{2}A(x)\mathbf{u}_x - \frac{1}{2}A(x)_x\mathbf{u} = \mathbf{0} \tag{4}$$

and define the scalar product and norm by

$$(\mathbf{u},\mathbf{v}) = \int_0^L \mathbf{u}^T\mathbf{v}\,dx \qquad ||\mathbf{u}||^2 = (\mathbf{u},\mathbf{u}). \tag{5}$$

A norm estimate is obtained if (4) is multiplied by $\mathbf{u}$ and integrated over $[0,L]$. We obtain

$$\frac{1}{2}\frac{d}{dt}||\mathbf{u}||^2 = -\frac{1}{2}(\mathbf{u},(A\mathbf{u})_x) - \frac{1}{2}(\mathbf{u},A\mathbf{u}_x) + \frac{1}{2}(A_x\mathbf{u},\mathbf{u}) = -\frac{1}{2}\mathbf{u}^TA\mathbf{u}|_0^L + \frac{1}{2}(A_x\mathbf{u},\mathbf{u}), \tag{6}$$

where the second equality is obtained from partial integration of $(\mathbf{u},(A\mathbf{u})_x)$, and from the symmetry of A which allows it to be moved between the arguments of the scalar product. If the boundary data are such that $\mathbf{u}^TA\mathbf{u}|_0^L \geq 0$, then the estimate

$$\frac{1}{2}\frac{d}{dt}||\mathbf{u}||^2 \leq \frac{1}{2}(A_x\mathbf{u},\mathbf{u}) \tag{7}$$

holds, which under the assumption that $max_x|A_x|$ is bounded leads to a stability estimate by use of Gronwall's lemma.

Let $x_j = j\Delta x$, $j = 0,\dots,N$ be a grid with spacing Δx, and let $\mathbf{u}_j(t)$ denote a numerical approximation of $\mathbf{u}(x_j,t)$. Consider the semi-discrete approximation of (4)

$$\frac{d}{dt}\mathbf{u}_j + \frac{1}{2}D(A(x_j)\mathbf{u}_j) + \frac{1}{2}A(x_j)D\mathbf{u}_j - \frac{1}{2}A(x_j)_x\mathbf{u}_j = \mathbf{0}, \tag{8}$$

where D is a centered finite difference operator approximating d/dx. Note that $A(x)$ is a given function, so that the exact derivative A_x can be used in (8). The discrete scalar product and norm are defined by

$$(\mathbf{u},\mathbf{v})_h = \sum_{j=0}^{N}\omega_j\mathbf{u}_j^T\mathbf{v}_j\Delta x \quad ||\mathbf{u}||_h = (\mathbf{u},\mathbf{u})_h.$$

where $\omega_j > 0$ are weights that are equal to one at most grid points, but are given special values near the boundaries $j = 0$ and $j = N$. The boundary modified norm weights together with special boundary modifications of D lead to the summation-by-parts property,

$$(\mathbf{u},D\mathbf{v})_h = -(D\mathbf{u},\mathbf{v})_h + \mathbf{u}_N^T\mathbf{v}_N - \mathbf{u}_0^T\mathbf{v}_0,$$

see [28] for details. Thanks to the summation-by-parts property, the same technique that led to the estimate (7) can be used to obtain the semi discrete estimate

$$\frac{1}{2}\frac{d}{dt}||\mathbf{u}||_h^2 \leq \frac{1}{2}(A_x\mathbf{u},\mathbf{u})_h. \tag{9}$$

The possible growth rate is determined by A_x in both (7) and (9), so that the discrete estimate will have the same growth rate as the estimate of the continuous problem.

Ducros et al. Type Conservative Splitting: For nonlinear systems, such as the Euler equations of gas dynamics, split approximations have been used for a long time see, e.g., Ducros et al and Blaisdel et al. [6, 2].

The split approximations makes use of (2) to rewrite different terms in the Euler equations as sums of three terms. The terms of the split form (2) are approximated by

$$\frac{1}{2}D(ab) + \frac{1}{2}D(a)b + \frac{1}{2}aD(b), \tag{10}$$

where D is a finite difference operator, and a and b are functions of x.

As shown in Ducros et al. [6], the approximation (10) can be written in conservation form. For example, with the second order operator $Du_j = (u_{j+1} - u_{j-1})/(2\Delta x)$, it holds that

$$\frac{1}{2}D(ab) + \frac{1}{2}D(a)b + \frac{1}{2}aD(b) = \frac{1}{4\Delta x}\Delta_+[(a_j + a_{j-1})(b_j + b_{j-1})], \tag{11}$$

where $\Delta_+ q_j = (q_{j+1} - q_j)$.

Equation (11) can be generalized to standard centered difference operators of $2p$-th order of accuracy,

$$D_p u_j = \frac{1}{\Delta x}\sum_{k=1}^{p} \alpha_k^{(p)} (u_{j+k} - u_{j-k}) \tag{12}$$

The coefficients $\alpha_k^{(p)}$ satisfy

$$\sum_{k=1}^{p} k\alpha_k^{(p)} = \frac{1}{2} \qquad \sum_{k=1}^{p} \alpha_k^{(p)} k^{2n+1} = 0, \quad n = 1, \ldots, p-1. \tag{13}$$

To derive the conservative form of the split approximation for an arbitrary operator, the right hand side of the algebraic identity

$$\begin{aligned} a_{j+k}b_{j+k} - a_{j-k}b_{j-k} + (a_{j+k} - a_{j-k})b_j + a_j(b_{j+k} - b_{j-k}) = \\ (a_{j+k} + a_j)(b_{j+k} + b_j) - (a_j + a_{j-k})(b_j + b_{j-k}) \end{aligned} \tag{14}$$

is written on conservative form by

$$\begin{aligned} &(a_{j+k} + a_j)(b_{j+k} + b_j) - (a_j + a_{j-k})(b_j + b_{j-k}) = \\ &\sum_{m=0}^{k-1}(a_{j-m} + a_{j+k-m})(b_{j-m} + b_{j+k-m}) - \sum_{m=0}^{k-1}(a_{j-1-m} + a_{j-1+k-m})(b_{j-1-m} + b_{j-1+k-m}) \end{aligned} \tag{15}$$

The conservative form of the split approximation becomes

$$\begin{aligned} &\frac{1}{2}D_p(ab) + \frac{1}{2}D_p(a)b + \frac{1}{2}aD_p(b) = \\ &\quad \frac{1}{\Delta x}\sum_{k=1}^{p}\frac{1}{2}\alpha_k^{(p)}\left((a_{j+k}b_{j+k} - a_{j-k}b_{j-k}) + a_j(b_{j+k} - b_{j-k}) + (a_{j+k} - a_{j-k})b_j\right) \\ &\quad = \frac{1}{\Delta x}\sum_{k=1}^{p}\frac{\alpha_k^{(p)}}{2}\left(\sum_{m=0}^{k-1}(a_{j-m} + a_{j+k-m})(b_{j-m} + b_{j+k-m})\right. \\ &\quad \left. - \sum_{m=0}^{k-1}(a_{j-1-m} + a_{j-1+k-m})(b_{j-1-m} + b_{j-1+k-m})\right) = \frac{1}{\Delta x}(h_{j+1/2} - h_{j-1/2}) \end{aligned} \tag{16}$$

where the numerical flux is defined by

$$h_{j+1/2} = \sum_{k=1}^{p} \frac{1}{2}\alpha_k^{(p)} \sum_{m=0}^{k-1} (a_{j-m} + a_{j+k-m})(b_{j-m} + b_{j+k-m}) \tag{17}$$

To simplify the formulas of the conservative form of split approximations for systems of equations, define

$$\Theta_{j+1/2}^{(p)}(a,b) = \sum_{k=1}^{p} \frac{1}{2}\alpha_k^{(p)} \sum_{m=0}^{k-1} (a_{j-m} + a_{j+k-m})(b_{j-m} + b_{j+k-m}). \tag{18}$$

For the three dimensional Euler equations of gas dynamics equations, the x-direction flux is

$$\mathbf{f} = (\rho u, \rho u^2 + p, \rho uv, \rho uw, (e+p)u)^T,$$

where the inviscid flux $\mathbf{f} = \mathbf{f}(\mathbf{u})$, the velocity vector $\mathbf{u} = (u(x,y,z), v(x,y,z), z(x,y,z))^T$ in the x-, y-, and z-directions, ρ denotes the density, p is the pressure, and e is the total energy. Denote $\mathbf{u}_j = (u_j, v_j, w_j)^T$ as the discretization at the j grid location with the y and z discretiztion indices suppress for simplicity. The flux components can be written as products of two factors in many different ways, leading to different split approximations. One Ducros et al. split-type approximation of the gas dynamics flux derivative that will be used in this study is given by

$$\mathbf{f}_x|_{x=x_j} \approx \begin{pmatrix} \frac{1}{2}D\rho_j u_j + \frac{1}{2}\rho_j D u_j + \frac{1}{2}u_j D\rho_j \\ \frac{1}{2}D\rho_j u_j^2 + \frac{1}{2}\rho_j u_j D u_j + \frac{1}{2}u_j D\rho_j u_j + Dp_j \\ \frac{1}{2}D\rho_j u_j v_j + \frac{1}{2}\rho_j v_j D u_j + \frac{1}{2}u_j D\rho_j v_j \\ \frac{1}{2}D\rho_j u_j w_j + \frac{1}{2}\rho_j w_j D u_j + \frac{1}{2}u_j D\rho_j w_j \\ \frac{1}{2}D u_j(e_j + p_j) + \frac{1}{2}u_j D(e_j + p_j) + \frac{1}{2}(e_j + p_j) D u_j \end{pmatrix}. \tag{19}$$

which by (17) can be written on conservative form with numerical flux function

$$\mathbf{h}_{j+1/2} = \frac{1}{2}\sum_{k=1}^{p} \alpha_k^{(p)} \sum_{m=1}^{k-1} \begin{pmatrix} (\rho_{j-m} + \rho_{j+k-m})(u_{j-m} + u_{j+k-m}) \\ (\rho_{j-m}u_{j-m} + \rho_{j+k-m}u_{j+k-m})(u_{j-m} + u_{j+k-m}) + p_{j-m} + p_{j+k-m} \\ (\rho_{j-m}v_{j-m} + \rho_{j+k-m}v_{j+k-m})(u_{j-m} + u_{j+k-m}) \\ (\rho_{j-m}w_{j-m} + \rho_{j+k-m}w_{j+k-m})(u_{j-m} + u_{j+k-m}) \\ (e_{j-m} + p_{j-m} + e_{j+k-m} + p_{j+k-m})(u_{j-m} + u_{j+k-m}) \end{pmatrix} \tag{20}$$

The more compact notation introduced in (18) allows (20) to be rewritten as

$$\mathbf{h}_{j+1/2} = \begin{pmatrix} \Theta_{j+1/2}^{(p)}(\rho, u) \\ \Theta_{j+1/2}^{(p)}(\rho u, u) + \Theta_{j+1/2}^{(p)}(p, 1) \\ \Theta_{j+1/2}^{(p)}(\rho v, u) \\ \Theta_{j+1/2}^{(p)}(\rho w, u) \\ \Theta_{j+1/2}^{(p)}(e + p, u) \end{pmatrix} \tag{21}$$

A Natural Nonconservative Splitting (not in the Ducros et al. type category):
The homogeneity property of the perfect gas dynamics inviscid flux implies that $\mathbf{f}(\mathbf{u}) = A(\mathbf{u})\mathbf{u}$, where $A(\mathbf{u})$ is the Jacobian of $\mathbf{f}(\mathbf{u})$. To make use of the homogeneity property, a non-conservative natural splitting is

$$\frac{1}{2}\mathbf{f}_x + \frac{1}{2}A\mathbf{u}_x + \frac{1}{2}A_x\mathbf{u} \tag{22}$$

IOP Conf. Series: Journal of Physics: Conf. Series **837** (2017) 012019 doi:10.1088/1742-6596/837/1/012019

where the discretization is

$$\frac{d}{dt}\mathbf{u}_j + \frac{1}{2}D_p\mathbf{f}_j + \frac{1}{2}A_jD_p\mathbf{u}_j + \frac{1}{2}D_p(A_j)\mathbf{u}_j = 0. \tag{23}$$

Here A_x and D_pA denote element wise application of differentiation and differencing respectively. The approximation (23) can be rewritten of conservative form with numerical flux

$$\mathbf{h}_{j+1/2} = \sum_{m=1}^{5} \begin{pmatrix} \Theta^{(p)}_{j+1/2}(A_{1,m}, u_m) \\ \Theta^{(p)}_{j+1/2}(A_{2,m}, u_m) \\ \Theta^{(p)}_{j+1/2}(A_{3,m}, u_m) \\ \Theta^{(p)}_{j+1/2}(A_{4,m}, u_m) \\ \Theta^{(p)}_{j+1/2}(A_{5,m}, u_m) \end{pmatrix}$$

where $A_{k,m}$ denotes element (k, m) of the matrix valued function $A(x)$, and u_m denotes the mth componen of the vector $\mathbf{u}$.

A Semi-conservative Entropy Splitting of the Euler Flux Derivatives:

Another splitting that gives entropy stability of the Euler equation of gas dynamics is by Olsson & Oliger, Gerritsen & Olsson, and Yee et al. [17, 11, 35]. They made used of Harten's symmetrizable form of the Euler equations in terms of the entropy variables [9] to obtain a semi-discrete splitting of the Euler equations with a discrete entropy stability by the summation-by-parts approach. During the computations, the entropy splitting is written in terms of the sum of a conservative portion for the interior scheme (interior grid points) and with a summation-by-parts for the boundary scheme (boundary points). Note that the Harten [9] and Gerritsen & Olsson entropy splitting form selects the un-physical branch of the inequality and was later corrected by Yee et al., hereafter referred to as the entropy splitting of the Euler equations. It is considered to be a semi-conservative splitting except at the boundary grid points. The entropy splitting of Olsson & Oliger, Gerritsen & Olsson, and Yee et al. [17, 11, 35] is a splitting which is of a form that is more suitable for the discrete stable energy norm estimate technique, including boundary scheme estimate for arbitrary order of central spatial schemes. See Yee et al. [36] for the formulation. For the 1D Euler equations the inviscid flux derivative $\mathbf{f}(\mathbf{u})_x$ for a perfect gas is split into the following via the entropy variables W discussed in Harten [9].

$$\mathbf{f}_x = \frac{\beta}{\beta+1}\mathbf{f}_x + \frac{1}{\beta+1}\mathbf{f}_W W_x, \quad \beta \neq -1 \tag{24}$$

$$W = [w_1, \quad w_2, \quad w_3, \quad w_3, \quad w_5]^T = \frac{p^*}{p}[e + \frac{\alpha-1}{\gamma-1}p, \quad -\rho u, -\rho v, \quad \rho]^T, \tag{25}$$

where

$$p^* = -(p\rho^{-\gamma})^{\frac{1}{\alpha+\gamma}} \tag{26}$$

and

$$\beta = \frac{\alpha+\gamma}{1-\gamma}, \quad \alpha > 0 \ or \ \alpha < -\gamma. \tag{27}$$

See Yee et al. [36, 35, 25] for the formulation, the choice for α and numerical examples.

Several split discretizations were compared in [10] where discretization by the entropy splitting form was shown by numerical experiments to be one of the best performing for smooth flows. For their skew-symmetric splitting extension to the ideal MHD, see Sjögreen & Yee and Sjögreen et al. [22].

IOP Conf. Series: Journal of Physics: Conf. Series **837** (2017) 012019 doi:10.1088/1742-6596/837/1/012019

3. MHD formulations

The equations of magnetohydrodynamics (MHD) is the system of conservation laws

$$\mathbf{u}_t + \mathbf{f}_x + \mathbf{g}_y + \mathbf{h}_z + \mathbf{e}\,\mathrm{div}\,\mathbf{B} = \mathbf{0}, \tag{28}$$

where the unknown field vector is

$$\mathbf{u} = (\rho,\ u,\ v,\ w,\ e,\ B^{(x)},\ B^{(y)},\ B^{(z)})^T$$

where ρ is density, $(u,\ v,\ w)$ are the velocities in the x, y and z directions, e the internal energy and $(B^{(x)},\ B^{(y)},\ B^{(z)})$ the magnetic field components. The x-direction flux is given by

$$\mathbf{f} = \begin{pmatrix} \rho u \\ \rho u^2 + p + \frac{1}{2}|\mathbf{B}|^2 - B^{(x)}B^{(x)} \\ \rho uv - B^{(x)}B^{(y)} \\ \rho uw - B^{(x)}B^{(z)} \\ u(e + p + \frac{1}{2}|\mathbf{B}|^2) - B^{(x)}\mathbf{u}^T\mathbf{B} \\ 0 \\ uB^{(y)} - vB^{(x)} \\ uB^{(z)} - wB^{(x)} \end{pmatrix},$$

and similar expressions hold for $\mathbf{g}$ and $\mathbf{h}$. Here p denotes the pressure

$$p = (\gamma - 1)(e - \frac{1}{2}\rho|\mathbf{w}|^2 - \frac{1}{2}|\mathbf{B}|^2),$$

where $\mathbf{w} = (u,\ v,\ w)$ and $\mathbf{B} = (B^{(x)},\ B^{(y)},\ B^{(z)})$. In the last term on the left hand side of (28)

$$\mathbf{e} = \mathbf{e}_G = (0,\ B^{(x)},\ B^{(y)},\ B^{(z)},\ \mathbf{w}^T\mathbf{B},\ u,\ v,\ w)^T \tag{29}$$

multiplies $\mathrm{div}\,\mathbf{B}$. This term could be removed, because $\mathrm{div}\,\mathbf{B} = 0$ from a physical standpoint. Numerically, if we can keep the discrete form of $\mathrm{div}\,\mathbf{B}$ to be very small, e.g., in the order of 10^{-6}, this non-conservative form is a symmetrizable system with a complete set of eigenvectors. Godunov [12] was the first to show that (28) can be symmetrized.

Some variants of (28) are to replace $\mathbf{e}$ by either

$$\mathbf{e}_J = (0,\ 0,\ 0,\ 0,\ 0,\ u,\ v,\ w)^T$$

as suggested in [13] or by

$$\mathbf{e}_B = (0,\ B^{(x)},\ B^{(y)},\ B^{(z)},\ 0,\ 0,\ 0,\ 0)^T$$

as suggested in [3]. The conservative form (28) is obtained by setting $\mathbf{e} = \mathbf{0}$ and denoted by $\mathbf{e}_0$

4. Ducros et al. type split approximations for MHD

There is no unique guiding principle for the Ducros splitting. Here, four different splitting schemes are considered. Since splitting (19) works well for non-MHD gas dynamics, the guiding principle we employed is to require that splitting for the MHD become (19) when the magnetic field is zero. The formulas show splitting of the x-direction flux, $\mathbf{f}$. Formulas for the splitting of **g** and **h** are similar. In the first MHD split scheme, denoted by DS1, the magnetic field components are split with density weighting,

$$\mathbf{f}_x|_{x=x_j} \approx \begin{pmatrix} \frac{1}{2}D\rho_j u_j + \frac{1}{2}\rho_j D u_j + \frac{1}{2}u_j D\rho_j \\ \frac{1}{2}D\rho_j u_j^2 + \frac{1}{2}\rho_j u_j D u_j + \frac{1}{2}u_j D\rho_j u_j + D(p_j + \frac{1}{2}|\mathbf{B}|^2) - \frac{1}{2}D((B^{(x)})^2) - B^{(x)}DB^{(x)} \\ \frac{1}{2}D\rho_j u_j v_j + \frac{1}{2}\rho_j v_j D u_j + \frac{1}{2}u_j D\rho_j v_j - \frac{1}{2}DB^{(x)}B^{(y)} - \frac{1}{2}B^{(x)}DB^{(y)} - \frac{1}{2}B^{(y)}DB^{(x)} \\ \frac{1}{2}D\rho_j u_j w_j + \frac{1}{2}\rho_j w_j D u_j + \frac{1}{2}u_j D\rho_j w_j - \frac{1}{2}DB^{(x)}B^{(z)} - \frac{1}{2}B^{(x)}DB^{(z)} - \frac{1}{2}B^{(z)}DB^{(x)} \\ f_5 \\ \frac{1}{2}DuB^{(x)} + \frac{1}{2}uDB^{(x)} + \frac{1}{2}B^{(x)}Du - \frac{1}{2}DuB^{(x)} - \frac{1}{2}\rho u D(B^{(x)}/\rho) - \frac{1}{2}B^{(x)}/\rho D\rho u \\ \frac{1}{2}DuB^{(y)} + \frac{1}{2}uDB^{(y)} + \frac{1}{2}B^{(y)}Du - \frac{1}{2}DvB^{(x)} - \frac{1}{2}\rho v D(B^{(x)}/\rho) - \frac{1}{2}B^{(x)}/\rho D\rho v \\ \frac{1}{2}DuB^{(z)} + \frac{1}{2}uDB^{(z)} + \frac{1}{2}B^{(z)}Du - \frac{1}{2}DwB^{(x)} - \frac{1}{2}\rho w D(B^{(x)}/\rho) - \frac{1}{2}B^{(x)}/\rho D\rho w \end{pmatrix}, \tag{30}$$

where the fifth component is

$$f_5 = \frac{1}{2}Du_j(e_j + p_j + \frac{1}{2}|\mathbf{B}|_j^2) + \frac{1}{2}u_j D(e_j + p_j + \frac{1}{2}|\mathbf{B}|_j^2) + \frac{1}{2}(e_j + p_j + \frac{1}{2}|\mathbf{B}|_j^2)Du_j$$
$$- \frac{1}{2}D\mathbf{u}^T\mathbf{B}B^{(x)} - \frac{1}{2}\rho\mathbf{u}^T\mathbf{B}DB^{(x)}/\rho - \frac{1}{2}B^{(x)}/\rho D(\rho\mathbf{u}^T\mathbf{B}). \tag{31}$$

The flux derivative approximation (30) can, by (17), be written on conservative form with numerical flux function

$$\mathbf{h}_{j+1/2} = \frac{1}{2}\sum_{k=1}^{p}\alpha_k^{(p)}\sum_{m=1}^{k-1}$$
$$\begin{pmatrix} (\rho_{j-m} + \rho_{j+k-m})(u_{j-m} + u_{j+k-m}) \\ h_2 \\ (\rho_{j-m}v_{j-m} + \rho_{j+k-m}v_{j+k-m})(u_{j-m} + u_{j+k-m}) - (B_{j-m}^{(x)} + B_{j+k-m}^{(x)})(B_{j-m}^{(y)} + B_{j+k-m}^{(y)}) \\ (\rho_{j-m}w_{j-m} + \rho_{j+k-m}w_{j+k-m})(u_{j-m} + u_{j+k-m}) - (B_{j-m}^{(x)} + B_{j+k-m}^{(x)})(B_{j-m}^{(z)} + B_{j+k-m}^{(z)}) \\ h_5 \\ (u_{j-m} + u_{j+k-m})(B_{j-m}^{(x)} + B_{j+k-m}^{(x)}) - (\rho_{j-m}u_{j-m} + \rho_{j+k-m}u_{j+k-m})(B_{j-m}^{(x)}/\rho_{j-m} + B_{j+k-m}^{(x)}/\rho_{j+k-m}) \\ (u_{j-m} + u_{j+k-m})(B_{j-m}^{(y)} + B_{j+k-m}^{(y)}) - (\rho_{j-m}v_{j-m} + \rho_{j+k-m}v_{j+k-m})(B_{j-m}^{(x)}/\rho_{j-m} + B_{j+k-m}^{(x)}/\rho_{j+k-m}) \\ (u_{j-m} + u_{j+k-m})(B_{j-m}^{(z)} + B_{j+k-m}^{(z)}) - (\rho_{j-m}w_{j-m} + \rho_{j+k-m}w_{j+k-m})(B_{j-m}^{(x)}/\rho_{j-m} + B_{j+k-m}^{(x)}/\rho_{j+k-m}) \end{pmatrix} \tag{32}$$

where

$$h_2 = (\rho_{j-m}u_{j-m} + \rho_{j+k-m}u_{j+k-m})(u_{j-m} + u_{j+k-m}) +$$
$$p_{j-m} + p_{j+k-m} + \frac{1}{2}|\mathbf{B}|_{j-m}^2 + \frac{1}{2}|\mathbf{B}|_{j+k-m}^2 - (B_{j-m}^{(x)} + B_{j+k-m}^{(x)})^2 \tag{33}$$

and

$$h_5 = (e_{j-m} + p_{j-m} + e_{j+k-m} + p_{j+k-m})(u_{j-m} + u_{j+k-m}) - (\rho_{j-m}\mathbf{u}_{j-m}^T\mathbf{B}_{j-m}$$
$$+ \rho_{j+k-m}\mathbf{u}_{j+k-m}^T\mathbf{B}_{j+k-m})(B_{j-m}^{(x)}/\rho_{j-m} + B_{j+k-m}^{(x)}/\rho_{j+k-m}). \tag{34}$$

IOP Conf. Series: Journal of Physics: Conf. Series **837** (2017) 012019 doi:10.1088/1742-6596/837/1/012019

The more compact notation introduced in (18) allows (30) to be rewritten as

$$\mathbf{h}_{j+1/2} = \begin{pmatrix} \Theta^{(p)}_{j+1/2}(\rho, u) \\ \Theta^{(p)}_{j+1/2}(\rho u, u) + \Theta^{(p)}_{j+1/2}(p + \frac{1}{2}|\mathbf{B}|^2, 1) - \Theta^{(p)}_{j+1/2}(B^{(x)}, B^{(x)}) \\ \Theta^{(p)}_{j+1/2}(\rho v, u) - \Theta^{(p)}_{j+1/2}(B^{(x)}, B^{(y)}) \\ \Theta^{(p)}_{j+1/2}(\rho w, u) - \Theta^{(p)}_{j+1/2}(B^{(x)}, B^{(z)}) \\ \Theta^{(p)}_{j+1/2}(e + p, u) - \Theta^{(p)}_{j+1/2}(\rho \mathbf{u}^T \mathbf{B}, B^{(x)}/\rho) \\ \Theta^{(p)}_{j+1/2}(u, B^{(x)}) - \Theta^{(p)}_{j+1/2}(\rho u, B^{(x)}/\rho) \\ \Theta^{(p)}_{j+1/2}(u, B^{(y)}) - \Theta^{(p)}_{j+1/2}(\rho v, B^{(x)}/\rho) \\ \Theta^{(p)}_{j+1/2}(u, B^{(z)}) - \Theta^{(p)}_{j+1/2}(\rho w, B^{(x)}/\rho) \end{pmatrix} \tag{35}$$

The notation in (35) will be used to describe the remaining three split forms. The second MHD split scheme is similar, but the magnetic field terms are split without density weighting. This splitting is denoted by DS2, and its numerical flux is given by

$$\mathbf{h}_{j+1/2} = \begin{pmatrix} \Theta^{(p)}_{j+1/2}(\rho, u) \\ \Theta^{(p)}_{j+1/2}(\rho u, u) + \Theta^{(p)}_{j+1/2}(p + \frac{1}{2}|\mathbf{B}|^2, 1) - \Theta^{(p)}_{j+1/2}(B^{(x)}, B^{(x)}) \\ \Theta^{(p)}_{j+1/2}(\rho v, u) - \Theta^{(p)}_{j+1/2}(B^{(x)}, B^{(y)}) \\ \Theta^{(p)}_{j+1/2}(\rho w, u) - \Theta^{(p)}_{j+1/2}(B^{(x)}, B^{(z)}) \\ \Theta^{(p)}_{j+1/2}(e + p + \frac{1}{2}|\mathbf{B}|^2, u) - \Theta^{(p)}_{j+1/2}(\mathbf{u}^T \mathbf{B}, B^{(x)}) \\ 0 \\ \Theta^{(p)}_{j+1/2}(u, B^{(y)}) - \Theta^{(p)}_{j+1/2}(v, B^{(x)}) \\ \Theta^{(p)}_{j+1/2}(u, B^{(z)}) - \Theta^{(p)}_{j+1/2}(w, B^{(x)}) \end{pmatrix} \tag{36}$$

The third splitting does not split the magnetic field components. It is denoted by DS3, and have numerical flux

$$\mathbf{h}_{j+1/2} = \begin{pmatrix} \Theta^{(p)}_{j+1/2}(\rho, u) \\ \Theta^{(p)}_{j+1/2}(\rho u, u) + \Theta^{(p)}_{j+1/2}(p + \frac{1}{2}|\mathbf{B}|^2 - (B^{(x)})^2, 1) \\ \Theta^{(p)}_{j+1/2}(\rho v, u) - \Theta^{(p)}_{j+1/2}(B^{(x)} B^{(y)}, 1) \\ \Theta^{(p)}_{j+1/2}(\rho w, u) - \Theta^{(p)}_{j+1/2}(B^{(x)} B^{(z)}, 1) \\ \Theta^{(p)}_{j+1/2}(e + p + \frac{1}{2}|\mathbf{B}|^2, u) - \Theta^{(p)}_{j+1/2}(\mathbf{u}^T \mathbf{B} B^{(x)}, 1) \\ 0 \\ \Theta^{(p)}_{j+1/2}(u B^{(y)}, 1) - \Theta^{(p)}_{j+1/2}(v B^{(x)}, 1) \\ \Theta^{(p)}_{j+1/2}(u B^{(z)}, 1) - \Theta^{(p)}_{j+1/2}(w B^{(x)}, 1) \end{pmatrix}. \tag{37}$$

The fourth splitting, denoted by DS4, differs from DS3 only in that the first term of the three magnetic field fluxes are split. The numerical flux of DS4 is given by

$$\mathbf{h}_{j+1/2} = \begin{pmatrix} \Theta^{(p)}_{j+1/2}(\rho, u) \\ \Theta^{(p)}_{j+1/2}(\rho u, u) + \Theta^{(p)}_{j+1/2}(p + \frac{1}{2}|\mathbf{B}|^2 - (B^{(x)})^2, 1) \\ \Theta^{(p)}_{j+1/2}(\rho v, u) - \Theta^{(p)}_{j+1/2}(B^{(x)}B^{(y)}, 1) \\ \Theta^{(p)}_{j+1/2}(\rho w, u) - \Theta^{(p)}_{j+1/2}(B^{(x)}B^{(z)}, 1) \\ \Theta^{(p)}_{j+1/2}(e + p + \frac{1}{2}|\mathbf{B}|^2, u) - \Theta^{(p)}_{j+1/2}(\mathbf{u}^T\mathbf{B}B^{(x)}, 1) \\ \Theta^{(p)}_{j+1/2}(u, B^{(x)}) - \Theta^{(p)}_{j+1/2}(uB^{(x)}, 1) \\ \Theta^{(p)}_{j+1/2}(u, B^{(y)}) - \Theta^{(p)}_{j+1/2}(vB^{(x)}, 1) \\ \Theta^{(p)}_{j+1/2}(u, B^{(z)}) - \Theta^{(p)}_{j+1/2}(wB^{(x)}, 1) \end{pmatrix}. \tag{38}$$

The performance of (30)–(38) will be evaluated by numerical experiments in Section 6.

5. Skew-Symmetric Splitting as Preprocessing Step in the Framework of the Nonlinear Filter Method of Yee & Sjögreen [33]

To make the discussion more self-contained, this section gives a brief overview of the high order nonlinear filter scheme of Yee et al. and Yee & Sjögreen [35, 34, 32, 33] for accurate computations of DNS and LES of compressible turbulence for a wide range of flow types by introducing as little shock-capturing numerical dissipation as possible.

Preprocessing Step by Skew-symmetric Splitting for Gas Dynamics: Before the application of a high-order non-dissipative spatial base scheme, a preprocessing step is employed to improve numerical stability. The inviscid flux derivatives of the governing equations are split into the following two ways, depending on the flow types and the desire for rigorous mathematical analysis or physical argument.

- Entropy splitting of [35] or the natural splitting described Section 3: These are non-conservative splittings and they are among some of the best in improving numerical stability for non-dissipative central schemes, especially for long time integration of shock-free turbulence.
- The Ducros et al. splitting [6] for systems (or variants of the conservative skew-symmetric splitting described earlier): These are conservative splitting and are suitable for problems with discontinuities.

Base Scheme Step Using the Preprocessing Step: A full time step is advanced using a high order non-dissipative (or very low dissipation) spatially central scheme on the split form of the governing partial differential equations (PDEs) (i.e., after the preprocessing step).

For the base scheme step, a full time step of high-order temporal discretization such as the third-order or fourth-order Runge-Kutta (RK3 or RK4) method is used. It is remarked that other temporal discretizations can be used for the base scheme step.

Post-Processing (Nonlinear Filter Step): To further improve the accuracy of the computed solution from the base scheme step, after a full time step of a non-dissipative high order spatial base scheme, the post-processing step is used to nonlinearly filter the solution by a dissipative portion of a high-order shock-capturing scheme with a local flow sensor. The flow sensor provides locations and amounts of built-in shock-capturing dissipation that can be further reduced or eliminated. At each grid point a local flow sensor is employed to analyze the regularity of the computed flow data. Only the strong discontinuity locations would receive the full amount of shock-capturing dissipation. In smooth regions no shock-capturing dissipation would be added, unless high frequency oscillations are developed, owning to the possibility of numerical instability in long time integrations of nonlinear governing PDEs. In regions with strong turbulence, if needed, a small fraction of the shock-capturing dissipation would be added to improve stability. See Yee & Sjögreen [33], Kotov et al. [15, 14] and Sjögreen & Yee [23, 24] for more details and the formula for corresponding numerical fluxes for the filter step (filter numerical fluxes). Filter numerical fluxes are not to be confused with the standard numerical fluxes for the full shock-capturing method.

The filter numerical fluxes for the filter step only involve the inviscid flux derivatives regardless if the flow is viscous or inviscid. If viscous terms are present, a matching high order central difference operator (as the inviscid difference operator) is included on the base scheme step. In viscous flow with non-periodic boundaries, for ease of summation-by-parts (SBP) numerical boundary closure [18] implementation for the viscous flux derivatives, the same inviscid central difference operator for the first derivative is employed twice for the viscous flux derivatives. This is due to the fact that if higher than second-order central base schemes are used, proper numerical boundary conditions in the vicinity of the non-periodic boundaries are needed for numerical stability and accuracy consideration. SBP numerical boundary closures are designed to help numerical stability and accuracy of the overall high order central scheme.

Remark For the gas dynamics the post-processing (nonlinear filter step) are applied to all of the equation set. For the MHD in uniform Cartesian grid, in order to obtain zero discrete div $\mathbf{B}$ error without any div $\mathbf{B}$ cleaning, the nonlinear filter step does not employ for the three magnetic field equations. See Yee & Sjögreen [34] for details.

The aforementioned high order nonlinear filter method is valid for the four forms of the MHD formulation and the four skew-symmetric splitting of the MHD to be used as the preprocessing step. In addition, the aforementioned high order nonlinear filter method is valid for the four forms of the MHD formulation and the **different high order entropy conservative numerical fluxes as the spatial base schemes** discussed in Sections 4 and 5 of Sjögreen & Yee [23, 24].

6. Numerical results

For smooth flow test cases, only the high order base schemes are activated. For problems containing shocks, the aforementioned nonlinear filter approach of Yee & Sjögreen is utilized. For all test cases we use the classical fourth-order Runge-Kutta time discretization. Here, for illustration purposes, only one smart flow sensor (among the many variants indicated in [33] and Kotov et al. [15, 14]) is chosen for the numerical experiment for the nonlinear filter approach. It is the third-order B-spline wavelet flow sensor developed in Sjögreen & Yee [20]. Due to a page limitation, many details of the development and extensive numerical testing with more representative test cases among the different formulations and the different governing equation sets will be reported in [22] for journal publication. For 3D DNS and LES gas dynamics computations include shock-free turbulence, turbulence with moderate and strong shocks are included in [15, 14, 24].

6.1. MHD test case for smooth flow - Alfvén wave

An Alfvén wave is a traveling transverse wave with phase $\phi = 2\pi(x+t)$,

$$\rho = 1, \;\; p = 1, \;\; \mathbf{v} = A(0, \; sin\phi, \; cos\phi), \;\; \mathbf{B} = A(1, \; sin\phi, \; cos\phi).$$

We will consider a two dimensional wave, traveling in the direction $\mathbf{n} = (\cos\alpha, \; \sin\alpha, \; 0)$, where $\alpha = 30°$. The size of the computational domain is $0 \le x \le 1/\cos\alpha$, and $0 \le y \le 1/\sin\alpha$

IOP Conf. Series: Journal of Physics: Conf. Series **837** (2017) 012019 doi:10.1088/1742-6596/837/1/012019

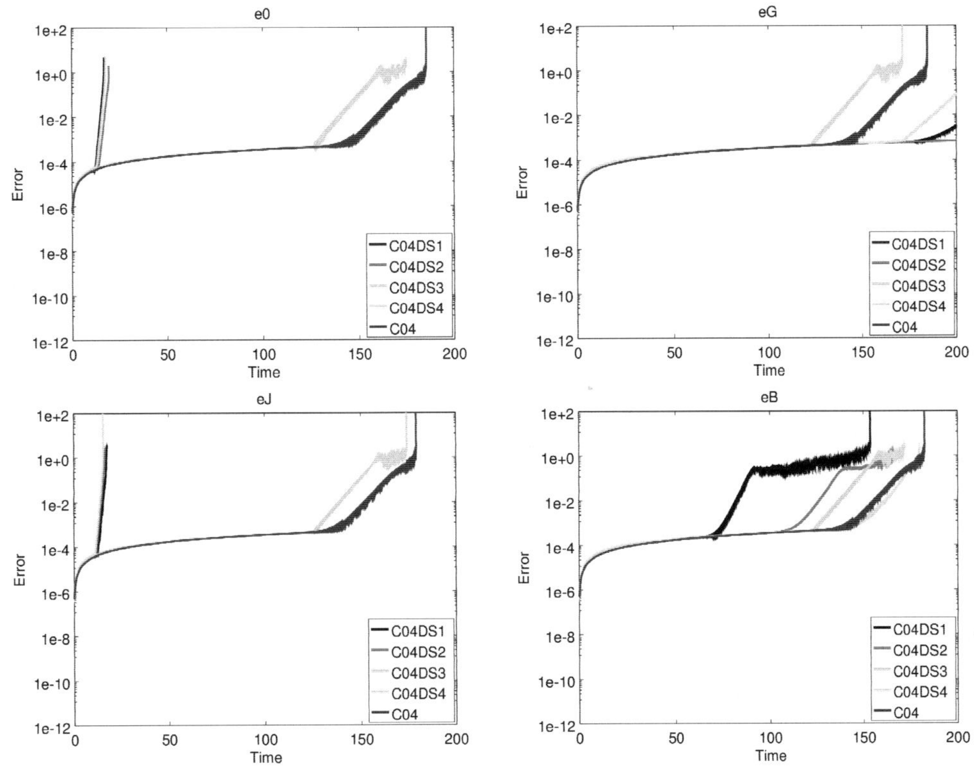

Figure 1. Alfvén wave, maximum norm errors vs. time using fourth-order central scheme in conjunction with the four Ducros et al. splittings. Top left $\mathbf{e} = \mathbf{e}_0$ (conservative equations), top right $\mathbf{e}_G$, bottom left $\mathbf{e} = \mathbf{e}_J$, bottom right $\mathbf{e} = \mathbf{e}_B$

with periodic boundary conditions. For this computation, $A = 0.1$. The grid spacing used was $\Delta x = \Delta y = 0.02$.

The magnetic field derivatives in the term $\mathbf{e}\,\mathrm{div}\,\mathbf{B}$ in (28), when present, are approximated by standard centered difference operators of the same order of accuracy as used to approximate the flux derivatives.

Figure 1 shows the evolution of the norm of the error for the four split forms DS1, DS2, DS3 and DS4 (30)–(38) inconjunction with the fourth-order central scheme. The comparison also include the standard fourth-order pure centered unsplit approximation (C04). The error is computed as the maximum norm of the pointwise error over x, y, and the eight solution components. The unsplit scheme is plotted in blue, C04DS1 in black (C04 in conjunction with DS1 split form), C04DS2 in red (C04 in conjunction with DS2 split form), C04DS3 in green (C04 in conjunction with DS3 split form), and C04DS4 in cyan color (C04 inconjunction with DS4 split form). The top left subplot of Fig. 1 depicts results obtained when $\mathbf{e} = \mathbf{e}_0$ (the conservative equations), and the top right subplot shows $\mathbf{e} = \mathbf{e}_G$, the bottom left subplot uses $\mathbf{e} = \mathbf{e}_J$, and the right subplot uses $\mathbf{e} = \mathbf{e}_B$. The unsplit scheme, C04, and the splitting with unsplit magnetic field variables, C04DS3, perform similarly for all three forms of the equations. The reason could be that the discretization of $\mathrm{div}\,\mathbf{B}$ is zero up to round-off errors for these two schemes, so that the exact form of $\mathbf{e}$ has a negligible effect on the result. The other schemes C04DS1, C04DS2, and C04DS4 have some split terms in the magnetic field equations, which prevents perfect discrete $\mathrm{div}\,\mathbf{B}$ conservation. Interestingly, the presence of $\mathbf{e}_G$ (top right subplot

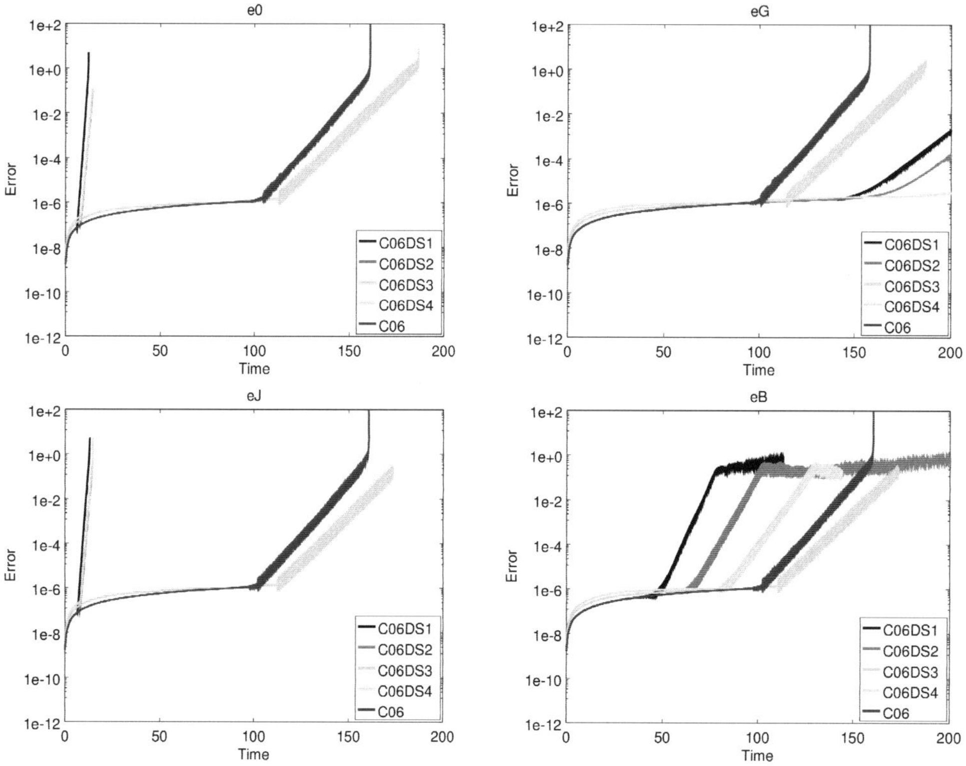

Figure 2. Alfvén wave, maximum norm errors vs. time using sixth-order central scheme in conjunction with the four Ducros et al. splittings. Top left $\mathbf{e} = \mathbf{e}_0$ (conservative equations), top right $\mathbf{e}_G$, bottom left $\mathbf{e} = \mathbf{e}_J$, bottom right $\mathbf{e} = \mathbf{e}_B$

of Fig. 1) makes the schemes C04DS1 and C04DS2 outperform C04 and C04DS3.

Figures 2–3 show the corresponding computations using the sixth-order and eight-order accurate schemes. These results are qualitatively similar to the fourth order accurate results in Fig. 1. As in the fourth-order case, the unsplit centered schemes C06 and C08 and the C06DS3 and C08DS3 split schemes are insensitive to the choice of source term, and show the longest stability for three of the source terms. However, with the source term, $\mathbf{e}_G$, the split schemes DS1,DS2, and DS4 all outperform DS3 and the centered schemes, also for the higher-order computations. The blow-up of the solution occurs at a somewhat earlier time with the higher-order accuracy. The conclusions here are only tested on one particular 2D smooth flow test case. It is possible that different performance would be obtained for multiscale smooth flows, flows with discontinuities, shock-free turbulent flows and turbulence with strong shocks.

Furthermore, only four choices of the Ducros et al. like splitting among many choices were chosen. Although it would not be hard to define other different split schemes, but there is no unique guiding principle for how this should be done to maximize the stability.

6.2. MHD test case with shocks - 2D Orzag-Tang vortex test case
The Orzag-Tang vortex starts from initial data

$$\rho = 25/9 \ \ (u, v, w) = (-\sin y, \sin x, 0) \ \ p = 5/3, \tag{39}$$

$$(B^{(x)}, B^{(y)}, B^{(z)}) = (-\sin y, \sin 2x, 0) \tag{40}$$

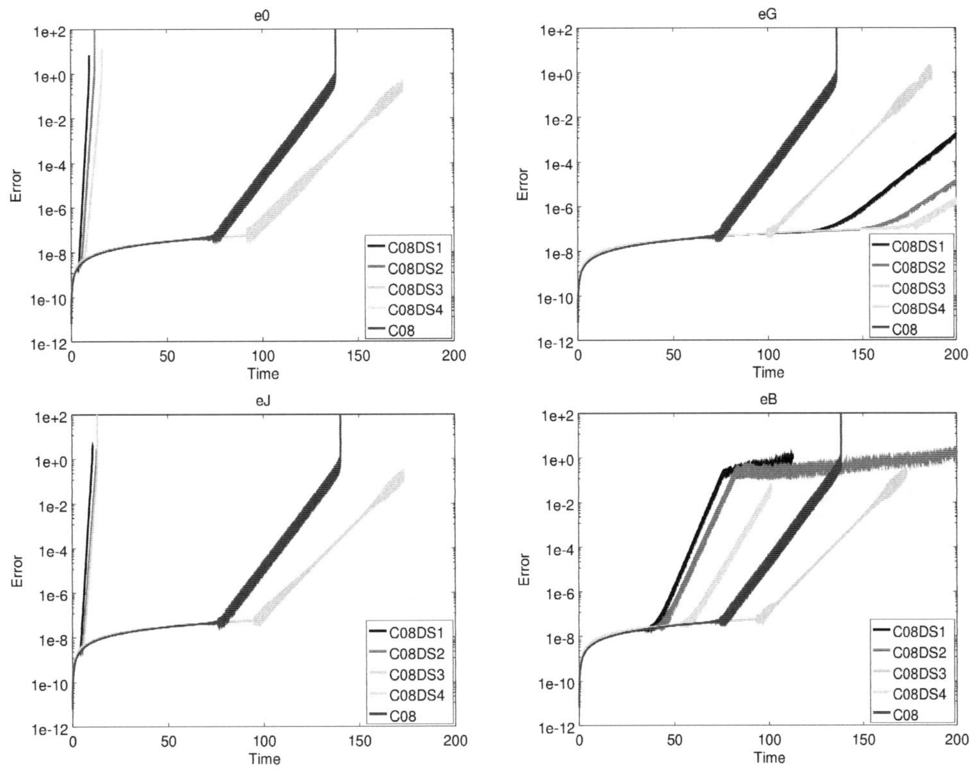

Figure 3. Alfvén wave, maximum norm errors vs. time using eighth-order central scheme in conjunction with the four Ducros et al. splittings. Top left $\mathbf{e} = \mathbf{e}_0$ (conservative equations), top right $\mathbf{e}_G$, bottom left $\mathbf{e} = \mathbf{e}_J$, bottom right $\mathbf{e} = \mathbf{e}_B$

and is solved in two space dimensions on a domain of size $2\pi \times 2\pi$ with periodic boundary conditions. The solution will be computed up to time 3.14, when the smooth initial data will have developed shock waves. Because shock waves appear, this problem requires nonlinear shock capturing filter as post processing to the non-dissipative Ducros et al. split form of the base scheme.

Results will be displayed as density contour levels at time 3.14 together with contours of $\operatorname{div}\mathbf{B}$ at the same time. All contour plots in this section use the same values of the contour levels. Furthermore, a logscale plot of the evolution of the norm of $\operatorname{div}\mathbf{B}$ will also be given. The computational domain was discretized by a grid with 100×100 grid points.

Figures 4–6 show results by the standard sixth-order centered base scheme filtered with the dissipation of the fifth-order WENO scheme after each time step (C06+WENO5fi), for the source term choices $\mathbf{e} = \mathbf{e}_0$, $\mathbf{e} = \mathbf{e}_G$, and $\mathbf{e} = \mathbf{e}_J$ respectively.

The centered base scheme preserves the discretized $\operatorname{div}\mathbf{B}$ perfectly as indicated on the middle subplots of Figs. 4–6 provided the nonlinear filter was not applied to the three magnetic field equations of the MHD. The empty middle subplots in Figs. 4–6 indicate an error-free discrete $\operatorname{div}\mathbf{B}$ to the order of the truncation error based on contour levels used to make the plot. It is noted that for uniform grid, when the base scheme solved for the full MHD equations (include all three magnetic field equations) and the nonlinear filter step is not applied on the three magnetic field equations, we do not need the standard $Div\mathbf{B}$ cleaning procedure. Due to a page limitation, see on expanded version of this paper for the error contours (not empty middle subplots) when

IOP Conf. Series: Journal of Physics: Conf. Series **837** (2017) 012019 doi:10.1088/1742-6596/837/1/012019

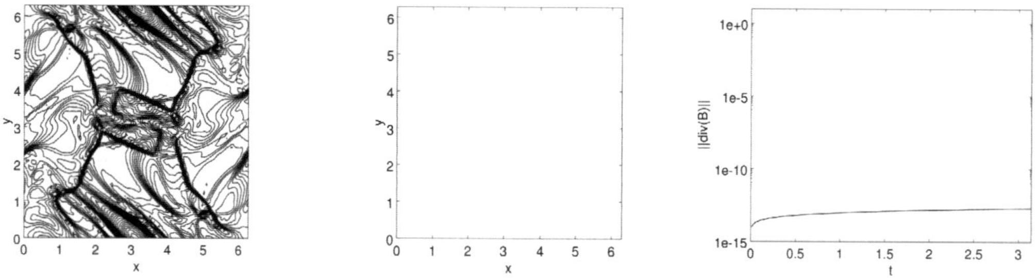

Figure 4. 2D Orzag-Tang vortex test case: C06+WENO5fi+e_0, density contours (left), div **B** contours (middle), norm of div **B** vs. time (right). No filter on the three magnetic field governing equations.

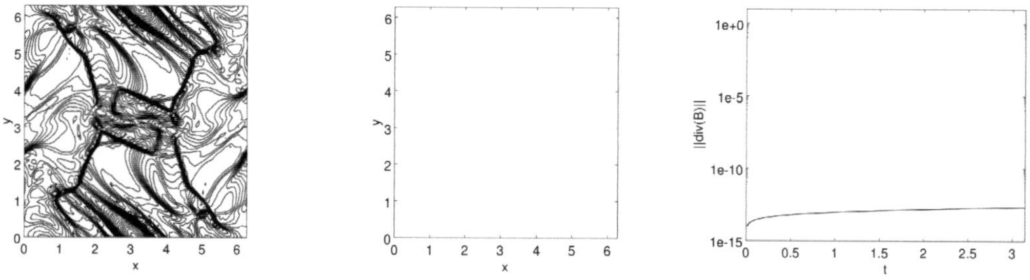

Figure 5. 2D Orzag-Tang vortex test case: C06+WENO5fi+e_G, density contours (left), div **B** contours (middle), norm of div **B** vs. time (right). No filter on the three magnetic field governing equations.

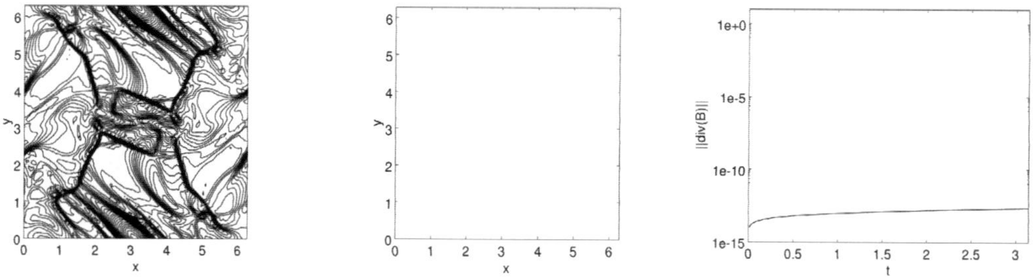

Figure 6. 2D Orzag-Tang vortex test case: C06+WENO5fi+e_J, density contours (left), div **B** contours (middle), norm of div **B** vs. time (right). No filter on the three magnetic field governing equations.

the filter step were employed on all three magnetic fields [22]. We also would like to mention that the entropy conserving schemes developed recently by Sjögreen & Yee [23, 24] under the same framework of the nonlinear filter approach might not have the perfect div **B** preservation property for the same test case for certain choice of MHD governing equation set and entropy numerical fluxes.

Warning: For magnetic field dominating flows, switching off the nonlinear filter on the magnetic field components might not be stable for problems where the shock waves are very strong. In this case, good standard $div(\mathbf{B})$ cleaning method is needed.

IOP Conf. Series: Journal of Physics: Conf. Series **837** (2017) 012019 doi:10.1088/1742-6596/837/1/012019

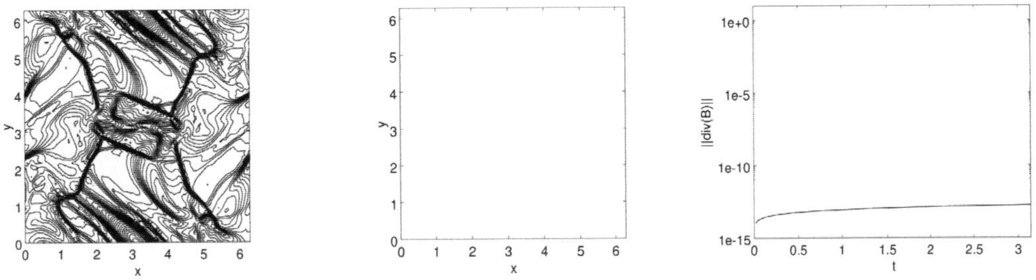

Figure 7. C06DS3+WENO5fi+e_0, density contours (left), div **B** contours (middle), norm of div **B** vs. time (right).

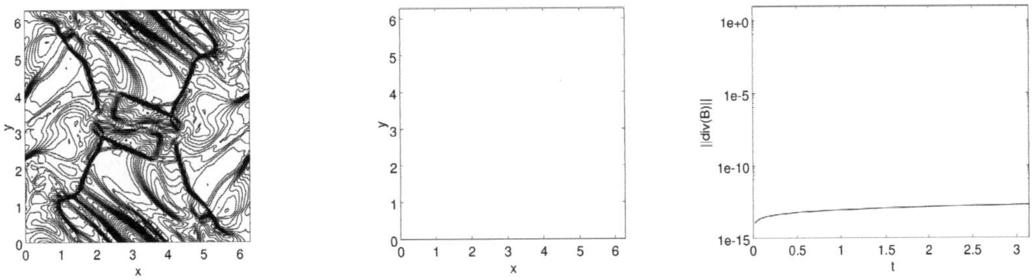

Figure 8. C06DS3+WENO5fi+e_G, density contours (left), div **B** contours (middle), norm of div **B** vs. time (right).

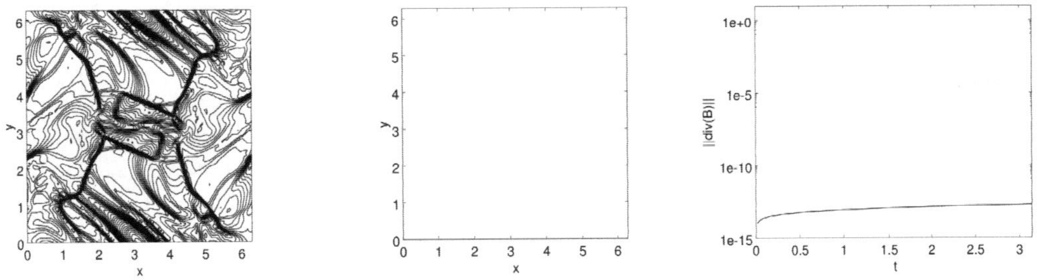

Figure 9. C06DS3+WENO5fi+e_J, density contours (left), div **B** contours (middle), norm of div **B** vs. time (right).

Figures 7–10 shows computations by the filter scheme with the sixth-order accurate Ducros et al. split scheme (37) as base scheme (C06DS3+WENO5fi) for the four different source terms. In these computations, the nonlinear filter was not applied to the magnetic field components, giving perfect div **B** preservation.

The Orzag-Tang problem does not involve long time integration, and therefore the long time stability studied in the Alfvén wave problem, is not an issue here. Any advantage of the split schemes over unsplit schemes deriving from the long time integration properties, would not be visible in this problem. The Orzag-Tang problem is solved here to show that the split base schemes are performing well also on problems where discontinuities are present.

DNS computations of 2D and 3D MHD turbulence to illustrate the stability and accuracy of our proposed schemes are a subject of our current work in progress.

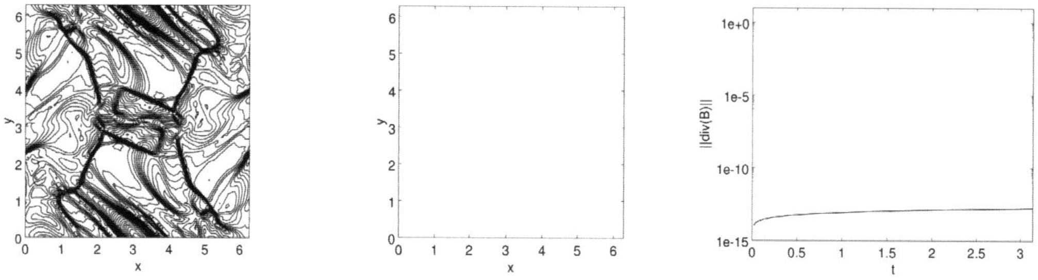

Figure 10. C06DS3+WENO5fi+e_B, density contours (left), div **B** contours (middle), norm of div **B** vs. time (right).

7. Conclusions

The Yee & Sjögreen and Kotov et al. [33, 15, 14] high order numerical method with the Ducros et al. skew-symmetric type of splitting for compressible gas dynamics is extended to ideal MHD. Four Ducros et al. type skew-symmetric splittings together with four formulations of the MHD are proposed and investigated. The proposed methods are able to stabilize non-dissipative high order central finite-difference methods for hyperbolic conservation laws. These methods are also able to maintain their high accuracy and stability for rapidly developing flows and also can improve the stability and accuracy of long time wave propagations and long time integration of turbulent flows. It was shown that the choice of the div **B** source term in the MHD equations has a large influence on the nonlinear stability. However, for problems where shock waves are present, without added dissipation the computed solutions exhibit oscillations. The nonlinear filter method by Yee & Sjögreen was demonstrated to maintain stability and give highly accurate computed solutions when using skew-symmetric splitting as a pre-processing step for the central base scheme.

The support of the DOE/SciDAC SAP grant DE-AI02-06ER25796 is acknowledged. The work was performed with the third author as a research scientist at the Bay Area Envirornmental Research Institute. Financial support from the NASA TTT/RCA program for the second author is gratefully acknowledged. The authors are grateful to Dr. Alan Wray of NASA Ames Research Center for the numerous valuable discussion throughout the course of this work.

References

[1] Arakawa, A.: Computational design for long-term numerical integration of the equations of fluid motion: two-dimensional incompressible flow, Part I. J. Comput. Phys. **1**, 119–143 (1966).

[2] Blaisdell, G.A., Spyropoulos, E.T., Qin, J.H.: The effect of the formulation of nonlinear terms on aliasing errors in spectral methods. Appl. Num. Math. **21**, 207–219 (1996).

[3] Brackbill, J.U., Barnes, D.C.: The effect of nonzero $\nabla \cdot \mathbf{B}$ on the numerical solution of the magnetohydrodynamics equations. J. Comput. Phys. **35**, 426–430 (1980).

[4] Bracket, M., Meiron, D, Orszag, S., Nickel, B., Morf, R., Frisch, U.: J. Fluid Mech. **130** 411 (1983).

[5] Chandrashekar, P., Klingenberg, C.: Entropy stable finite volume scheme for ideal compressible MHD on 2-D Cartesian Meshes. SIAM J. Numer. Anal. **54**, 1313–1340 (2016).

[6] Ducros, F., Laporte, F., Soulères, T., Guinot, V., Moinat, P., Caruelle, B.: High-order fluxes for conservative skew-symmetric-like schemes in structured meshes: application to compressible flows: J. Comput. Phys. **161**, 114–139 (2000).

[7] Ducros, F., Ferrand, V., Nicoud, F., Weber, C. Darracq, D., Gacherieu, C., Poinsot, T.: J. Comput. Phys. **1152**, 517-549 (1999).

[8] Harten, A: The artificial compression method for computations of shocks and contact discontinuities: III. Self-adjusting hybrid schemes: J. comput. Phys. **32**, 363-389 (1978).

[9] Harten, A: On the symmetric form of systems for conservation laws with entropy: J. Comput Phys. **49**, 151 (1983).

[10] Honein A.E., Moin, P.: Numerical aspects of compressible turbulence simulations. Report No. TF-92, Dept. of Mech. Eng., Stanford University, (2005).

[11] Gerritsen, M, Olsson, P.: Designing an efficient solution strategy for fluid flows. I. A stable high order finite difference scheme and sharp shock resolution for the Euler equations. J. Comput. Phys. **129**, 245 (1996).

[12] Godunov, S.K.: The symmetric form of magnetohydrodynamics equation. Num. Meth. Mech. Cont. Media **1**, 26–34 (1972).

[13] Janhunen, P.: A Positive conservative method for MHD based on HLL and Roe methods. J. Comput. Phys. **160**, 649–661 (2000).

[14] Kotov, D.V., Yee, H.C., Wray, A.A., Sjögreen, B., Kritsuk, A.G.: Numerical disipation control in high order shock-capturing schemes for LES of low speed flows. J. Comput. Phys. **307**, 189–202 (2016).

[15] Kotov, D.V., Yee, H.C., Wray, A.A., Sjögreen, B.: High order numerical methods for dynamic SGS model of turbulent flows with shocks. Commun. Comput. Phys. **19**, 273–300 (2016).

[16] Kreiss, H.O.: Numerical Methods for Solving Time-Dependent Problems for Partial Differential Equations. Les Presses de l'Université de Montreéal (1978).

[17] Olsson, P., Oliger, J.: Energy and maximum norm estimates for nonlinear conservation laws. RIACS Technical Report 94.01, (1994).

[18] Olsson, P. Summation by Parts, Projections and Stability II. Math. Comp. **64**, 212 (1995).

[19] Pirozzoli, S.: Generalized conservative approximations of split convective derivative operators. J. Comput. Phys., **219**, 7180-90 (2010).

[20] Sjögreen, B., Yee, H.C.: Multiresolution wavelet based adaptive numerical dissipation control for high order methods. J. Scientific Computing, **20**, 211-255 (2004).

[21] Sjögreen, B., Yee, H.C.: On skew-symmetric splitting and entropy conservation schemes for the Euler equations. In: proc. of ENUMATH09, June 29- July 2, Uppsala University, Sweden (2009).

[22] Sjögreen, B., Yee, H.C.: Improving Numerical Stability via Skew-Symmetric Splitting in Multiscale Gas Dynamics and MHD Turbulence Flows. Extended version of the Proceeding of ASTRONUM-2016, June 6-10, 2016, Monterey, CA, USA, in preparation for submission to J. Comput. Phys., 2016.

[23] Sjögreen, B., Yee, H.C.: On high order entropy conserving numerical flux for multiscale gas dynamics and MHD turbulent simulations. Proceeding of ICOSAHOM-2016, June 27-July 1, 2016, Rio de Janeiro, Brazil.

[24] Sjögreen, B., Yee, H.C.: Construction of high order entropy conserving numerical flux for multiscale gas dynamics and MHD turbulent simulations. Extended version of the Proceeding of Proceeding of ICOSAHOM-2016, June 27-July 1, 2016, Rio de Janeiro, Brazil, submitted to J. Comput. Phys., 2016.

[25] Sandham, N.D., Li, Q., Yee, H.C.: Entropy splitting for high-order numerical simulation of compressible turbulence. J. Comput. Phys. **23**, 307-322 (2002).

[26] Shu, C.-W., Osher, S.J.: Efficient implementation of essentially non-oscillatory shock capturing schemes. J. Comput. Phys., **83**, 32–78 (1989).

[27] Tadmor, E.: Numerical viscosity and the entropy condition for conservative difference schemes. Math. Comput. **43**, 369–381 (1984).

[28] Strand, B.: Summation by parts for finite difference approximations for d/dx. J. Comput. Phys. **110**, 47–67 (1994).

[29] Taylor, G., Green, A., Proc. R. Soc. Lond. A **158**, 499 (1937).

[30] Tauber, E., Sandham, N.D.,: Comparison of three large-eddy simulatitons of shock-induced trubulent separation bubbles: Shock waves, **19**, 469-478 (2009).

[31] Winters, A.R., Gassner, G.J.: Affordable, entropy conserving and entropy stable flux functions for the ideal MHD equations. J. Comput. Phys. **304**, 72–108 (2016).

[32] Yee, H. C., Sjögreen, B.: Development of low dissipative high order filter schemes for multiscale Navier-Stokes MHD systems. J. Comput. Phys., **225** 910-934 (2007).

[33] Yee, H. C., Sjögreen, B.: High Order Filter Methods for Wide Range of Compressible flow Speeds. Proceedings of the ICOSAHOM09, June 22-26, 2009, Trondheim, Norway.

[34] Yee, H. C., Sjögreen, B.: Efficient low dissipative high order schemes for multiscale MHD flows, II: Minimization of $\nabla \cdot B$ numerical error. J. Sci. Comput. **29**, 115–164 (2006).

[35] Yee, H. C., Vinokur, M, Djomehri, M. J.: Entropy Splitting and Numerical Dissipation. J. Comp. Phys. **162**, 33–81 (2000).

[36] Yee, H. C., Sandham, N.D., Djomehri, M. J.: Low-Dissipative High Order Shock-Capturing Methods Using Characteristtic-Based Filters. J. Comput. Phys., **150**, 199-238 (1999).

IOP Conf. Series: Journal of Physics: Conf. Series **837** (2017) 012020 doi:10.1088/1742-6596/837/1/012020

Factorized Runge–Kutta–Chebyshev Methods

Stephen O'Sullivan

School of Mathematical Sciences, Dublin Institute of Technology, Kevin Street, Dublin 8, Ireland

E-mail: `stephen.osullivan@dit.ie`

Abstract. The second-order extended stability Factorized Runge–Kutta–Chebyshev (**FRKC2**) explicit schemes for the integration of large systems of PDEs with diffusive terms are presented. The schemes are simple to implement through ordered sequences of forward Euler steps with complex stepsizes, and easily parallelised for large scale problems on distributed architectures. Preserving 7 digits for accuracy at 16 digit precision, the schemes are theoretically capable of maintaining internal stability for acceleration factors in excess of 6000 with respect to standard explicit Runge-Kutta methods. The extent of the stability domain is approximately the same as that of **RKC** schemes, and a third longer than in the case of **RKL2** schemes. Extension of **FRKC** methods to fourth-order, by both complex splitting and Butcher composition techniques, is also discussed.

A publicly available implementation of **FRKC2** schemes may be obtained from `maths.dit.ie/frkc`

1. Introduction

Factorized Runge–Kutta–Chebyshev (FRKC) methods are well suited to the numerical integration of problems where diffusion limits the efficiency of standard explicit techniques. In general, such systems of PDEs may be presented as semi-discrete ordinary differential equations of the form

$$w' = f(t,\, w). \tag{1}$$

Here, the associated Jacobian is assumed to have negative eigenvalues lying close to the real axis, a good approximation for many systems of interest in astrophysical contexts.

The main use of extended stability Runge–Kutta (ESRK) methods is to fill the gap between unconditionally stable but operationally complex implicit methods, and simply implemented explicit schemes which suffer from stability constraints for stiff problems. ESRK methods are particularly useful for problems involving diffusion, where the work required by standard explicit techniques goes as the inverse square of the mesh spacing, while for extended stability methods it goes as the inverse mesh spacing. ESRK explicit schemes may be broadly divided into factorized and recursive types.

Factorized ESRK methods are particularly straightforward to implement at second-order, consisting solely of forward Euler steps. At orders above two, splitting methods or, alternatively, additional finishing stages are required for nonlinear problems. First considered by [1, 2, 3], factorized ESRK methods fell out of common usage for some time, until revived in 1996 as SuperTimeStepping [4] at first-order, and later extended to second-order applications in astrophysical simulations by means of Richardson extrapolation [5, 6]. DUMKA methods exist at third- and fourth-order [7].

The perennial problem with the factorized formalism has been that, when a very large number of stages is used, the internal amplification factors can easily drown numerical precision. Factorized methods have, as a result, largely taken a back seat to recursive methods which manage internal stability by mapping the three-term recurrence relations for orthogonal polynomials to the stability polynomials [8]. Recursive methods have been developed up to fourth-order [9, 10, 11, 12, 13, 14, 15].

In the following, a formulation of factorized methods is presented which has high internal stability, and is more straightforward to implement than recursive methods, and demonstrates comparable efficiency.

2. FRKC stability polynomials

Stability polynomials form the backbone of ESRK numerical schemes and encapsulate the linear stability properties. Linearising a system of semi-discrete ODEs

$$w' = Aw, \tag{2}$$

the FRKC stability polynomial [16] is obtained by seeking a polynomial of degree L which yields a forward Euler scheme of order N for linear problems through its roots via

$$W^L = W^0 + T\sum_{l=1}^{L} a_l f(W^{l-1}) \quad W^0 = w^n,\, w^{n+1} = W^L. \tag{3}$$

The M-th stability polynomial of order N, with $L = MN$, determined via $w^{n+1} = R^N_M(TA)w^n$, must match the first $N+1$ terms in the Taylor expansion of the evolution operator

$$R^N_M(TA) = \mathrm{e}^{TA} + \mathcal{O}(TA)^{N+1}. \tag{4}$$

Equivalently, the linear order conditions may be expressed as constraints on the derivatives of the stability polynomial evaluated at zero:

$$R^{N\,(n)}_M(0) = 1, \quad n = 1, \cdots, N. \tag{5}$$

In addition, stability requires that the polynomial is bounded according to

$$|R^N_M(z)| \leq 1 \quad \forall \quad z = T\lambda. \tag{6}$$

The objective is to determine a closed form for the polynomial such that the extent of the stability domain along the negative real axis β is as great as possible.

It is shown in [16] that the FRKC stability polynomial of rank N, and degree L, is a sum of Chebyshev polynomials of the first kind given by

$$B^N_M(z) = d^N_0 + 2\sum_{k=1}^{N} d^N_k C_{kM}(z), \tag{7}$$

where C_{kM} is the Chebyshev polynomial of the first kind of degree kM, and the coefficients d_k are determined through the linear order conditions given by Equation 5. The resultant scheme follows immediately from the roots of the polynomial, ζ_l, with coefficients given by

$$a_l = \frac{1}{M^2\alpha_M}\frac{1}{1-\zeta_l}. \tag{8}$$

The dependency of β on L is presented in Figure 1 at second-order ($N = 2$). While the optimal value for $\beta/L^2\beta_{RK2}$ is 0.41 [17], where β_{RK2} is the conventional second order Runge–Kutta stability limit, values of 0.330, 0.333, and 0.25 are obtained for **FRKC2**, **RKC** [8], and **RKL2** [15] respectively.

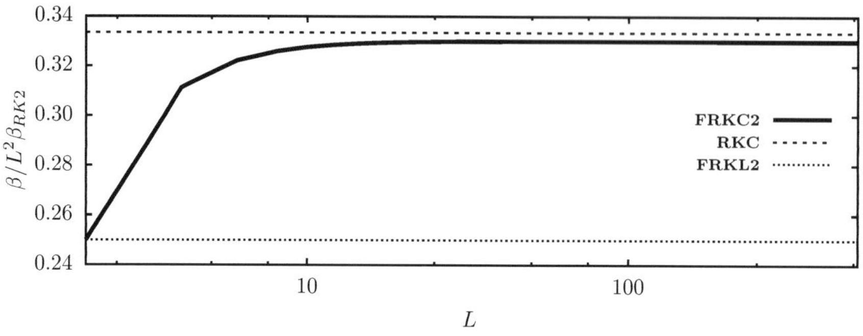

Figure 1: Extent of real stability domain at second-order for **FRKC2** (solid line). For large values of stage-number L, the scheme has a stability domain which is $0.33L^2$ times the extent of the reference second-order Runge–Kutta scheme. Also shown are the corresponding values for **RKC** (dashed line) and **RKL2** (dotted line).

2.1. Damping

Along the real axis on the interior of the stability domain of the stability polynomial, there are points which are marginally stable, as shown in Figure 2. This is remedied by introducing the damping parameter $\nu \equiv \nu_0/N$ via

$$a_l = \frac{1}{(1-\nu)M^2\alpha_M}\,\frac{1-\mu_l}{1-(1-2\mu_l)\zeta_l}, \tag{9}$$

and again enforcing the order conditions given by Equation 5 via Newton-Raphson iteration over the parameters μ_l, which consist of N distinct values, each repeated M times. As a result, the marginally stable maxima in $|R|$ along the real axis are scaled by $\sim (1-\nu_0)$ at the expense of reducing the extent of the stability domain β along the real axis by approximately $(1-\nu)$.

The damping process may also be used to make the scheme applicable to problems with small hyperbolic components, with Péclet numbers $Pe \lesssim 10$. For the case $Pe \approx 10$, illustrated in Figure 2, there is a 25% loss in the extent of the stability domain observed.

2.2. Internal stability

Internal instability may be caused when the product of any of the possible sub-sequences of steps act to generate large values which drown out numerical precision. Following the idea of Lebedev [18, 19], but with a more effective approach, the timesteps are ordered to *approximately* minimise $\mathcal{Q}$, where

$$\mathcal{Q} \equiv \max(\mathcal{Q}_{j,k}(x)), \quad 1 \le j \le k \le L, \quad x \in [-\beta, 0], \tag{10}$$

is the maximum over the internal amplification factors defined by

$$\mathcal{Q}_{j,k}(x) = \prod_{l=j}^{k} |1 + a_l x|. \tag{11}$$

While the optimal value of $\mathcal{Q}$ is approximately L^2, for the purposes of constructing a stabilization algorithm, a practical upper bound of $10L^2$ is chosen. In order to maintain this bound, the estimated maximum amplification factor $\overline{\mathcal{Q}}$ is held to a minimum while l runs from 1 to L, where

$$\overline{\mathcal{Q}} \equiv \left\| \max\left(\prod_{j=1}^{l} v_{j,k}, \prod_{j=l+1}^{L/2} v_{j,k} \right) \right\|_1 . \tag{12}$$

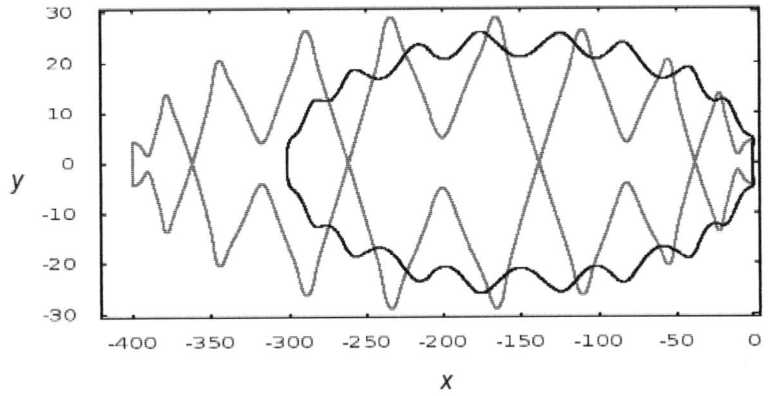

Figure 2: Stability domain $|R(x+Iy)| = 1$ for **FRKC** polynomial without damping (grey line) and with sufficient damping applied for mixed hyperbolic-parabolic problems with $Pe \approx 10$ (black line). The extent of the stability domain along the real axis is contracted by approximately 25% in the latter case.

The amplification factors $v_{j,k}$ are defined by $v_{j,k} = |1 + a_j x_k|$, where $x_k \in [\overline{\beta}, 0]$ are L uniformly distributed values over the reduced range $\overline{\beta} = (1 - nC)\beta$, with $C = 10^{-4}$. Initially, $n = 1$, however, in a limited number of cases, the process is repeated with n incremented until the required bound is satisfied. In this work, the mean value of n for the second-order schemes was found to be 1.5. Figure 3 shows the realised values of $\mathcal{Q}$ obtained via the stabilization algorithm for second-order schemes. Preserving 7 digits for accuracy, a scheme consisting of 10^4 stages is therefore theoretically viable in a numerical integration carried out to 16 digit precision.

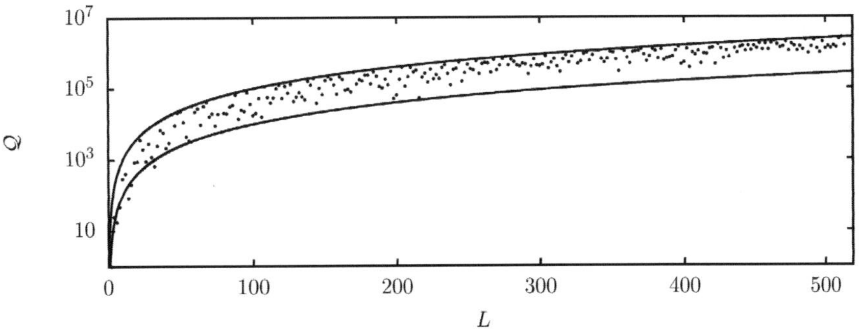

Figure 3: The maximum realised internal amplification factor $\mathcal{Q}$ as a function of L for second-order schemes. Guidelines are at L^2 and $10L^2$.

2.3. Convex Monotone Property

The convex monotone property (CMP), relevant to problems with spatially varying diffusion coefficients, was recently considered by [15]. Figure 4 shows solutions obtained for a problem describing two materials at different temperatures and with different diffusion coefficients placed into contact. For Chebyshev polynomial-based schemes such as **RKC** and **FRKC2**, features in the solution associated with failure to meet the CMP are evident if the schemes are forced to take large steps initially (Figures 4a and 4c). However, with error control implemented as discussed in Section 3, the features do not appear (Figure 4d). The **RKL2** scheme maintains the CMP naturally (Figure 4b).

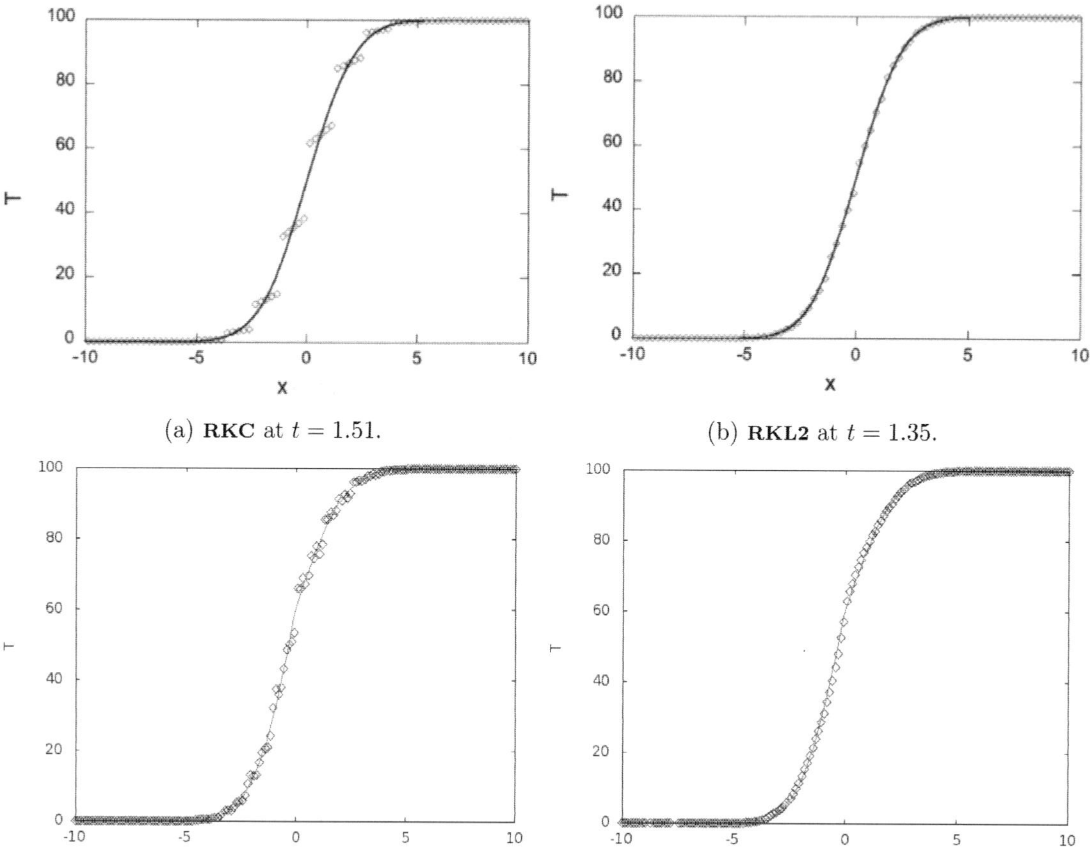

(a) **RKC** at $t = 1.51$.

(b) **RKL2** at $t = 1.35$.

(c) **FRKC2** at $t = 1.24$ without error control.

(d) **FRKC2** at $t = 1.24$ with error control.

Figure 4: Heat conduction between materials of different thermal conductivity illustrating influence of the convex monotone property (CMP) [15].

3. FRKC2 schemes

A system $w' = f(w)$ is assumed such that the associated Jacobian has an eigenvalue of maximum magnitude $|\lambda|_{\max}$. Then, given a numerical solution w^n at some time index n, $L = 2M$ stages W^l (for $l = 0, \cdots, L$) are evaluated such that $W^0 = w^n$ and a second-order accurate solution $w^{n+1} = W^L$ is obtained a time T later. The intermediate stages of the scheme are determined via the Euler steps

$$W^{l+1} = W^l + Ta_l f(W^l). \tag{13}$$

Error control is straightforward since a first-order solution is available at no additional cost in function evaluations. This first-order solution $\widehat{W}^L$ is obtained by considering only the real parts of a_l and $f(W^l)$. Setting $\widehat{W}^0 = W^0$,

$$\widehat{W}^{l+1} = \widehat{W}^l + T\,\mathrm{Re}(a_l)\,\mathrm{Re}(f(W^l)). \tag{14}$$

The error, scaled to a specified tolerance TOL, is estimated using

$$\|err\| = \left\| \frac{|W^{l+1} - \widehat{W}^{l+1}|}{TOL(1 + \max(|W^{l+1}|,\, |\widehat{W}^{l+1}|)} \right\|. \tag{15}$$

If $\|err\| > 1$, the step is rejected and retried with T scaled by $SAFE/\sqrt{\|err\|}$. Otherwise, a predictive controller is used to determine the subsequent timestep calibrated to the required tolerance via

$$T^{n+1} = \left(\frac{SAFE}{\sqrt{\|err^n\|}}\right)\left(\frac{T^n}{T^{n-1}}\right)\sqrt{\frac{\|err^{n-1}\|}{\|err^n\|}}, \tag{16}$$

using values of T and $\|err\|$ from previous timesteps. $SAFE$ is a safety factor chosen with a value 0.8 here. The reader is referred to [20] for further details of error control procedures.

3.1. **FRKC2** *public code*

A freely available C implementation of the second-order **FRKC2** schemes may be accessed at `maths.dit.ie/frkc`. The files frkc2core.c and frkc2user.c provide the code for internal calculations required by the **FRKC2** scheme and the code specific to the particular problem respectively. For a given value of M, up to 257, the extent of the stability domain along the real axis, $\overline{\beta}$, and the maximum realised internal amplification factor $\mathcal{Q}$ (see also Figure 3) are given on line $3M - 2$ of frkc2arks.dat. The real and imaginary parts of a_l are recorded on lines $3M - 2$ and $3M$ respectively.

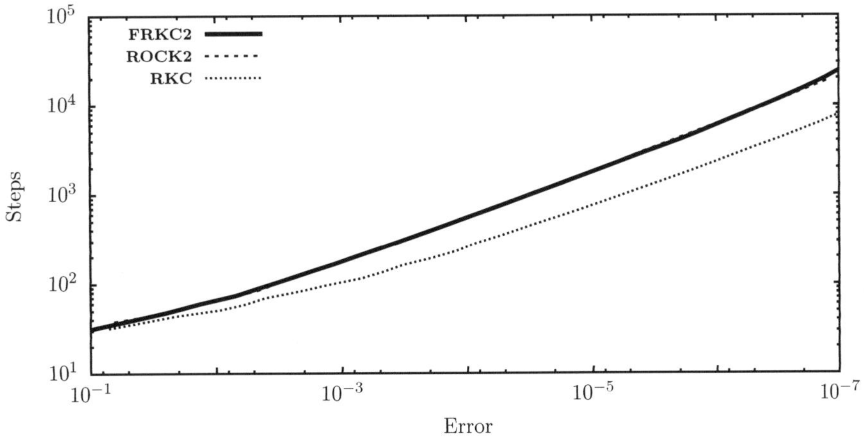

Figure 5: Efficiency comparison for the two-dimensional Brusselator problem for **FRKC2**, **ROCK2**, and **RKC**. (The lines for **FRKC2** and **ROCK2** are almost coincident.)

The default problem provided in frkc2user.c is a two-species reaction diffusion Brusselator

$$\begin{aligned}
\partial v/\partial t &= 0.02\left(\partial^2 v/\partial {x_1}^2 + \partial^2 v/\partial {x_2}^2\right) + 1 - 4v + wv^2, \\
\partial w/\partial t &= 0.02\left(\partial^2 w/\partial {x_1}^2 + \partial^2 w/\partial {x_2}^2\right) + 3v - v^2 w, \\
v(0,\, x) &= 1 + \sin(2\pi x) \quad w(0,\, x) = 3 + \cos(2\pi y),
\end{aligned} \tag{17}$$

which possesses a spectral radius of approximately 6400 for a 200×200 mesh. The initial state is a perturbation of the equilibrium solution [21]which is given by $v = 1$, $w = 3$. Figure 5 shows the number of steps required to attain a given error in the solution for **FRKC2**, **ROCK2** [22], and **RKC**. It is evident that, for a given precision, there is little difference in the number of steps required by **FRKC2** and **ROCK2**. At higher degrees of acceleration (fewer steps), the difference between the three schemes is negligible.

4. FRKC4 scheme

4.1. Complex splitting

Above second-order, nonlinear order conditions are present which require additional consideration. One approach, given a semi-linear parabolic (reaction-diffusion) equation of the form $w' = Aw + f_B(w)$, is to split the nonlinear part $f_B(w)$, which is typically easily integrated, from the linear diffusion terms Aw. For orders above two, this requires complex timesteps [23, 24] and may be prescribed in the form

$$w^{n+1} = \mathrm{e}^{T_{k_J}B}\mathrm{e}^{T_{k_{J-1}}A} \cdots \mathrm{e}^{T_{k_3}B}\mathrm{e}^{T_{k_2}A}\mathrm{e}^{T_{k_1}B}w^n. \quad (18)$$

Figure 6 shows the split **FRKC4s** scheme is competitive with **ROCK4** [12]. However, in support of the splitting approach, it may be noted that **ROCK4** suffers significantly from internal stability issues arising from the finishing stages required for the nonlinear order conditions (discussed further in Section 4.2) which effectively limits the scheme to about 150 stages.

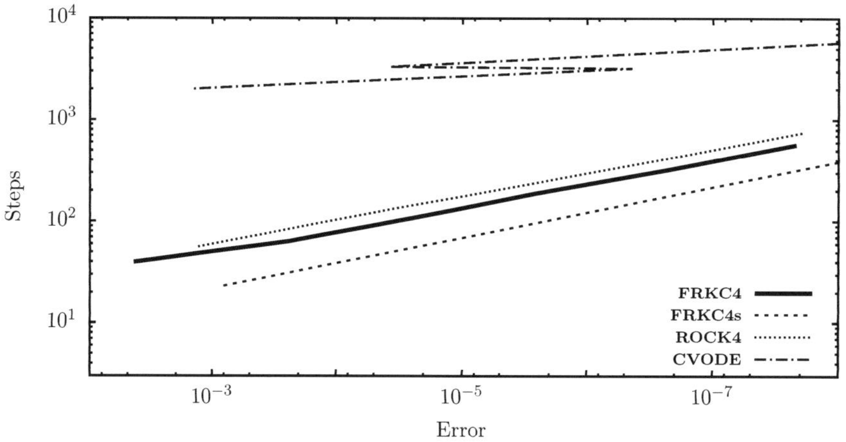

Figure 6: Performance results derived from the estimated error for the Brusselator problem. Shown are data for the fourth-order ESRK schemes **FRKC4s**, **FRKC6s**, **ROCK4**, and the fifth-order implicit **CVODE** scheme.

4.2. Butcher composition

An alternative to complex splitting is Butcher composition. At fourth-order, as illustrated in Table 1, $L - 4$ forward Euler steps are adopted from the **FRKC** stability polynomial with $N = 4$ as the scheme a, and appropriate finishing stages for the scheme $\overline{a}$ are subsequently derived.

According to a theorem of Hairer & Wanner [25], given the B-series a, b, the composite scheme ab is determined via

$$a\overline{a}(t) = \frac{1}{\alpha(t)} \sum \left(\sum_{i=0}^{q} \binom{q}{i} \overline{a}(s_i(t)) \prod_{z \in d_i(t)} a(z) \right), \quad (19)$$

0						
c_2	a_1					
c_3	a_1	a_2				
c_4	a_1	a_2	a_3			
$\vdots$	$\vdots$	$\vdots$	$\vdots$	$\ddots$		
c_L	a_1	a_2	a_3	$\cdots$	a_{L-5}	
	a_1	a_2	a_3	$\cdots$	a_{L-5}	a_{L-4}

0				
$\overline{c}_2$	$\overline{a}_{21}$			
$\overline{c}_3$	$\overline{a}_{31}$	$\overline{a}_{32}$		
$\overline{c}_4$	$\overline{a}_{41}$	$\overline{a}_{42}$	$\overline{a}_{43}$	
	$\overline{a}_1$	$\overline{a}_2$	$\overline{a}_3$	$\overline{a}_4$

Table 1: **FRKC4** order composition tableaux.

where rooted trees t are used to represent derivatives in Taylor series. The first summation in Equation 19 is over all $\alpha(t)$ different labelings of t, $s_i(t)$ is the subtree formed by the first i indices, and $\{z \in d_i(t)\}$ is the difference set of subtrees formed by the remaining indices. The eight order conditions at fourth order are then given by

$$
\begin{aligned}
a\overline{a}(\bullet) &= a(\bullet) + \overline{a}(\bullet),\\
a\overline{a}([\bullet]) &= 2a(\bullet)\overline{a}(\bullet) + a([\bullet]) + \overline{a}([\bullet]),\\
a\overline{a}([[\bullet]]) &= 3a(\bullet)^2\overline{a}(\bullet) + 3a(\bullet)\overline{a}([\bullet]) + a([[\bullet]]) + \overline{a}([[\bullet]]),\\
a\overline{a}([\bullet,\bullet]) &= 3a(\bullet)\overline{a}([\bullet]) + 3a([\bullet])\overline{a}(\bullet) + a([\bullet,\bullet]) + \overline{a}([\bullet,\bullet]),\\
a\overline{a}([\bullet,\bullet,\bullet]) &= 4a(\bullet)^3\overline{a}(\bullet) + 6a(\bullet)^2\overline{a}([\bullet]) + 4a(\bullet)\overline{a}([[\bullet]]) + a([\bullet,\bullet,\bullet]) + \overline{a}([\bullet,\bullet,\bullet]),\\
a\overline{a}([[\bullet],\bullet]) &= 4a(\bullet)^2\overline{a}([\bullet]) + 4a([\bullet])a(\bullet)\overline{a}(\bullet) + (8/3)a(\bullet)\overline{a}([[\bullet]]) + (4/3)a(\bullet)\overline{a}([\bullet,\bullet]),\\
&\quad + 2a([\bullet])\overline{a}([\bullet]) + a([[\bullet],\bullet]) + \overline{a}([[\bullet],\bullet]),\\
a\overline{a}([[\bullet,\bullet]]) &= 6a(\bullet)^2\overline{a}([\bullet]) + 4a(\bullet)\overline{a}([\bullet,\bullet]) + 4a([[\bullet]])\overline{a}(\bullet) + a([[\bullet,\bullet]]) + \overline{a}([[\bullet,\bullet]]),\\
a\overline{a}([[[\bullet]]]) &= 4a(\bullet)\overline{a}([\bullet,\bullet]) + 6a([\bullet])\overline{a}([\bullet]) + 4a([\bullet,\bullet])\overline{a}(\bullet) + a([[[\bullet]]]) + \overline{a}([[[\bullet]]]).
\end{aligned}
\tag{20}
$$

Hence, for given scheme a, imposing the required order conditions on $a\overline{a}$ yields equations for $\overline{a}$, which are in turn easily solved for $\overline{a}$. The reader is referred to [25] for further details.

Figure 6 shows a comparison of the **FRKC4** scheme based on composition methods with other schemes. The reference solution is provided by a fifth-order implicit preconditioned BDF solver from the **CVODE** numerical package [26]. The number of steps required for a given precision is comparable for **ROCK4** and **FRKC4** and somewhat greater than the split **FRKC4s** scheme. This difference may be ascribed to a loss of precision due to the accumulation of errors over the finishing stages [27].

5. Conclusions

FRKC extended stability Runge–Kutta (ESRK) schemes are shown to well-suited to the integration of large-scale problems governed by systems of PDEs where the explicit timescale is constrained by diffusion. An implementation of the **FRKC2** second-order schemes, publicly available at `maths.dit.ie/frkc`, is presented. The fourth-order **FRKC4** schemes are also presented with nonlinear order conditions addressed via both splitting and composition methods. These schemes are shown to be competitive with alternative ESRK methods.

5.1. Acknowledgments

I am grateful to organisers of Astronum2016 for the invitation to present this work in Monterey and to an anonymous referee for helpful comments.

6. References

[1] Saulev V 1960 *Fizmatgiz, Moscow*
[2] Guillou A and Lago B 1960 *Ier Congr. Ass. Fran. Calcul., AFCAL* 43–56
[3] Gentzsch W and Schluter A 1978 *Z. Angew. Math. Mech* **58** T415–T416
[4] Alexiades V, Amiez G and Gremaud P 1996 *Communications in numerical methods in engineering* **12** 31–42
[5] O'Sullivan S and Downes T P 2006 *Monthly Notices of the Royal Astronomical Society* **366** 1329–1336
[6] O'Sullivan S and Downes T P 2007 *Monthly Notices of the Royal Astronomical Society* **376** 1648–1658
[7] Medovikov A A 1998 *BIT Numerical Mathematics* **38** 372–390
[8] van Der Houwen P J and Sommeijer B P 1980 *ZAMM - Journal of Applied Mathematics and Mechanics / Zeitschrift fr Angewandte Mathematik und Mechanik* **60** 479–485
[9] Verwer J G 1996 *Applied Numerical Mathematics* **22** 359–379
[10] Sommeijer B, Shampine L and Verwer J 1998 *Journal of Computational and Applied Mathematics* **88** 315–326
[11] Abdulle A and Medovikov A A 2001 *Numerische Mathematik* **90** 1–18
[12] Abdulle A 2002 *SIAM Journal on Scientific Computing* **23** 2041–2054
[13] Martin-Vaquero J and Janssen B 2009 *Computer Physics Communications* **180** 1802–1810
[14] Meyer C D, Balsara D S and Aslam T D 2012 *Monthly Notices of the Royal Astronomical Society* **422** 2102–2115
[15] Meyer C D, Balsara D S and Aslam T D 2014 *Journal of Computational Physics* **257** 594–626
[16] O'Sullivan S 2015 *Journal of Computational Physics* **300** 665 – 678
[17] Van Der Houwen P J 1977 *Construction of integration formulas for initial value problems* (North Holland)
[18] Lebedev V I 2000 *Computational mathematics and mathematical physics* **40** 1729–1740
[19] Lebedev V 1994 *Numerical methods and applications* 45–80
[20] Hairer E and Wanner G 1996 *Springer series in computational mathematics* **14**
[21] Nicolis G and Prigogine I 1977 *Self-organization in nonequilibrium systems : from dissipative structures to order through fluctuations* (New York: Wiley)
[22] Abdulle A 2001 *Chebyshev methods based on orthogonal polynomials* Ph.D. thesis
[23] Castella F, Chartier P, Descombes S and Vilmart G 2009 *BIT Numerical Mathematics* **49**(3) 487–508
[24] Blanes S, Casas F, Chartier P and Murua A 2013 *Mathematics of Computation* **82** 1559–1576
[25] Hairer E, Nørsett S and Wanner G 1993 *Springer series in computational mathematics* **8**
[26] Cohen S D and Hindmarsh A C 1996 *Computers in physics* **10** 138–143
[27] Hundsdorfer W and Verwer J G 2013 *Numerical solution of time-dependent advection-diffusion-reaction equations* vol 33 (Springer Science & Business Media)

IOP Conf. Series: Journal of Physics: Conf. Series **837** (2017) 012021 doi:10.1088/1742-6596/837/1/012021

Spatial Multiphysics Models of the Radiation Gas Dynamics of Super Orbital Re-Entry Space Vehicles

S.T. Surzhikov

Institute for Problems in Mechanics of the Russian Academy of Sciences, Russia, Moscow,119526

e-mail: surg@ipmnet.ru

Abstract. The multiphysics models of non equilibrium physical and chemical processes that should be considered together with radiative gas dynamic models for a description of the aero physics of super-orbital space vehicles are described.
As an example of application of the multiphysics approach a three-dimensional numerical simulation analysis of radiative gas dynamics of Apollo-17 command module is analyzed. The results of calculations are compared with other numerical predictions, analytical correlations, and available experimental data.

1. Introduction

A variety of multiphysics models developed for a description of large amount of related processes have found a wide application in modern computational physics, especially in basic problems of astrophysics [1,2], physical mechanics [3,4], plasma physics [5–7], theory of combustion, flame, and explosion [8, 9], electromagnetics [10] *et al.*

The classic examples of the multi-physics problems are the relevant tasks arisen in the radiative gas dynamics (RadGD) of re-entry space vehicles. For a reliable description of the processes developed near and inside the streamlined surface at extremely high velocities it is necessary to include in consideration the following branches of science:

- gas dynamics and continuum mechanics of solid materials,
- thermodynamics and statistical physics,
- applied quantum mechanics, quantum chemistry, and quantum statistics,
- physical and chemical kinetics of gases and heat protective materials,
- radiation heat transfer,
- computational fluid dynamics (CFD).

Solution to such a multiphysical problem appears to be very complicated due to the fact that the physical processes under consideration are interdependent and appropriate partial differential equations of mathematical physics turn out to be strongly nonlinear. These are some examples of such an interrelation:

- gas heating in shock layer and attendant chemical kinetics,
- nonequilibrium dissociation and ionization, as well as attendant relaxation processes,
- thermo-chemistry of heat-protection materials and the conditions in gas phase in surrounding boundary layers,

- physical kinetics of the processes of filling and depletion of quantum states in atoms and radiation heat transfer,
- coupling physical-chemical and radiative processes with gas dynamics.

Investigation of various multiphysical problems of radiative gas dynamics is directly connected with creating and developing of new theoretical and computational models. Wide spectrum of physical-chemical, quasi-classical and quantum models are used for a description of nonequilibrium processes, such as:

- quasi-classical theory of velocities of chemical reactions,
- quantum-chemistry models of chemical transformations,
- collisional-radiative (CR) models,
- quantum-kinetic models,
- multi-temperature kinetic models.

As for the interaction between radiation and gas dynamic processes, one can also use various models of such a coupling, for example:

- strong radiative-gas-dynamic interaction,
- weak radiative-gas-dynamic interaction,
- radiative gas dynamics with a strong reabsorbtion of heat radiation,
- volume emission approximation (RadGD of optically thin gas),
- RadGD of optically dense gas,
- RadGD with weak and strong influence of fine structure o atomic lines.

The principal feature of modern models of radiation gasdynamics is a consideration of the spatial nature of the radiation heat transfer. To take into account this feature, two-dimensional and three-dimensional computer models of computational physics should be used. Non-local nature of the selective heat radiation makes these models extremely labour- and time- consuming.

A general problem of the fundamental aero physics is the reliable computational and theoretical predictions of the aerothermodynamic and electro-physical parameters of compressed layers near the space vehicle surfaces. Up to now this problem has not been solved the despite a significant progress made in recent decades.

Depending on flight conditions of spacecrafts, substantially different mechanisms for elementary physical and chemical processes are realized in the heated gas of the shock layer. The diversity of these processes, together with uncertainty in a knowledge of mechanisms of corresponding reactions, often leads to different results of the kinetic modeling not only in the values of reaction rates, but also in their nomenclature.

Another unresolved issue is the need to take into account physical and chemical non-equilibrium behind the shock front due to rarefaction of gas flows. Absence of thermalization of internal degrees of freedom leads to the necessity of using different models of chemical and physical kinetics, among which the models of multimode relaxation are the most applicable. These models are based on the Born-Oppenheimer approximation and on the various kinds of the resonant energy exchange between quantum states.

Differences in translational, vibrational and electronic temperatures require for a modification of the models of chemical kinetics through the use of models of nonequilibrium dissociation. Such models have been developed, however, the lack of reliable criteria for the use of different models leads to the fact that in practice each research has to look for the most suitable model in particular situation.

Prediction of the radiative heating of entering space vehicles and study of coupled radiative gas dynamic processes in shock layers at orbital and especially for the super-orbital conditions of return into Earth atmosphere has revived a new interest due to modern plans for a return from Moon and closest planets of the Solar system. In spite on the fact that radiation heat transfer processes, as applied to the problems of aerothermodynamics of re-entry space vehicles, have been considered more than forty years, unanswered questions still remaining a lot up to now. The scientific tasks followed from these questions have to be resolved to provide reliable aerothermodynamic prediction of creating space vehicles of the new generation.

This paper is dedicated to the analyze of spatial radiative heating of the Apollo command modules. This paper is dedicated to the analyze of spatial radiative heating of the Apollo command modules. It is well known that significant resources for testing and quantification of the aerothermodynamic characteristics of the re-entering command modules were spent within the framework of the Apollo program [11-13]. An engineering-level predictions of the expected flight heating rates were created on the basis of the ground-based tests in low enthalpy wind tunnels, shock tunnels and the free-flight facilities. Two flight tests of the Apollo command modules were conducted at orbital (Apollo-09,11) and at super-orbital speeds (Apollo–17,20) [13]. Table 1, reproduced from Ref.13, lists the relevant entry conditions for each of the four Apollo test flights. In table 1 V_∞ is the entry velocity at atmospheric interface, α is the angle of attack, and $Q_{w,an}$ is the maximum theoretical heating rate determined by engineering correlations at a spherical stagnation point. Entry trajectories of these flight tests are presented in figure 1.

Table 1. Apollo descent modules entry conditions [13]

Spacecraft Number	V_∞ **, km/s (nominal)**	α **, deg**	$Q_{w,an}$ **, W/cm^2**
009	7.67	20	186
011	8.29	18	91
020	10.73	25	237
017	9.60	25	488

In the present the numerical investigation bases on some previously published works [14-18]. Detailed analysis of radiative gas dynamics of experimental space vehicles Fire-II and Stardust was presented in [14,15]. Several recent publications were dedicated to studying the real re-entry conditions of super-orbital capsule [16,17]. Some analytical correlations for intensity of convective and radiative heating, based on a large number of numerical simulation results, were recently suggested in [18].

In our previous papers [19–22] we analyzed radiative aerothermodynamics of segmental-conical space vehicle of large size (the Prospective Transport Vehicle – PTV) that is intended for flight of six astronauts. Start of this investigation was made in papers [19, 20], where the radiative aerothermodynamics of the orbital re-entry space vehicles similar to PTV was studied.

Some problems of radiative gas dynamics of re-entry space vehicle of large size and super orbital speed, as well as of coupled radiative gas dynamic interaction and non-equilibrium dissociation for large-scale returned space vehicles were considered in [21, 22].

The paper presents numerical results of the heating prediction for the surface of the Apollo command modules at orbital and super orbital velocities with a use of the author's computer codes.

Computer platform **NERAT-3D** (**N**on-**E**quilibrium **R**adiation **A**ero-**T**hermodynamics) + **ASTEROID** [19–24] together with various models of chemical kinetics of ionized air is used for such numerical study. The computational fluid dynamic code **NERAT-3D** is based on groups of governing equations that reflect multiphysical nature of the problem under consideration. These are: (*a*) the Navier-Stokes and continuity equations, (*b*) the equations of mass conservation of separate species, (*c*) the equation of energy conservation, including total and vibrational energy of molecules N_2, O_2 and NO, and (*d*) the equation of radiation heat transfer in the multi-group approximation.

All equations are formulated in curvilinear coordinate system related with the Apollo geometry, which was borrowed from [25] (see figure 2). Calculations were carried out with using a multi block and multi grid computational technology in three-dimensional geometry. The finite-difference approach was used.

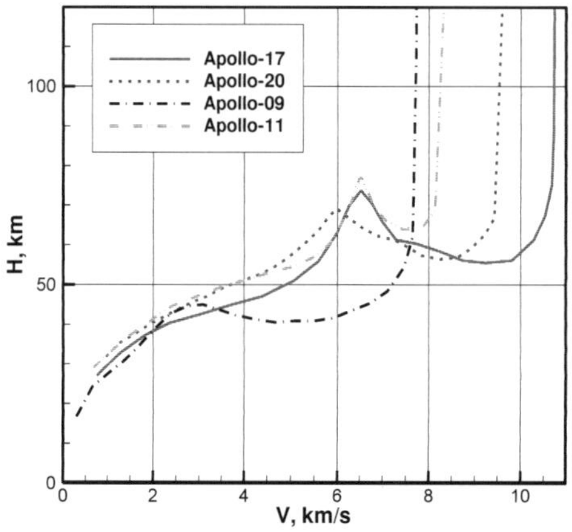

Figure 1. Trajectories of Apollo command modules.

Three dimensional investigations of radiative gas dynamic processes are extremely time-consuming. Therefore multi-block and multi-grid numerical techniques were used for investigation of the set of governing equations by the time-relaxation method, whereas the multi-group spectral approximation was used for solving the selective radiation heat transfer equation. Computer code **NERAT-3D** was used together with the following databases of thermodynamic, transport, kinetic and spectral optical properties:

- Database of thermodynamic properties of individual species [26],
- Database of chemical kinetics [27],
- Database of parameters of intermolecular interaction [28],
- Database of physical kinetics [29],
- Database of spectral optical properties is based on the code **ASTEROID** [30].

Radiation emission from the shock layer was calculated with the use of vibrational temperatures, therefore their behavior in shock layer is very significant for the surface radiative heating determination, as well as for the radiative signatures of space vehicles prediction. The dominant optically active components in the shock layer are the following: N_2, O_2, NO, N, O, N^+, O^+ and N_2^+. General features of these distributions are the following: large mass fractions of molecular N_2 and O_2 and low mass fractions of atomic N and O in the vicinity of the command module surfaces. Such behaviors of mass fractions are in full agreement with a used assumption concerning catalytic properties of the surface. Obtained distributions of translational and vibrational temperatures will be analyzed below. The vibrational temperatures are introduced in the so called multi-temperature models to characterize the energy accumulated in vibrational degree of freedoms of the diatomic molecules. In this case one uses an assumption concerning Boltzmann's energy distribution inside some separated degrees of freedom. To verify and validate obtained numerical simulation results several well known calculation and experimental data were used (Fire-II and Stardust) [23,24].

Two calculation cases are presented in this paper. In the first case, the non-equilibrium and quasi-equilibrium regimes at maximal velocity of the space vehicle (V_∞>9 km/s) were investigated. This case directly corresponds to super-orbital entry.

The second case characterizes the second entry of the command module into atmosphere (V_∞ <7 km/s), and actually corresponds to orbital entry, since the trajectory parameters of the second re-entry actually coincide with corresponding parameters of regular orbital entry.

The calculation data presented in this paper allow to get a full representation concerning specific properties of the flow field and the aerothermodynamics of super-orbital entering of Apollo command modules in the non-equilibrium and quasi-equilibrium regimes. These data include gas dynamic functions of the flow field at the angle of attack α=25^0, distributions of translational and vibrational temperatures, as well as mass fractions of gas species. Presented data on spectral radiation heat fluxes and corresponding cumulative functions give a representation about most significant radiative processes which form integral radiative heating of the space vehicle.

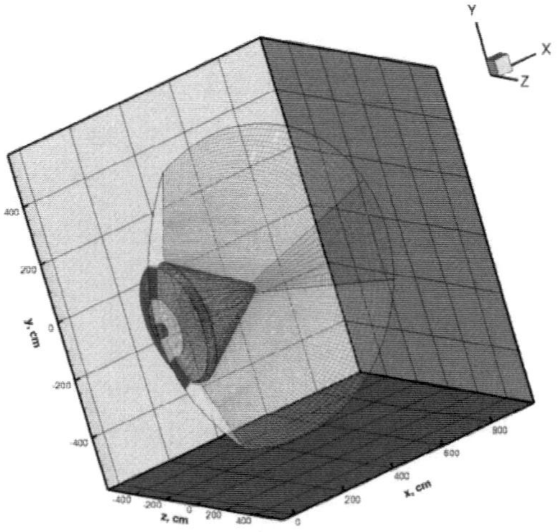

Figure 2. Multi-block calculation grid.

Finally, the numerical simulation results obtained within the framework of the three-dimensional approach will be compared with the data of two-dimensional calculations and engineering correlations [18].

2. Computational multiphysical model of radiative gas dynamics of re-entry command module

At every time step the following groups of governing equations are integrated successively:

$$\frac{\partial \mathbf{W}}{\partial t}+\frac{\partial \mathbf{F}_x}{\partial x}+\frac{\partial \mathbf{F}_y}{\partial y}+\frac{\partial \mathbf{F}_z}{\partial z}=0, \tag{1}$$

$$\mathbf{W}=\begin{bmatrix}\rho\\ \rho u\\ \rho v\\ \rho w\end{bmatrix},\quad \mathbf{F}_x=\begin{bmatrix}\rho u\\ \rho u^2+p-\tau_{xx}\\ \rho uv-\tau_{xy}\\ \rho uw-\tau_{xz}\end{bmatrix},\quad \mathbf{F}_y=\begin{bmatrix}\rho v\\ \rho uv-\tau_{xy}\\ \rho v^2+p-\tau_{yy}\\ \rho vw-\tau_{yz}\end{bmatrix},\quad \mathbf{F}_z=\begin{bmatrix}\rho w\\ \rho uw-\tau_{xz}\\ \rho vw-\tau_{yz}\\ \rho w^2+p-\tau_{zz}\end{bmatrix}, \tag{2}$$

$$\rho c_p\frac{\partial T}{\partial t}+\rho c_p\mathbf{V}\,\mathrm{grad}\,T=\mathrm{div}\left(\lambda\,\mathrm{grad}\,T\right)+\frac{\partial p}{\partial t}+\mathbf{V}\,\mathrm{grad}\,p+\Phi_\mu+Q_{vib}-\mathrm{div}\,\mathbf{q}_R-\sum_{i=1}^{N_s}h_i\dot{\varpi}_i+$$

$$+\sum_{i=1}^{N_s}\rho c_{p,i}D_i\left(\mathrm{grad}\,Y_i\cdot\mathrm{grad}\,T\right), \tag{3}$$

$$\frac{\partial \rho_i}{\partial t} + \text{div}\,\rho_i \mathbf{V} = -\text{div}\,\mathbf{J}_i + \dot{\varpi}_i, \quad i = 1,2,\ldots,N_s, \tag{4}$$

$$\frac{\partial \rho_m e_{\text{v},m}}{\partial t} + \text{div}\left(\rho_m \mathbf{V} e_{\text{v},m}\right) = \dot{e}_{\text{v},m}, \; m = 1,2,\ldots,N_V, \tag{5}$$

$$\mathbf{\Omega}\frac{\partial J_\omega(\mathbf{r},\mathbf{\Omega})}{\partial \mathbf{r}} + \kappa_\omega(\mathbf{r}) J_\omega(\mathbf{r},\mathbf{\Omega}) = j_\omega(\mathbf{r}), \tag{6}$$

$$p = \rho \frac{R_0}{M_\Sigma} T = \rho R_0 T \sum_i^{N_s} \frac{Y_i}{M_i}, \tag{7}$$

where p and ρ are the pressure and density; $\tau_{ij} = \mu\left[\left(\frac{\partial u_i}{\partial x_j} + \frac{\partial u_j}{\partial x_i}\right) - \frac{2}{3}\delta_{ij}\frac{\partial u_k}{\partial x_k}\right]$, $i,j,k = 1,2,3$ are the viscous components of the total stress tensor $\Pi_{ij} = -p\delta_{ij} + \tau_{ij}$; δ_{ij} is the Kronecker delta symbol; $u_i, u_j, u_k \to u, v, w$ are the projections of gas velocity $\mathbf{V}$ on axis of the Cartesian coordinate system; t is the time; T is the translational temperature; μ and λ are the viscosity and heat conductivity coefficients, c_p is the specific heat capacity of gas mixture; $c_p = \sum_i^{N_s} Y_i c_{p,i}$; Y_i is the mass fraction of species i; $c_{p,i}$ and h_i are the specific heat capacity at constant pressure and specific enthalpy of species i; $\dot{\varpi}_i$ is the reaction rate for species i; D_i is the effective diffusion coefficient of species i; ρ_i and $\mathbf{J}_i$ are the density and mass diffusion flux for species i; $\mathbf{J}_i = -\rho D_i\,\text{grad}\,Y_i$; N_s is the number of species; $e_{\text{v},m}$ is the specific vibrational energy of m-th mode of species i; $\rho_{i(m)}$ is the density of ith species having m^{th} mode of vibration; $e^0_{\text{v},m} = e_{\text{v},m}(T_V = T)$ is the equilibrium specific vibrational energy of mth mode of species i; $R_0 = 8.314\times10^7$ erg/(K·mole) is the universal gas constant; $T_{V,m}$ is the vibrational temperature defining energy distribution for separate vibrational mode (which is called as the mth mode) of ith species;

$$\dot{e}_{\text{v},m} = \rho_{i(m)}\frac{e^0_{\text{v},m} - e_{\text{v},m}}{\tau_m} - e_{\text{v},m}\dot{\varpi}_{i(m)}, \quad e_{\text{v},m} = \frac{R_0\theta_m}{M_{i(m)}\left[\exp(\theta_m/T_{\text{V},m}) - 1\right]}, \tag{8}$$

$$\Phi_\mu = \mu\left[2\left(\frac{\partial u}{\partial x}\right)^2 + 2\left(\frac{\partial v}{\partial y}\right)^2 + 2\left(\frac{\partial w}{\partial z}\right)^2 + \left(\frac{\partial v}{\partial x} + \frac{\partial u}{\partial y}\right)^2 + \left(\frac{\partial w}{\partial y} + \frac{\partial v}{\partial z}\right)^2 + \left(\frac{\partial u}{\partial z} + \frac{\partial w}{\partial x}\right)^2 - \frac{2}{3}\left(\frac{\partial u}{\partial x} + \frac{\partial v}{\partial y} + \frac{\partial w}{\partial z}\right)^2\right]$$

is the dissipative function. Other variables are defined below.

Equation (5) describes vibrational relaxation processes in non-equilibrium gas flows. The term Q_{vib} in Eq.(3) is the energy source due to vibrational relaxation in multi component gas mixture (vibrational source turn); N_V is the number of vibrational modes (in the case under consideration $N_V = 3$: $m = 1$ for N_2 vibration mode, $m = 2$ for O_2 vibration mode, $m = 3$ for the NO vibration modes).

In the case under consideration thermodynamic properties of gas mixture cannot be described using a single temperature because there are relaxation zones with non-equilibrium (i.e. non

IOP Conf. Series: Journal of Physics: Conf. Series **837** (2017) 012021 doi:10.1088/1742-6596/837/1/012021

Boltzmann's) occupation of internal degrees of freedom of molecules and atoms. The most simple and popular non-equilibrium thermodynamic models are based on introduction of various translational and vibrational temperatures. Such models are often used for a description of relaxation processes behind shock wave fronts formed due to collisional processes and in expansion parts of gas flows. In such cases the Kirchhoff low is applied to averaged vibrational temperature or to separate temperatures.

Using a symbolic notation for the n-th chemical reaction

$$\sum_{j=1}^{N_s} a_{j,n}\left[X_j\right] = \sum_{j=1}^{N_s} b_{j,n}\left[X_j\right],\ n = 1,2,\ldots,N_r,$$

the rate of formation of i-th gas species (in mole/cm^3·s) can be written for n^{th} chemical reaction as follows:

$$\left(\mathrm{d}X_i/\mathrm{d}t\right)_n = k_{f,n}\left(b_{i,n} - a_{i,n}\right)\prod_{j=1}^{N_s} X_j^{a_{i,n}} - k_{r,n}\left(b_{i,n} - a_{i,n}\right)\prod_{j=1}^{N_s} X_j^{b_{i,n}} = \left(b_{i,n} - a_{i,n}\right)\left(S_{f,i}^n - S_{r,i}^n\right),\ n = 1,2,\ldots,N_r \quad (9)$$

where $a_{i,n}, b_{i,n}$ are the stoichiometric coefficients of the nth reaction; $\left[X_j\right]$ are the chemical symbols of reagents and products of reactions; N_s and N_r are the number of gas species and number of the chemical reactions; $k_{f,n}$, $k_{r,n}$ are the rate constants of the forward and reverse reactions; $S_{f,i}^n$, $S_{r,i}^n$ are the rates of the forward and reverse reactions.

The rate of formation of mole per cm^3 for i^{th} gas species is written in the following form:

$$W_i = \sum_{n=1}^{N_r}\left(b_{i,n} - a_{i,n}\right)\left(S_{f,i}^n - S_{r,i}^n\right),\ \text{mol/(cm}^3\text{·s)}, \quad (10)$$

and the mass formation rate of species i can be calculated as follows:

$$\dot{\varpi}_i = M_i W_i,\ \text{g/(cm}^3\text{·s)}. \quad (11)$$

To calculate the rates of formation it is necessary to know the forward and reverse reaction constants for each n-th reaction

$$k_{f(r),n} = A_{f(r),n} T^{n_{f(r),n}} \exp\left(-E_{f(r),n}/kT\right), \quad (12)$$

where $A_{f(r),n}, n_{f(r),n}, E_{f(r),n}$ are the approximation coefficients for the forward (f) and reverse (r) reaction rates. Kinetic model, presented in [27] was applied. Note, that the equilibrium constant for each n-th chemical reaction is determined as follows

$$K_n = k_{f,n}/k_{r,n}, \quad (13)$$

that in turn are determined by thermodynamic data [26].

Thermodynamic properties of individual chemical species were calculated by the polynomial approximations [26]. This data base contains tables of the approximation constants for each individual species in the temperature range of 100 – 20 000 K. The Gibbs function can be calculated as follows:

$$\Phi_j = \varphi_{1,j} + \varphi_{2,j}\ln x + \varphi_{3,j}x^{-2} + \varphi_{4,j}x^{-1} + \varphi_{5,j}x + \varphi_{6,j}x^2 + \varphi_{7,j}x^3, \quad (14)$$

$$\left(\frac{\mathrm{d}\Phi}{\mathrm{d}x}\right)_j = \left(\varphi_{2,j} - 2\varphi_{3,j}x^{-2} - \varphi_{4,j}x^{-1} + \varphi_{5,j}x + 2\varphi_{6,j}x^2 + 3\varphi_{7,j}x^3\right)\frac{1}{x},$$

$$\left(\frac{\mathrm{d}^2\Phi}{\mathrm{d}x^2}\right)_j = \left(-\varphi_{2,j} + 6\varphi_{3,j}x^{-2} + 2\varphi_{4,j}x^{-1} + 2\varphi_{6,j}x^2 + 6\varphi_{7,j}x^3\right)\frac{1}{x^2}.$$

General properties of thermodynamic potentials allow predict the enthalpy, heat capacity at constant pressure and equilibrium constants as following:

$$h_j = xT\left(\frac{\mathrm{d}\Phi}{\mathrm{d}x}\right)_j + \varphi_{8,j}\times 10^3,\ \text{J/mole},$$

$$c_{p,j} = 2x\left(\frac{\mathrm{d}\Phi}{\mathrm{d}x}\right)_j + x^2\left(\frac{\mathrm{d}^2\Phi}{\mathrm{d}x^2}\right)_j,\ \text{J/mole·K},$$

$$\ln K_n\left(p/p_*\right) = \frac{1}{R_0 T}\sum_{j=1}^{N_s}\left(a_{j,n} - b_{j,n}\right)\left(-T\Phi_j + \varphi_{8,j}\times 10^3\right), \tag{15}$$

where $p_* = 101325$ Pa; $x = T/10000$.

The radiation heat transfer equation (6) is formulated in a general form. The solution of this equation allows calculate a total radiation heat flux that can be presented in the following form:

$$\mathbf{q}_R = \mathbf{q}_R(\mathbf{r}) = \int_{4\pi}\mathrm{d}\Omega \int_{\Delta\omega_{tot}} J_\omega(\mathbf{r},\mathbf{\Omega})\mathbf{\Omega}\mathrm{d}\omega = \int_{\Delta\omega_{tot}} \mathbf{q}_{R,\omega}(\mathbf{r})d\omega,$$
$$\mathbf{q}_{R,\omega}(\mathbf{r}) = \int_{4\pi} J_\omega(\mathbf{r},\mathbf{\Omega})\mathbf{\Omega}\mathrm{d}\Omega, \qquad q_R = (\mathbf{q}_R\cdot\mathbf{n}), \tag{16}$$

where $J_\omega(\mathbf{r},\mathbf{\Omega})$ is the spectral intensity of heat radiation; $\kappa_\omega(\mathbf{r})$ is the spectral absorption coefficient; $j_\omega(\mathbf{r})$ is the spectral emission coefficient calculated in the case of local thermodynamic equilibrium (LTE) by the Kirchhoff low

$$j_\omega(\mathbf{r}) = \kappa_\omega(\mathbf{r})J_{b,\omega}(\mathbf{r}), \tag{17}$$

where $J_{b,\omega}(\mathbf{r})$ is the black body spectral intensity (the Planck function); $\mathbf{r}$ is the coordinate of a point; $\mathbf{\Omega}$ is the directional vector; $\Delta\omega_{tot}$ is the total spectral range of heat radiation (in the given case $\Delta\omega_{tot}$ =1000 – 200 000 cm^{-1}).

Viscosity and heat conductivity of gas mixtures are calculated by the following formulas [31]

$$\mu = \frac{1}{\sum_{i=1}^{N_c}\left(Y_i/\mu_i\right)}, \quad \lambda = \frac{1}{2}\left[\sum_{i=1}^{N_c} x_i\lambda_i + 1\Big/\sum_{i=1}^{N_c}\left(x_i/\lambda_i\right)\right], \tag{18}$$

$$\mu_i = 2.67\times 10^{-5}\frac{\sqrt{M_i T}}{\sigma_i^2\Omega_i^{(2,2)*}},\ \text{g/cm·s}, \quad \lambda_i = 8330\sqrt{\frac{T}{M_i}}\frac{1}{\sigma_i^2\Omega_i^{(2,2)*}},\ \text{erg/cm·K}, \tag{19}$$

where σ_i is the effective collision diameter, A; $\Omega_i^{(2,2)*} = f\left(T_i\right)$ is the collision integral; $T_i = kT/\varepsilon_i$.

IOP Conf. Series: Journal of Physics: Conf. Series **837** (2017) 012021 doi:10.1088/1742-6596/837/1/012021

The effective diffusivity for species i is calculated by the Wilke formula [32].

$$D_i = \frac{1-x_i}{\sum_{j\neq i}^{N_c}\left(x_j/D_{ij}\right)},\ \text{cm}^2/\text{s},\quad D_{i,j} = 1.858\times10^{-3}\sqrt{T^3\frac{M_i+M_j}{M_iM_j}}\frac{1}{p\sigma_{i,j}^2\Omega_{i,j}^{(1,1)*}}. \tag{20}$$

The collision integrals are calculated by the Anfimov approximation

$$\Omega_i^{(2,2)*} = 1.157T_i^{-0.1472},\quad \Omega_{i,j}^{(1,1)*} = 1.074T_{i,j}^{-0.1604}, \tag{21}$$

where [27]

$$T_{i,j} = \frac{kT}{\varepsilon_{i,j}},\quad \varepsilon_{i,j} = \sqrt{\varepsilon_i\varepsilon_j},\quad \sigma_{i,j} = \frac{1}{2}\left(\sigma_i+\sigma_j\right). \tag{22}$$

Boundary conditions for pseudo-catalytic surface are used in the following form:

$$\left(Y_i\right)_w = Y_{i,\infty},\quad i = 1,2,3,\ldots,N_s, \tag{23}$$

where $Y_{i,\infty}$ are the mass fractions in the oncoming gas flow.

The surface temperature is calculated using the following assumption:

$$\varepsilon\sigma T_w^4 = Q_w, \tag{24}$$

where ε is the surface emissivity factor; $\sigma = 5.67\cdot10^{-12}$ W/(cm^2·K^4) is the Stephan-Boltzmann constant; Q_w is the total heating of the surface (the sum of the convective and radiation heat fluxes).

The ray tracing method was used for calculation of radiation heat fluxes to the surface of Apollo command module. To calculate radiation heat flux to any element of the space vehicle surface, it is necessary to introduce a local spherical coordinate system with an origin at this point and the z-axis of which is directed along a local normal to the surface, and next to integrate the radiation heat transfer equation along each ray emitted from the chosen point. .

In this local coordinate system each ray $\mathbf{\Omega}$ is defined by two angular coordinates, namely, the latitude angle $\theta\in\left[0,\pi/2\right]$ and the azimuth angle $\varphi\in\left[0,2\pi\right]$.

The spectral radiation heat flux is defined by the following formula:

$$q_{R,\omega}\left(\mathbf{r}_j\right) = \int_0^{2\pi}\mathrm{d}\varphi\int_0^{\pi/2}J_\omega\left(\mathbf{r}_j,\mathbf{\Omega}\right)\cos\theta\sin\theta\mathrm{d}\theta, \tag{25}$$

where $\mathbf{r}_j$ is the radius-vector of j-th point on the surface in the laboratory frame of reference; $J_v\left(\mathbf{r}_j,\mathbf{\Omega}\right)$ is the spectral radiation intensity at the point $\mathbf{r}_j$ corresponding to a direction of the unit vector $\mathbf{\Omega}$.

Introduction of finite-difference mesh of angular directions allows to integrate equation (6) over angular variables and to calculate the spectral radiation flux:

$$W_v(\mathbf{R}_j) = \sum_{m=1}^{N_\varphi-1}\left(\varphi_{m+1}-\varphi_m\right)\sum_{n=1}^{N_\theta-1}J_v\left(\mathbf{\Omega}_{m,n}\right)\times\frac{1}{2}\left(\sin\theta_{n+1}\cos\theta_{n+1}+\sin\theta_n\cos\theta_n\right)\left(\theta_{n+1}-\theta_n\right) =$$

$$= \sum_{m=1}^{N_\varphi-1}\left(\varphi_{m+1}-\varphi_m\right)\sum_{n=1}^{N_\theta-1}J_v\left(\mathbf{\Omega}_{m,n}\right)\frac{\left(\cos^2\theta_{n+1}-\cos^2\theta_n\right)}{2}, \tag{26}$$

where N_φ is the number of rays emitted in the azimuthal direction; N_θ is the number of rays emitted in the latitude direction.

The directing cosines

$$\mathbf{\Omega}_{m,n} = (\omega_x)_{m,n}\mathbf{i} + (\omega_y)_{m,n}\mathbf{j} + (\omega_z)_{m,n}\mathbf{k} \tag{27}$$

are calculated by the following formulas:

$$(\omega_x)_{m,n} = \sin\theta_n \cos\varphi_m, \ (\omega_y)_{m,n} = \sin\theta_n \sin\varphi_m, \ (\omega_z)_{m,n} = \cos\theta_n. \tag{28}$$

To calculate spectral radiation intensity the radiation heat transfer equation should be integrated over an inhomogeneous optical path. For this purpose the following solution to the radiation heat transfer equation can be used

$$J_\nu(\tau_\nu) = \int_0^{\tau_\nu} B_\nu(\tau'_\nu)\exp\left[-(\tau_\nu - \tau'_\nu)\right]d\tau'_\nu, \tag{29}$$

where $\tau_\nu = \int_0^s \kappa_\nu ds'$ is the optical depth; $B_\nu(T) = \frac{2h\nu^3}{c^2}\left[\exp(h\nu/kT) - 1\right]^{-1}$ is the black body radiation intensity; s is the coordinate along the ray, $s = 0$, $s = L$ are the initial (on the surface) and final (on the outer surface of the calculation range or on the space vehicle conjugate surface) coordinates of the segment with directing vector $\mathbf{\Omega}_{m,n}$. The finite-difference scheme for spatial variables s is formulated for each ray $\mathbf{\Omega}_{m,n}$.

Numerical integration in equation (29) is realized as follows:

$$J(\tau_{NS}, \mathbf{\Omega}_{m,n}) = \exp(-\tau_{NS})\left\{\sum_{k=1}^{NS-1} \tilde{B}_k \exp(\tau_k)\left[\exp(\tau_{k+1} - \tau_k) - 1\right]\right\}, \tag{30}$$

where $\tilde{B}_k = \frac{(B_k + B_{k+1})}{2}$, $\tau_l = \sum_{i=1}^{l-1} \frac{(\kappa_i + \kappa_{i+1})}{2}(s_{i+1} - s_i)$, $l = 1, \ldots, NS$.

The set of governing equations was integrated up to convergence of the velocity projections, temperature, pressure, density, and chemical species concentrations with relative error of about $10^{-3} - 10^{-5}$. The gas dynamic equations were integrated separately from other equations with the second order of accuracy in space without the use of additional numerical limits and of smoothing operators [33].

The energy conservation equation was integrated by the implicit finite-different method of the second order in space and time. Continuity equations for chemical species were integrated also by an implicit method of the second order of accuracy.

As was mentioned above, the spectral and group optical properties of gases of complex chemical composition were on-line calculated by the author's code **ASTEROID**.

3. Spatial radiative gas dynamics of super-orbital Apollo command module

The aerothermodynamics of Apollo command module was numerically investigated for four trajectory points shown in table 2. In present paper we will analyze in detail the calculated data for the first point. This trajectory point corresponds to extremely nonequilibrium RadGD regime of super-orbital. It should be taken into account that in these cases the incoming gas is very rare. This means that the using of the continuum model of flow field can give rise doubts concerning its justification. Nevertheless, large dimension of the command module and the acceptable correspondence to experimental and flight data demonstrated in previous works under similar conditions, actually justify the use of this model.

Two-dimensional images of the three dimensional gas dynamic fields around Apollo command module at the angle of attack of $\alpha=25^0$ are shown in figures 3 and 4. Figures and 4 show, correspondingly, the pressure, longitudinal velocity and Mach number, the translational temperature of all heavy particles and vibrational temperatures of N_2, O_2, and nitrogen oxide.

Table 2. Trajectory parameters used as initial data for calculations

#	*t*, s	*H*, km	V_∞, km/s	p_∞, erg/cm^3	ρ_∞, g/cm^3	Particles concentration, 1/cm^3	Mean free pass, *L*, cm
1	40	76.2	10.7	19.5	$0.330\cdot10^{-07}$	$0.687\cdot10^{+15}$	0.246
2	53	67.3	10.5	78.0	$0.120\cdot10^{-06}$	$0.249\cdot10^{+16}$	0.068
3	220	73.2	6.5	31.8	$0.523\cdot10^{-07}$	$0.109\cdot10^{+16}$	0.156
4	390	61.0	5.9	192.	$0.273\cdot10^{-06}$	$0.568\cdot10^{+16}$	0.029

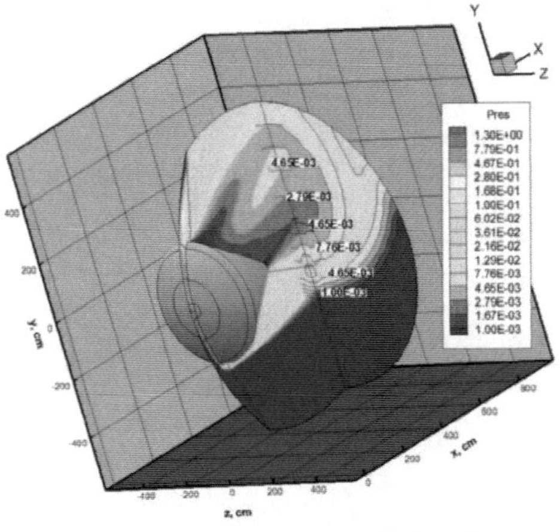

(*a*)

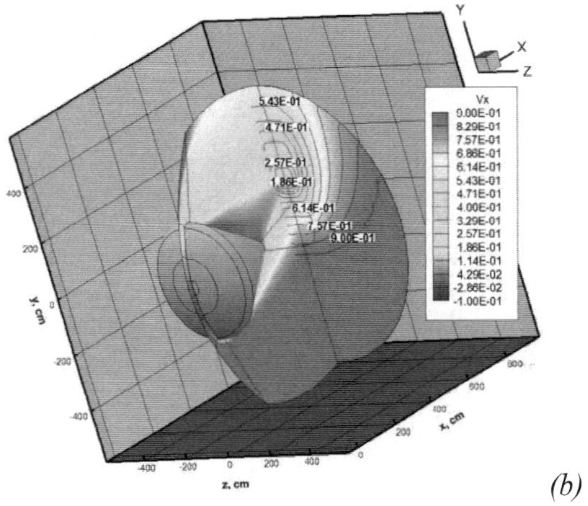

(*b*)

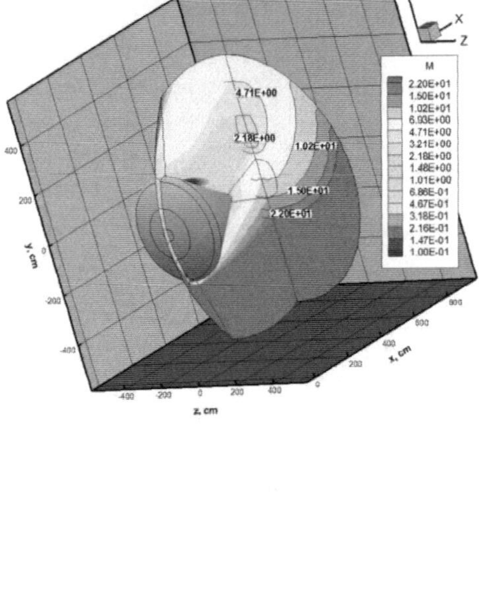

(*c*)

Figue 3. Pressure Pres= p/p_0 (*a*), where p_0 is the brake pressure in erg/cm^3, longitudinal velocity $V_x=u/V_\infty$ (*b*), and Mach number (*c*) for the 1st trajectory point.

The functions of a flow field presented in figure 3 show general peculiarities of hypersonic flows around entering command module. These are: an emergence of high pressure shock layer above aerodynamic front shield and domains of subsonic flow in shock layer and above back side surface where a vortex gas movement of large scale is observed. High temperature shock layer and total asymmetry of the flow field due to the angle of attack are also clearly seen there. The flow field in a wake is characterized by higher vibrational temperatures as compared to translational ones.

Mass fractions of atoms N and O and charged particles are presented in figures 5,6. These figures give general representation concerning a degree of dissociation of diatomic molecules calculated with applying the models of non-equilibrium dissociation [7].

The distributions of ionized atoms and electrons presented in figures 7 show that under the considered conditions the general sources of electrons are the collisions of nitrogen atoms and electrons. The nitrogen oxide generally appear in the domain of shock wave front.

Distributions of pressure (figure 3), mass fractions of atomic nitrogen and oxygen (figure 5), as well as molecular and ionized species allow understand general peculiarities in formation of the radiation heat fluxes to surface.

General results of numerical simulations are presented in figure 7, where distributions of convective heat fluxes, as well as the integral radiation heat fluxes along the surface in meridional cross sections are shown. In the first case, the radiation heat flux exceeds the convective heat flux on the front shield surface, especially in the domain of leeward part.

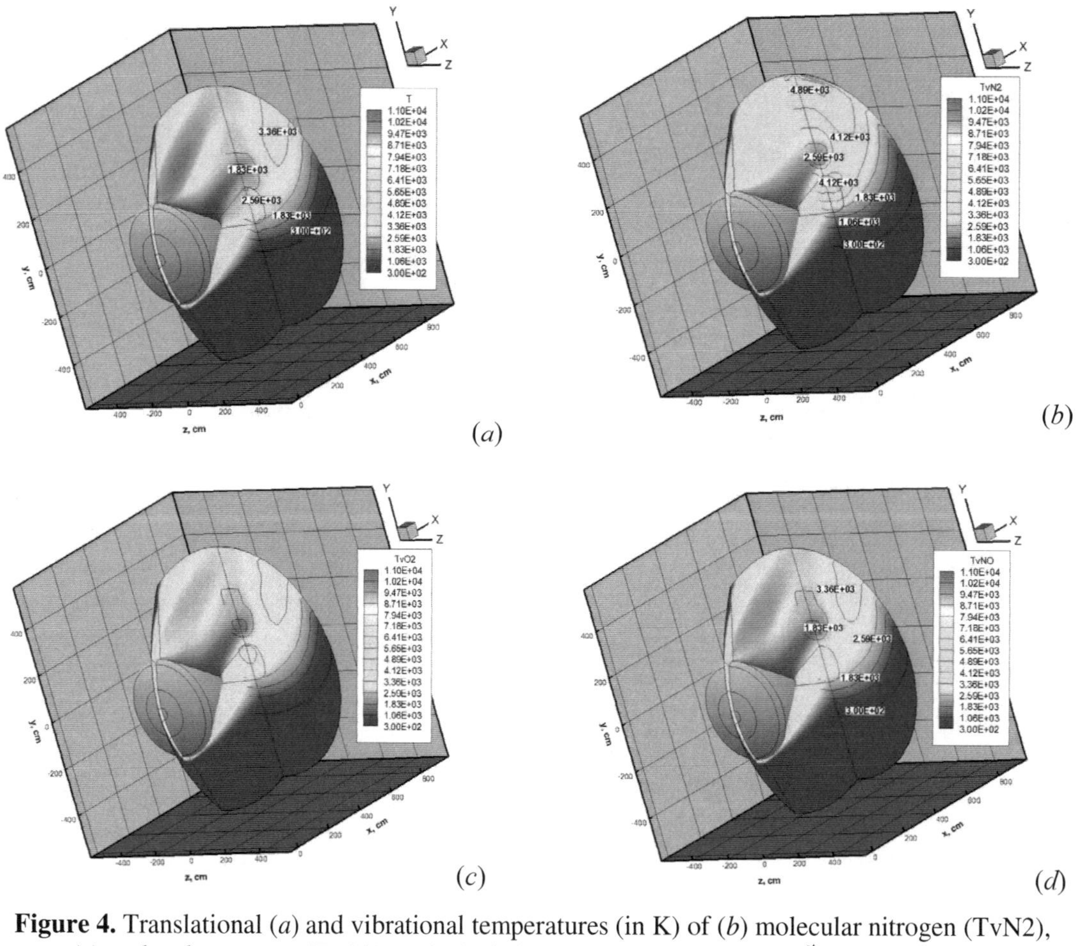

Figure 4. Translational (*a*) and vibrational temperatures (in K) of (*b*) molecular nitrogen (TvN2), (*c*) molecular oxygen (TvO2), and (*d*) nitride oxide (TvNO) for the 1st trajectory point.

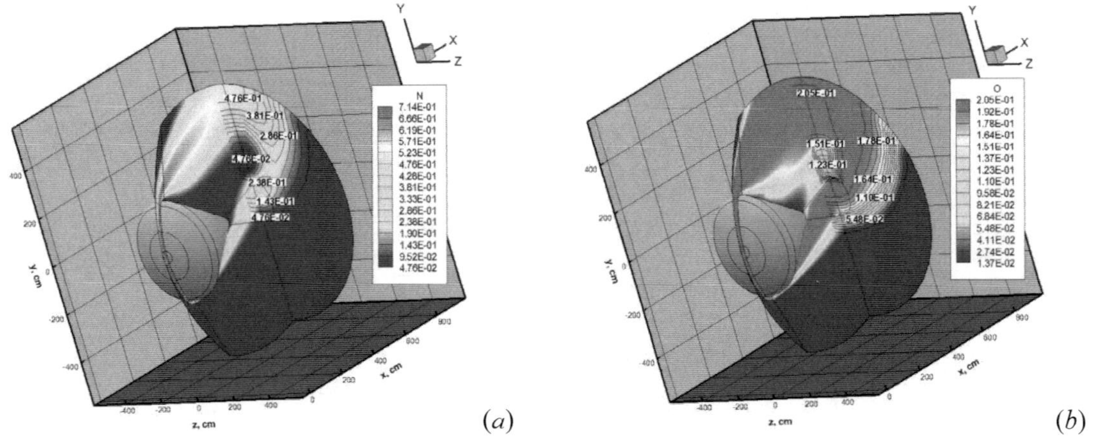

(*a*) (*b*)

Figure 5. Mass fractions of N (*a*) and O (*b*) for the 1st trajectory point.

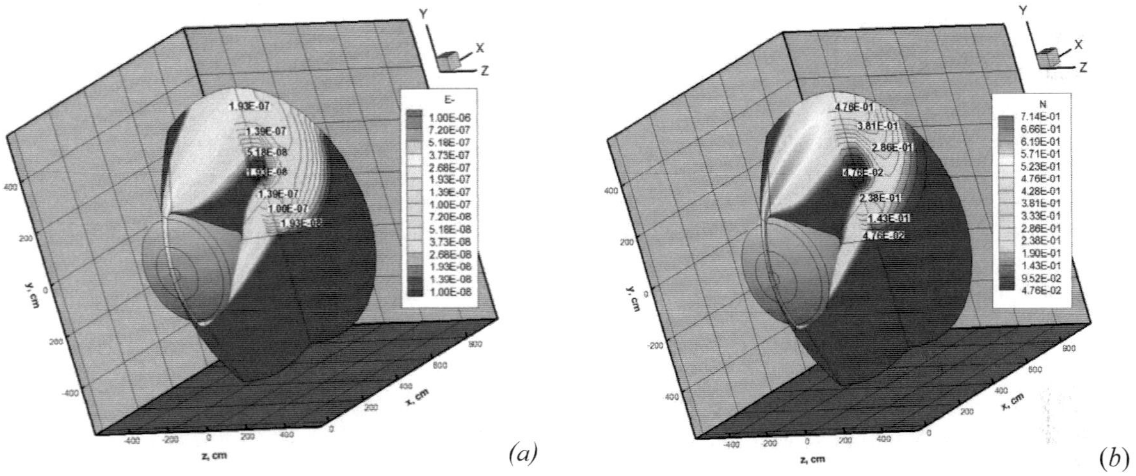

(*a*) (*b*)

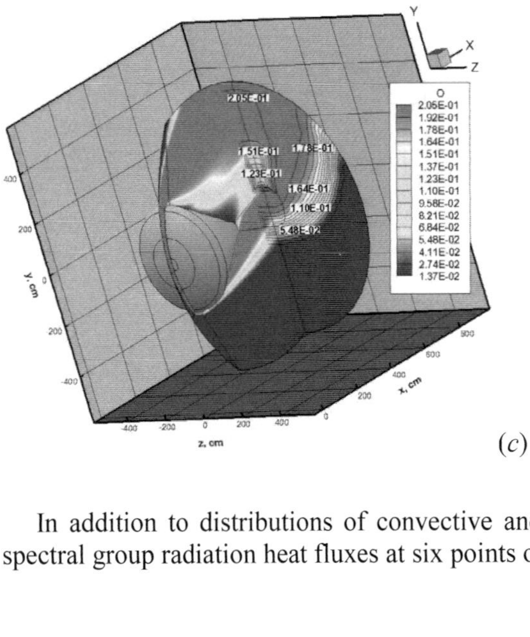

(*c*)

Figure 6. Mass fractions of charged particles electrons (*a*), N^{+}(*b*), and O^{+}(*c*) for the 1st trajectory point.

In addition to distributions of convective and integral radiation heat fluxes along a surface, the spectral group radiation heat fluxes at six points on the surface in the meridional plane are presented in figure 8. Coordinates of these points can be calculated from relations presented in these figures, where s_1,s_2, ... are the longitudinal coordinates along the surface measured from the point #1 in figure 8, and S_{surf} is the total length. From these figures one can see that the radiative heating of the front shield significantly exceeds radiative heating for back surface.

As mentioned above, the investigated calculation cases correspond to a rare gas flow. This immediately follows that significant part of ultraviolet (UV) radiation (for wavenumbers of heat radiation of ω>50 000 cm^{-1}) achieves a surface.

Finally we compare the numerical simulation results obtained in present work with available data on preliminary prediction [12,34,35] and analytical correlations [18]. This comparison (figure 9) demonstrates acceptable agreement with previous investigations.

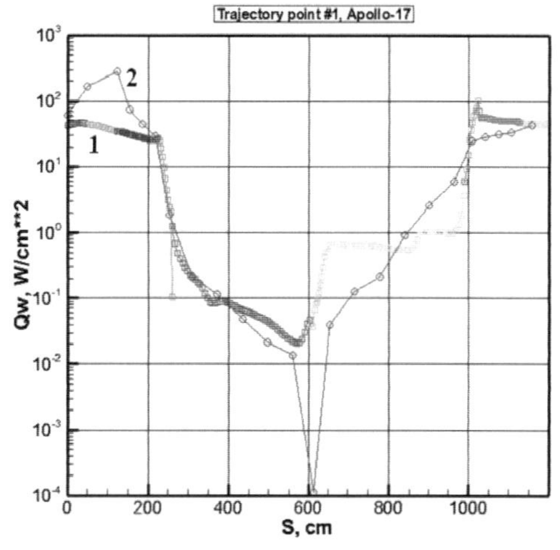

(*a*)

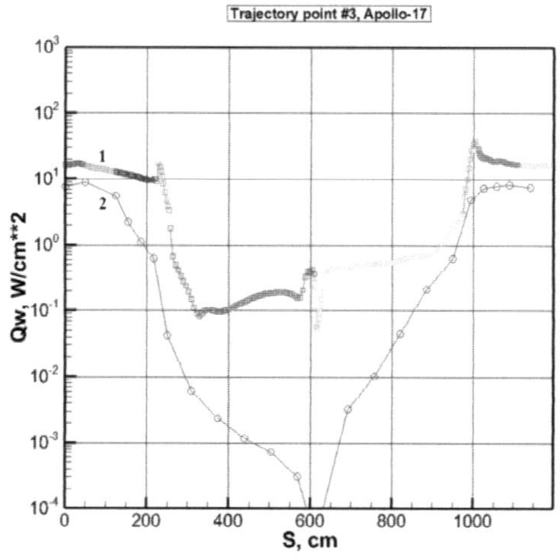

(*b*)

Figure 7. Disribution of the convective (1) heat flux ($Q_{w,tot} = Q_{w,c} + Q_{w,dif}$, $Q_{w,c}$ is the conductive part of the convective heat flux, $Q_{w,dif}$ is the diffusive part) and integral radiation (2) heat fluxes on the front shield with catalytic surface. Trajectory point #1 (*a*) and #3 (*b*).

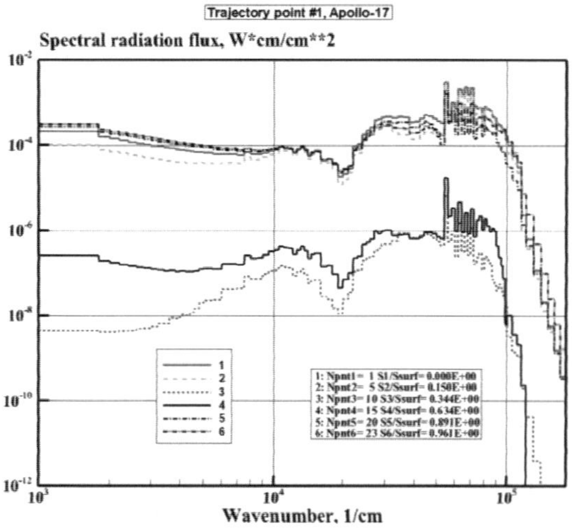

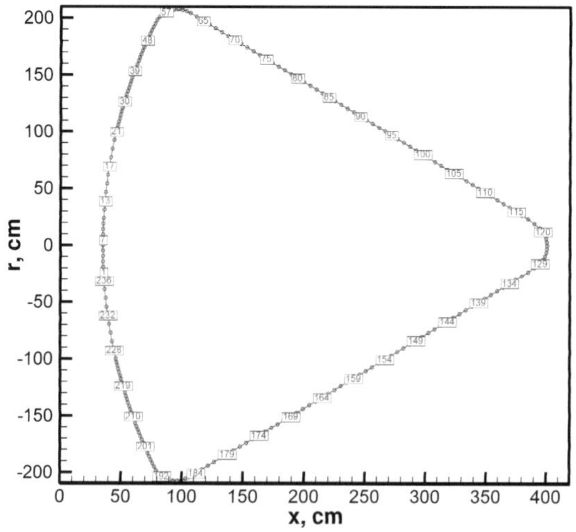

Figure 8. Radiation heat fluxes at six points on the surface in meridional plane of the Apollo-17 for the 1st trajectory points (left picture). Grid for prediction of radiation heat fluxes on the surface. Order of the points specifies the order of bypass the surface (right picture).

4. Conclusion

The examination of urgent multiphysics problems that should be considered for reliable prediction of aerothermodynamics of super-orbital space vehicles has been presented.

Numerical analysis of three-dimensional radiative aerothermodynamics of super-orbital Apollo command module based on the numerical simulation data obtained with using the author computer codes, has been presented. Distinguishing feature of presented data is a comparison of convective and radiative heating of the landing module surfaces, as well as a comparison of obtained calculation data on convective heating with correlation data for heating of front aerodynamic shield.

The author's three dimensional computational fluid dynamic code NERAT(3D)+ASTEROID has been used for investigation of radiative gas dynamics of super-orbital landing module at entering Earth atmosphere under angles of attack. Three dimensional calculations of convective and radiative heating have been performed. It has been shown that for some trajectory points the radiative heating exceeds the convective heating.

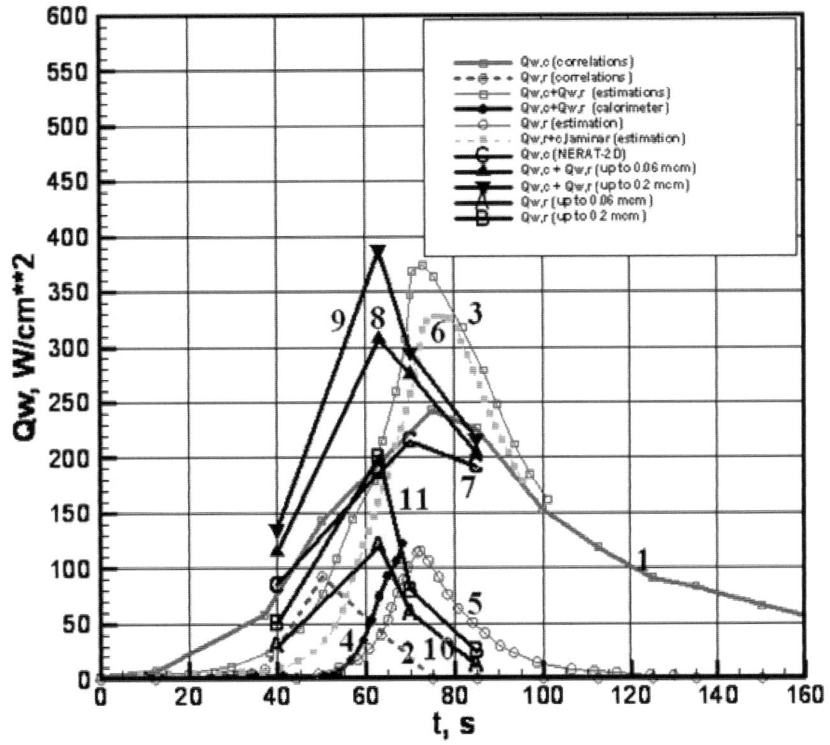

Figure 9. Data on convective and radiative heating of stagnation point of Apollo-17 along trajectory, which were measured in flight [12], and were predicted in given work and with different methods [18,34,35]:

1 – convective heat flux [18], 2 – integral radiative heat flux [18], 3,6 – prediction of the total heat flux [12], 4 – calorimeter flight data [12], 5 – prediction of the integral radiation heat fluxes [34,35], 6 – convective heat flux (two-dimensional calculations) [36], 8 – total heat flux with radiation heat flux including UV region, 9 – total heat flux without radiation heat flux in UV region, 10 – radiation heat flux including UV region, 11 – radiation heat flux without UV region.

Acknowledgments
This work was supported by the Russian academy of sciences within the framework of the Program of basic research, and RFBR grant #16-01-00379.

References

[1] Pringle G E and King A R 2007 *Astrophysical Flows*. (Cambridge University Press) 206 p.
[2] LeVeque R J, Mihalas D, Dorfi E A and Muller E 1998 *Computational methods for astrophysical fluid flows* (Springer Verlag Berlin, Heidelberg)
[3] Bond J W Jr, Watson K M and Welch J A Jr 1965 *Atomic Theory of Gas Dynamics* (Reading, Massachusetts: Addison-Wessley)
[4] Park C. 1990 *Nonequilibrium Hypersonic Aerothermodynamics* (John Wiley & Sons)
[5] Capitelli M, Bruno D and Laricchiuta A 2013 Fundamental Aspects of Plasma Chemical Physics *Springer Series on Atomic, Optical, and Plasma Physics* **74**
[6] Bittencourt J A 2004 *Fundamentals of Plasma Physics* (Springer)
[7] Surzhikov S T 2013 *Computational Physics of Electric Discharges in Gas Flows* (Berlin/Boston: Walter de Gruyter GmbH)
[8] Oran E S and Boris J P 2001 *Numerical Simulation of Reactive Flow* (Cambridge University Press)
[9] Kessler D A, Oran E S and Kaplan C R 2010 Towards the Development of a Multiscale, Multiphysics Method for the Simulation of Rarefied Gas Flows *J. Fluid Mech.* **661**
[10] Shang Joseph J S 2016 *Computational Electromagnetic-Aerodynamics* (John Wiley & Sons.Inc.)
[11] Lee D B, Bertin J J and Goodrich W D October 1970 Heat-Transfer Rate and Pressure Measurements Obtained During Apollo Orbital Entries *NASA TN D-6028*
[12] Lee D B and Goodrich W D, April 1972 The Aerothermodynamic Environment of the Apollo Command Module During Superorbital Entry *NASA TN D-6792*
[13] Lee D B January 1972 Apollo Experience Report: Aerothermodynamics Evaluation *NASA TN D– 6843*
[14] Olynick D R, Henline W D, Hartung L C and Candler G V 1994 Comparison of Coupled Radiative Navier-Stokes Flow Solutions with the Project Fire-II Flight Data *AIAA 94-195*
[15] Olynick D R, Chen Y K and Tauber M E 1999 Aerodynamics of the Stardust Sample Return Capsule *J. Spacecraft and Rockets*, **36** 3 442
[16] Johnston C O, Hollis B R and Sutton K 2008 Spectrum Modeling for Air Shock-Layer Radiation at Lunar-Return Conditions *J Spacecraft and Rockets* **45** 5 865
[17] Johnston C O, Hollis B R and Sutton K 2008 Nonequilibrium Stagnation-Line Radiative Heating for Fire-II *J Spacecraft and Rockets* **45** 6 1185
[18] Brandis A M and Johnston C O 2014 Characterization of Stagnation-Point Heat Flux for Earth Entry *AIAA 2014–2374*
[19] Djadkin A, Beloshitsky A, Shuvalov M and Surzhikov S 2011 Nonequilibrium Radiative Gasdynamics of Segmental-Conical Space Vehicle of Large Size *AIAA 2011–0453*
[20] Djadkin A, Beloshitsky A, Shuvalov M, and Surzhikov S 2013 Uncertainties in Heating Predictions of Segmental-Conical Space Vehicle Resulting From Data on Chemical and Physical Kinetics *AIAA 2013–1056*
[21] Surzhikov S T 2015 Radiative Gasdynamics of Re-entry Space Vehicle of Large Size with Superorbital Velocity *AIAA 2015–0980*
[22] Surzhikov S T, 2012 Coupled radiative gasdynamic interaction and non-equilibrium dissociation for large-scale returned space vehicles *J. Chem Phys* **398** pp.56–63
[23] Surzhikov S T 2016 Radiative Gas Dynamics of the Fire-II Superorbital Space Vehicle *Technical Physics* **61** 3 349

[24] Surzhikov S T and Shang J S 2012 Coupled Radiation-Gasdynamic Model for Stardust Earth Entry Simulation *J. of Spacecraft and Rockets* **49** 5 875
[25] Wright M J, Prabhu D K and Martinez E R 2004 Analysis of Afterbody Heating Rates on the Apollo Command Modules, Part 1: AS–202, *AIAA 2004–2456*
[26] Gurvich L V, Veitc I V, Medvedev V A *et al* 1978 *Thermodynamic Properties of Individual Substances* (HandBook Moscow: «Nauka»)
[27] Park C, Jaffe R L and Partridge H 1993 Chemical-Kinetic Parameters of Hyperbolic Earth Entry *J. of Thermophysics and Heat Transfer* **15** 1 76
[28] Svehla R A 1962 Estimated Viscosities and Thermal Conductivities of Gases at High Temperatures *NASA TR-R-132*
[29] Millikan R, White D 1963 Systematics of Vibrational Relaxation *J of Chem. Phys.* **39** 12 3209
[30] Surzhikov S T 2000 Computing System for Mathematical Simulation of Selective Radiation Transfer *AIAA Paper 2000-2369*
[31] Bird R B, Stewart W E, Lightfoot E N 1965 *Transport Phenomena* (John Wiley & Sons. Inc.)
[32] Wilke C R 1950 *Chem. Engn. Progr.* **46** 95
[33] Edwards J R , Liou M - S 1998 Low-Diffusion Flux-Splitting Methods for Flow at all Speeds. *AIAA Journal* **36** 9 1610
[34] Page W A, Compton D L, Borucki W J, Ciffone D L and Cooper D M 1968 Radiative Transport in Inviscid Nonadiabatic Stagnation Region Shock Layers *AIAA 3rd Thermophysics Conf. Los Angeles*
[35] Detra R W, Kemp N H and Riddel F R 1957 Addendum to Heat Transfer to Satellite Vehicles Re-Entering the Atmosphere *Jet Propulsion* **27** 12 1256
[36] Surzhikov S T 2016 Radiative Gas Dynamics of Apollo Command Module at Angle of Attack *AIAA paper 2016-4022*